Basic Mathematics
for Calculus

Basic Mathematics
for Calculus

THIRD EDITION

Dennis G. Zill

Jacqueline M. Dewar

Warren S. Wright

Loyola Marymount University

Wadsworth Publishing Company
Belmont, California
A Division of Wadsworth, Inc.

Mathematics Publisher: Kevin Howat
Editorial Assistant: Sally Uchizono
Developmental Editor: Anne Scanlan-Rohrer
Assistant Editor: Barbara Holland
Managing Designer: MaryEllen Podgorski
Print Buyer: Karen Hunt
Illustrator: Carl Brown
Compositor: Polyglot Pte. Ltd.

Printed in the United States of America   34

2 3 4 5 6 7 8 9 10——92 91 90 89 88

Library of Congress Cataloging-in-Publication Data

Zill, Dennis G., 1940–
  Basic mathematics for calculus.

  Includes index.
  1. Mathematics—1961–   . I. Dewar, Jacqueline M.
II. Wright, Warren S.   III. Title.
QA39.2.Z54   1988   512′.1   87-23014
ISBN 0-534-08682-9

Contents

Functions and Graphs 112

Exponential and Logarithmic Functions 179

Trigonometric Functions 217

5 Trigonometric Applications   285

6 Conic Sections   315

Appendix   377

Tables 398

Answers to Odd-Numbered Problems   409

Index   443

Preface to the Third Edition

An instructor is always faced with the dilemma of too much material and too little time. In the vast precalculus market we see texts that cover everything from topics in elementary algebra to topics in linear and abstract algebra. Some texts are at such a sophisticated level that they appear to be oriented toward students who do not need a course in precalculus mathematics. Others include too much review or are paced too slowly to adequately prepare students for calculus in a single course. Thus, we feel there is a great need for a text that quickly gets to the heart of the matter, a text that presents only those basic topics that will be of direct and immediate use in most calculus courses. It has been our experience that a student, unable to solve a problem, may seek clarification from the examples in the text. Therefore, we have taken care to include many examples clearly delineated from the textual discussion. It is also our philosophy to illuminate an example, discussion, or problem with a figure whenever possible.

Here are some of the changes that appear in the third edition.

- The chapters have been rearranged so that essential precalculus material is easier to cover: exponential and logarithmic functions are treated earlier; discussion of the conic sections is delayed until Chapter 6; and the algebra review, now presented in Chapter 0, is optional.

- The number of exercises has been substantially increased. Problems that require greater analytical or geometrical insight have been added, as well as problems that focus on skills needed in calculus.

- The material on graphing has been substantially revised. Shifting, reflecting, and stretching of graphs is now covered in detail. A greater emphasis is placed on the interpretation of graphs.

- All examples are now numbered for easier reference.

- Use of the calculator has been emphasized throughout the text. The treatment of logarithms and trigonometry reflects the ready access that students now have to scientific calculators.

- Review problem sets have been added. These problem sets appear at the end of the second section and each succeeding section in every chapter. The additional problems enable students to practice and maintain new skills acquired in previous sections as they proceed through the chapter.

Throughout the revision we have adhered to our original goal of producing a "no-nonsense" approach to precalculus mathematics. Since we have deliberately kept the number of topics reasonable, most of the material can easily be covered in a one-quarter or a one-semester course. Our style is informal, intuitive, and straightforward; we avoid a formal theorem/proof format.

In conclusion we would like to thank the many users of the first two editions, the editorial and production staffs at Wadsworth Publishing Company, and the following reviewers for their helpful comments and suggestions: Hugh B. Easler, College of William and Mary; Louis F. Hoezle, Bucks County Community College; Steve Scarborough, Loyola Marymount University; Ralph Bean, Stockton State College; Sandy Wager, Penn State University; Robert P. Boner, Western Maryland College; Johnny A. Johnson, University of Houston; Sandra Schrader, University of Richmond; and Dorothy Barrett, Trinity College.

Dennis G. Zill
Jacqueline M. Dewar
Warren S. Wright
Los Angeles, California

Basic Mathematics
for Calculus

0 Review of Algebra

0.1 Real Numbers

A student of precalculus should be familiar with most of the material discussed in this section.

Numbers: Integer, Rational, and Irrational

Recall that the set* of counting numbers, or **positive integers**, consists of

$$N = \{1, 2, 3, 4, \ldots\}.$$

The set N is a subset of the set of **integers**

$$I = \{\ldots, -3, -2, -1, 0, 1, 2, 3, \ldots\}.$$

The set I includes both the positive and negative integers and the number zero, which is neither positive nor negative. In turn, the set of integers is a subset of the set of **rational numbers**,

$$Q = \left\{ \frac{p}{q} \,\middle|\, p \text{ and } q \text{ are integers, } q \neq 0 \right\}.$$

This latter set Q consists of all fractions, such as

$$-\frac{1}{2}, \quad \frac{17}{5}, \quad \frac{4}{6}, \quad \frac{10}{-2} = -5, \quad \frac{6}{1} = 6, \quad \frac{0}{8} = 0, \quad \frac{-22}{-22} = 1.$$

Note that an expression such as $9/0$ is undefined since $q \neq 0$ in p/q.

The set of rational numbers is not sufficient to solve some of the more elementary algebraic or geometric problems.[†] For example, there is no rational number p/q for

---

* A brief review of set theory is given in Appendix A-1. The counting numbers are also called the **natural numbers**.

[†] The discovery of this fact (about 400 B.C.) caused a very serious problem for the group of Greek mathematicians known as the Pythagoreans. These individuals believed that rational "numbers constitute the entire heaven."

1

Figure 0.1

which

$$\left(\frac{p}{q}\right)^2 = 2.$$

(See Problem 58). Hence rational numbers cannot be used to describe the length of a diagonal of a unit square such as shown in Figure 0.1. By the Pythagorean theorem, the length of the diagonal d must satisfy

$$d^2 = (1)^2 + (1)^2 = 2.$$

We write $d = \sqrt{2}$ and call d "the square root of 2." As we have just indicated, $\sqrt{2}$ is not a rational number; rather it belongs to the set of **irrational numbers**. The numbers in this set cannot be expressed as a quotient of integers. Other examples of irrational numbers are $\sqrt{3}$, π, $\sqrt{5}/4$. When we study decimal representations, we shall discuss the distinction between rational and irrational numbers in greater detail.

If we denote the set of irrational numbers by the symbol H, then the set of **real numbers** R can be written as the union of two disjoint sets

$$R = Q \cup H.$$

We should also note that the set of real numbers R can be written as the union of three disjoint sets,

$$R = R^- \cup \{0\} \cup R^+,$$

where R^- is the set of **negative real numbers** and R^+ is the set of **positive real numbers**. Elements of the set

$$\{0\} \cup R^+$$

are called **nonnegative real numbers**.

Decimals   Every real number can be written in **decimal form**. For example,

$$\tfrac{1}{4} = 0.25$$

$$\tfrac{25}{7} = 3.571428571428\ldots$$

$$\tfrac{7}{3} = 2.3333\ldots$$

$$\pi = 3.141592654\ldots$$

$$\tfrac{131}{99} = 1.323232\ldots$$

$$\sqrt{2} = 1.41421356\ldots$$

Numbers such as 0.25 and 1.6 are said to be **terminating decimals**. We say that numbers such as

$$1.\overset{\text{repeats}}{\overbrace{32}}\ \overbrace{32}\ \overbrace{32}\ldots \quad \text{and} \quad 3.\overset{\text{repeats}}{\overbrace{571428}}\ \overbrace{571428}\ldots$$

are **repeating decimals**. It can be shown that every rational number has either a repeating or a terminating decimal representation. Conversely, every repeating or terminating decimal is a rational number. It is also a basic fact that every decimal number is a real number. Hence, it follows that the set of irrational numbers consists of all decimals that neither terminate nor repeat. Thus, π and $\sqrt{2}$ have nonrepeating and nonterminating decimal representations.

Example 1   Write $n = 0.9999\ldots$ as a quotient of integers.

Solution   We use a simple "trick." Since $n = 0.9999\ldots$, we have $10n = 9.9999\ldots$. Subtracting n from $10n$ gives

$$10n = 9.9999\ldots$$
$$\frac{n = 0.9999\ldots}{9n = 9},$$

which implies that $n = \frac{9}{9} = 1$.   ∎

In Example 1 note that we can also write the number $n = 1$ as $n = 1.0000\ldots$. This simply indicates that certain rational numbers do not possess unique decimal representations.

Example 2   Write $n = 5.46565\ldots$ as a quotient of integers.

Solution   First find multiples of n that have the same decimal part. We can use

$$1000n = 5465.6565\ldots \quad \text{and} \quad 10n = 54.6565\ldots.$$

Note that subtracting $10n$ from $1000n$ cancels the common decimal part. Thus

$$990n = 5411 \quad \text{and so} \quad n = \frac{5411}{990}.$$ ∎

Less Than and Greater Than   If we are given any finite set of real numbers, we are intuitively aware that these numbers can always be arranged in increasing order. For example, the reader should be able to arrange the numbers

$$\tfrac{16}{3}, 7, \sqrt{2}, \pi, -1, -3, 0$$

in order as

$$-3, -1, 0, \sqrt{2}, \pi, \tfrac{16}{3}, 7.$$

The idea that -1 is smaller than 7 is expressed by "-1 is less than 7" and is written symbolically as

$$-1 < 7.*$$

We may also say that "7 is greater than -1" and write

$$7 > -1.$$

For any two distinct real numbers a and b, there is always a third real number between them, for example their average, $(a + b)/2$. If $a = -2$ and $b = 6$, then the number

$$\frac{a + b}{2} = \frac{-2 + 6}{2} = 2$$

satisfies the relation

$$-2 < 2 < 6.$$

That is, the number 2 is simultaneously greater than -2 and less than 6. Similarly, for any two distinct points A and B on a straight line, there is always a third point between them. As shown in Figure 0.2, the midpoint M of the segment AB is halfway between the given points. These and other similarities between the set of real numbers and the set of points on a straight line suggest using a line to "picture" the set of real numbers. This can be done as follows.

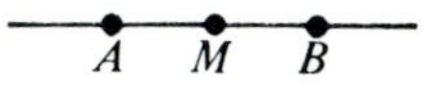

Figure 0.2

The Real Number Line

Given any horizontal straight line, we choose a point on the line to represent the number 0. This particular point is called the **origin**. If we now select a line segment of unit length as shown in Figure 0.3, each positive real number x can be represented by the point at a distance x units to the right of the origin. Similarly, each negative real number $-x$ can be represented by the point at a distance x units to the left of the origin. This association results in a one-to-one correspondence between the set of real numbers and the set of points on a straight line, called the **real number line** or **coordinate line**. For any given point P on the number line, the number p, which corresponds to this point, is called the **coordinate** of P.

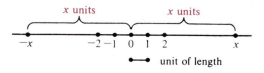

Figure 0.3

In general we shall not distinguish between a point on the number line and its coordinate. For example, we shall simply refer to the point on the line with coordinate 5 as "the point 5."

* A detailed discussion of inequalities is presented in Section 0.9.

Absolute Value   As shown in Figure 0.4, the distance from the point 3 to the origin is 3 units and the distance from the point -3 to the origin is 3 or $-(-3)$ units.

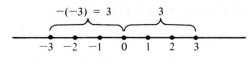

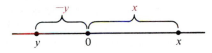

Figure 0.4

It follows from our discussion of the number line that, in general, the distance from any number to the origin is the "unsigned value" of that number. More precisely, as shown in Figure 0.5, for any positive real number x the distance from the point x to the origin is x units, but for any negative number y, the distance from the point y to the origin is $-y$ units.*

Figure 0.5

Of course, for $x = 0$, the distance to the origin is 0. The concept of the distance from a point on the number line to the origin is described by the **absolute value** of a number.

> ***Definition 0.1***   *For any real number a, the **absolute value** of a, denoted by $|a|$, is*
>
> $$|a| = \begin{cases} a & \textit{if a is nonnegative} \\ -a & \textit{if a is negative.} \end{cases} \qquad (1)$$

Example 3   **a.** Since 3 and $\sqrt{2}$ are positive numbers,

$$|3| = 3 \quad \text{and} \quad |\sqrt{2}| = \sqrt{2}.$$

b. But since -3 and $-\sqrt{2}$ are negative numbers,

$$|-3| = -(-3) = 3 \quad \text{and} \quad |-\sqrt{2}| = -(-\sqrt{2}) = \sqrt{2}. \qquad \blacksquare$$

Example 4   **a.** $|2 - 2| = |0| = 0$
b. $|2 - 5| = |-3| = -(-3) = 3$
c. $|2| - |-5| = 2 - [-(-5)] = 2 - 5 = -3 \qquad \blacksquare$

---

* We emphasize that y represents a *negative* number in this discussion, such as $y = -3$, and hence $-y$ is a *positive* number *even though y is preceded by a minus sign.*

Example 5   Find $|\sqrt{2} - 3|$.

Solution   To find $|\sqrt{2} - 3|$ we must first determine whether $\sqrt{2} - 3$ is positive or negative. Since $\sqrt{2} \approx 1.4$, we see that $\sqrt{2} - 3$ is a negative number. Hence from (1),

$$|\sqrt{2} - 3| = -(\sqrt{2} - 3) = -\sqrt{2} + 3 = 3 - \sqrt{2}.$$ ∎

For any real number x and its negative, $-x$, the distance to the origin is the same. In other words,

$$|x| = |-x|. \tag{2}$$

Distance Between Points   The concept of absolute value not only describes the distance from a point to the origin, it is also useful in defining the distance between two points on the number line.

> **Definition 0.2**   If a and b are two points on the real number line, *the distance from a to b is*
>
> $$d(a, b) = |b - a|. \tag{3}$$

Example 6   The distance from -5 to 2 is

$$d(-5, 2) = |2 - (-5)| = 7.$$ ∎

Example 7   The distance from 3 to $\sqrt{2}$ is

$$d(3, \sqrt{2}) = |\sqrt{2} - 3| = 3 - \sqrt{2}.$$ ∎

Example 8   The distance from $\sqrt{2}$ to 3 is

$$d(\sqrt{2}, 3) = |3 - \sqrt{2}| = 3 - \sqrt{2}.$$ ∎

Note that in general the distance $d(a, b)$ from a to b is the same as the distance $d(b, a)$ from b to a, since, by equation (2),

$$|b - a| = |-(b - a)| = |a - b|.$$

See Figure 0.6.

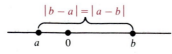

Figure 0.6

Coordinate of the Midpoint

In the following example we use the distance formula (3) to show that the coordinate m of the midpoint of a line segment is the average

$$m = \frac{a + b}{2}$$

of the coordinates of the endpoints of the segment. See Figure 0.7.

Example 9 The endpoints of the line segment AB shown in Figure 0.7 have coordinates a and b, respectively. Verify that the coordinate m of the midpoint M is $(a + b)/2$.

Solution Since M is the midpoint of AB, the distance from M to A equals the distance from M to B. So, from (3), the coordinate m satisfies

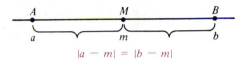

$$|a - m| = |b - m|, \qquad (4)$$

Figure 0.7

as shown in Figure 0.7. Because each point on the number line has a unique coordinate, we need only show that the equality (4) holds for $m = (a + b)/2$. We substitute $(a + b)/2$ for m and find that

$$|a - m| = \left| a - \frac{a + b}{2} \right| = \left| \frac{2a - (a + b)}{2} \right| = \left| \frac{a - b}{2} \right|$$

and

$$|b - m| = \left| b - \frac{a + b}{2} \right| = \left| \frac{2b - (a + b)}{2} \right| = \left| \frac{b - a}{2} \right|.$$

Since

$$\frac{a - b}{2} = -\left(\frac{b - a}{2} \right),$$

we have by (2),

$$\left| \frac{a - b}{2} \right| = \left| \frac{b - a}{2} \right|,$$

as desired. ∎

Example 10 The midpoint of the line segment joining the points 5 and -2 is

$$\frac{5 + (-2)}{2} = \frac{3}{2}.$$

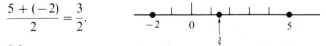

See Figure 0.8.

Figure 0.8

Exercise 0.1

1. Construct a number line and locate on it the points

$$0, -\tfrac{1}{2}, 1, -1, 2, -2, \tfrac{4}{3}, 2.5.$$

2. Construct a number line and locate on it the points

$$0, 1, -1, \sqrt{2}, -3, -\sqrt{2}, \sqrt{2} + 1.$$

[*Hint:* With a ruler and compass you can locate $\sqrt{2}$ as shown in Figure 0.9.]

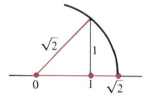

Figure 0.9

In Problem 3–10, use long division to find a decimal expression for the given fraction.

3. $\frac{7}{6}$ **4.** $\frac{11}{9}$ **5.** $\frac{5}{7}$ **6.** $\frac{13}{250}$

7. $\frac{5}{12}$ **8.** $\frac{74}{99}$ **9.** $\frac{27}{8}$ **10*.** $-\frac{15}{29}$

In Problems 11–18, write the given decimal as a quotient of two integers.

11. $0.181818\ldots$ **12.** $0.3333\ldots$ **13.** $1.24999\ldots$ **14.** $0.77777\ldots$

15. 7.512 **16.** $35.79252525\ldots$ **17.** 0.692 **18.** $-2.347347\ldots$

In Problems 19–36, find the given absolute value.

19. $|-7|$ **20.** $|7|$ **21.** $|22|$

22. $|-\frac{4}{3}|$ **23.** $|-x|$, x positive   **24.** $|\sqrt{5}|$

25. $|0.13|$ **26.** $|-\sqrt{5}|$ **27.** $|-|-3||$

28. $|7 - 2|$ **29.** $|2 - 6|$ **30.** $||6| - |2||$

31. $|\sqrt{5} - 3|$ **32.** $|-6| - |-2|$ **33.** $-(|9| - |-4|)$

34. $|\sqrt{6} - 5|$ **35.** $|a - b|$, $a < b$ **36.** $|h|$, h negative.

In Problems 37–44, (a) find the distance between the given points; (b) find the coordinate of the midpoint of the line segment joining the given points.

37. $2, 5$ **38.** $7, 3$ **39.** $-100, 255$ **40.** $0.6, 0.8$

41. $6, -4.5$ **42.** $-5, -8$ **43.** $-\frac{1}{4}, \frac{7}{4}$ **44.** $\frac{3}{2}, -\frac{3}{2}$

* The repeating part of a decimal can be very long.

In Problems 45–52, let m be the midpoint of the line segment whose left endpoint is a and whose right endpoint is b. Find the indicated quantities from the given data.

45. $m = 6$, $d(a, b) = 6$; a, b

46. $m = -8$, $d(a, b) = 12$; a, b

47. $m = 1$, $d(a, m) = 4$; a, b

48. $m = 0$, $d(m, b) = 6$; a, b

49. $a = 8$, $m = 14$; $b, d(a, b)$

50. $b = -2$, $d(a, m) = 10$; a, m

51. $a = 6$, $d(m, b) = \sqrt{2}$; b, m

52. $b = 18$, $m = 9$; $a, d(a, m)$

Calculator Problems   *Scientific calculators have a key which gives a decimal approximation to the number π. In Problems 53–56, use a calculator as an aid in finding the given absolute value.*

53. $|\pi - \frac{22}{7}|$

54. $\left|\dfrac{\pi}{2} - 1.57\right|$

55. $|6.28 - 2\pi|$

56. $\left|4.71 - \dfrac{3\pi}{2}\right|$

57. Is the product of two irrational numbers necessarily irrational? The quotient?

58. Show that $\sqrt{2}$ cannot be written as a quotient of integers. [*Hint:* Assume that there is a fraction, p/q, reduced to lowest terms, such that $(p/q)^2 = 2$. This simplifies to $p^2 = 2q^2$, which implies that p^2 and hence p is an even integer, say $p = 2r$. Make this substitution and consider $(2r/q)^2 = 2$. You should arrive at a contradiction of the fact that p/q was reduced to lowest terms.]

59. Show that the sum of an irrational number and a rational number is irrational. [*Hint:* Assume that it is a rational number. Use the fact that the sum and the difference of two rational numbers are rational.]

0.2 **Exponents and Radicals**

Integer Exponents

For any real number x and any positive integer n, the symbol x^n, read "x to the nth power," represents the product of n factors of x. Thus,

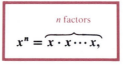

$$x^n = \overbrace{x \cdot x \cdots x,}^{n\ \text{factors}}$$

for any positive integer n. The number n is called the **exponent** of x and x is called the **base**. For example,

$$5^2 = 5 \cdot 5 = 25,$$

$$\left(-\frac{1}{2}\right)^5 = \left(-\frac{1}{2}\right)\left(-\frac{1}{2}\right)\left(-\frac{1}{2}\right)\left(-\frac{1}{2}\right)\left(-\frac{1}{2}\right) = -\frac{1}{32},$$

and

$$y^3 = y \cdot y \cdot y.$$

Also for any positive integer n, we define

$$x^{-n} = \frac{1}{x^n}, \qquad x \neq 0.$$

For example,

$$2^{-3} = \frac{1}{2^3} = \frac{1}{8},$$

$$\left(-\frac{1}{10}\right)^{-4} = \frac{1}{(-\frac{1}{10})^4} = \frac{1}{\frac{1}{10,000}} = 10,000,$$

and

$$(0.25)^{-2} = \frac{1}{(0.25)^2} = \frac{1}{0.0625} = 16.$$

For any $x \neq 0$, the **reciprocal of** x is

$$\frac{1}{x} = x^{-1}.$$

For example, the reciprocal of 3 is $\frac{1}{3}$ and the reciprocal of $-\frac{1}{4}$ is $\frac{1}{-\frac{1}{4}} = -4$.

For any nonzero base x it is convenient to define

$$x^0 = 1.$$

For example,

$$2^0 = 1, \qquad (-\tfrac{7}{2})^0 = 1, \quad \text{and} \quad (\sqrt{2} + \sqrt{3})^0 = 1.$$

Note: The expression 0^0 is not given any definition.

Laws of Exponents The student has probably seen the following laws of integer exponents. If m and n are integers and x and y are arbitrary numbers, then

$$x^m x^n = x^{m+n} \tag{1}$$

$$(x^m)^n = x^{mn} \tag{2}$$

$$(xy)^n = x^n y^n \tag{3}$$

$$\left(\frac{x}{y}\right)^n = \frac{x^n}{y^n} \tag{4}$$

$$\frac{x^m}{x^n} = x^{m-n}, \tag{5}$$

where it is assumed that x (or y) $\neq 0$, if x (or y) is in a denominator.

Example 1 **a.**
$$a^5 a^4 = a^{5+4} \qquad [\text{by } (1)]$$
$$= a^9.$$

b.
$$a^5 a^{-4} = a^{5+(-4)} \qquad [\text{by } (1)]$$
$$= a^1$$
$$= a.$$

Example 2 **a.**
$$(b^3)^2 = b^{3(2)} \qquad [\text{by } (2)]$$
$$= b^6.$$

b.
$$(b^3)^{-2} = b^{3(-2)} \qquad [\text{by } (2)]$$
$$= b^{-6}$$
$$= \frac{1}{b^6}.$$

Example 3
$$(5x)^3 = 5^3 x^3 \qquad [\text{by } (3)]$$
$$= 125x^3.$$

Example 4 **a.**
$$\left(\frac{y}{4}\right)^5 = \frac{y^5}{4^5} \qquad [\text{by } (4)]$$
$$= \frac{y^5}{1024}.$$

b.
$$\left(\frac{y}{4}\right)^{-5} = \frac{y^{-5}}{4^{-5}} \qquad [\text{by } (4)]$$
$$= \frac{\dfrac{1}{y^5}}{\dfrac{1}{4^5}}$$
$$= \frac{4^5}{y^5}$$
$$= \frac{1024}{y^5}.$$

Example 5 **a.**
$$\frac{a^5}{a^3} = a^{5-3} \qquad [\text{by } (5)]$$
$$= a^2.$$

b.
$$\frac{a^{-5}}{a^{-3}} = a^{-5-(-3)} \qquad [\text{by (5)}]$$
$$= a^{-2}$$
$$= \frac{1}{a^2}.$$

■

As shown in the following examples, the laws of exponents are useful in simplifying algebraic expressions.

Example 6
$$(5a^5b^3)(7a^{-2}b^4) = 5(7)a^{5+(-2)}b^{3+4}$$
$$= 35a^3b^7.$$

■

Example 7
$$\frac{(-6xy^2)^3}{x^2y^5} = \frac{(-6)^3x^3(y^2)^3}{x^2y^5}$$
$$= \frac{-216x^3y^{2(3)}}{x^2y^5}$$
$$= -\frac{216x^3y^6}{x^2y^5}$$
$$= -216x^{3-2}y^{6-5}$$
$$= -216xy.$$

■

Example 8
$$\left(\frac{x^3y^{-2}}{y^{-3}}\right)^{-1} = (x^3y^{-2-(-3)})^{-1}$$
$$= (x^3y^1)^{-1}$$
$$= \frac{1}{x^3y}.$$

■

Scientific Notation   Recall that a number such as 257 means
$$(2 \times 100) + (5 \times 10) + (7 \times 1) = 2(10^2) + 5(10^1) + 7(10^0)$$
$$= 10^2[2 + 5(10^{-1}) + 7(10^{-2})]$$
$$= 10^2(2 + 0.5 + 0.07)$$
$$= 2.57 \times 10^2.$$

Every positive real number N can be written as a multiple of a power of 10,

$$N = n \times 10^c,$$

where n is a decimal number satisfying $1 \le n < 10$ and c is an integer. A number written in this form is said to be in **scientific notation**. For example,

$$6{,}900{,}000 = 6.9 \times 10^6 \quad \text{and} \quad 0.0000000537 = 5.37 \times 10^{-8}.$$

The positive exponent 6 in the form 6.9×10^6 means that we move the decimal point six places to the *right*, whereas the negative exponent -8 in the form 5.37×10^{-8} means that we move the decimal point eight places to the *left*.

Example 9 The distance to the Andromeda galaxy (Messier 31), located in the direction of the constellation Andromeda, is 2,200,000 light years from our own Milky Way galaxy. A light year is a measure of distance equivalent to (approximately) 6 million million miles. Thus, in the compact form of scientific notation the distance to the Andromeda galaxy is

$$(2{,}200{,}000) \times (6{,}000{,}000{,}000{,}000) = (2.2 \times 10^6)(6 \times 10^{12})$$

$$= 13.2 \times 10^{18}$$

$$= 1.32 \times 10^{19} \text{ miles.} \qquad \blacksquare$$

Radicals For any positive integer p and any real number r, we call r a **pth root** of a number x if

$$r^p = x,$$

and we write

$$r = \sqrt[p]{x}.$$

The symbol $\sqrt[p]{x}$ is called a **radical**; the number p is the **index** of the radical; and x is called the **radicand**.

If p is odd, then for any value of x there is exactly *one* real pth root of x. For example,

$$\sqrt[3]{125} = 5, \quad \sqrt[5]{-32} = -2, \quad \text{and} \quad \sqrt[3]{\tfrac{1}{8}} = \tfrac{1}{2}.$$

If p is even and x is positive, there are *two* real pth roots of x. However, it is common practice to reserve the symbol $\sqrt[p]{x}$ for the positive pth root and to denote the negative pth root by $-\sqrt[p]{x}$. For example,

$$\sqrt[4]{\tfrac{1}{81}} = \tfrac{1}{3} \quad \text{and} \quad -\sqrt[4]{\tfrac{1}{81}} = -\tfrac{1}{3},$$

$$\sqrt[6]{1{,}000{,}000} = 10 \quad \text{and} \quad -\sqrt[6]{1{,}000{,}000} = -10.$$

If the index p is 2, it is usually omitted from the radical and $\sqrt{x}$ is called the **square root** of x. For example, $\sqrt{4} = 2$. Since square roots are often misunderstood, it is worth repeating that

$$\sqrt[p]{x} \text{ is a positive number when } p \text{ is even and } x \text{ is positive.}$$

Thus, $\sqrt{4}$ is *not* ± 2. This is the same convention used in all calculators; when a positive number is entered, the $\boxed{\sqrt{}}$ key will yield a positive answer. If we want the negative square root of, say, 4, we must write a minus sign in front of the radical: $-\sqrt{4} = -2$. Finally, if p is even and x is negative, there is no real pth root of x.*

It should be noted that for any index p, $\sqrt[p]{0} = 0$.

Rules for Radicals The basic **rules for radicals** are: For positive integers m and n,

$$(\sqrt[n]{x})^n = x \tag{6}$$

$$\sqrt[n]{x}\,\sqrt[n]{y} = \sqrt[n]{xy} \tag{7}$$

$$\frac{\sqrt[n]{x}}{\sqrt[n]{y}} = \sqrt[n]{\frac{x}{y}} \tag{8}$$

$$\sqrt[m]{\sqrt[n]{x}} = \sqrt[mn]{x}, \tag{9}$$

where it is assumed that the value of x (or y) is such that these radicals are real numbers.

It can also be shown that

$$\sqrt[n]{x^n} = x, \qquad \text{if } n \text{ is odd}, \tag{10}$$

$$\sqrt[n]{x^n} = |x|, \qquad \text{if } n \text{ is even}. \tag{11}$$

It is a common mistake to simplify $\sqrt{x^2}$ as x. This is valid only for *nonnegative x*. For example, if $x = -3$, we see that

$$\sqrt{(-3)^2} = \sqrt{9} = 3 \neq -3.$$

The correct result is given by (11):

$$\sqrt{(-3)^2} = |-3| = 3.$$

In the following examples, we assume that each variable is positive so that the pth roots are always defined.

Example 10

$$\sqrt[3]{40} = \sqrt[3]{8 \cdot 5}$$
$$= \sqrt[3]{8} \cdot \sqrt[3]{5} \qquad [\text{by (7)}]$$
$$= 2\sqrt[3]{5}. \qquad \blacksquare$$

---

* An even root of a negative number, for example $\sqrt{-5}$, is a complex number. Complex numbers are discussed in Appendix A-3.

Example 11

$$\sqrt[4]{160,000} = \sqrt[4]{16(10,000)}$$
$$= \sqrt[4]{16} \cdot \sqrt[4]{10,000} \qquad [\text{by } (7)]$$
$$= 2 \cdot 10$$
$$= 20.$$

∎

Example 12

$$\sqrt[3]{\frac{27y^7}{x^3}} = \frac{\sqrt[3]{27y^7}}{\sqrt[3]{x^3}} \qquad [\text{by } (8)]$$
$$= \frac{\sqrt[3]{27} \cdot \sqrt[3]{y^7}}{\sqrt[3]{x^3}} \qquad [\text{by } (7)]$$
$$= \frac{3 \cdot \sqrt[3]{y^6} \cdot \sqrt[3]{y}}{x} \qquad [\text{by } (7)]$$
$$= \frac{3y^2 \sqrt[3]{y}}{x}.$$

∎

Rationalizing the Denominator

When the denominator of an expression contains one or more radicals, it is customary to simplify the expression by eliminating these radicals. This process is called **rationalizing the denominator**. For example,

$$\frac{1}{\sqrt{5}} = \frac{1}{\sqrt{5}} \cdot \frac{\sqrt{5}}{\sqrt{5}}$$
$$= \frac{\sqrt{5}}{5}$$

and

$$\frac{1}{\sqrt[3]{2}} = \frac{1}{\sqrt[3]{2}} \cdot \frac{(\sqrt[3]{2})^2}{(\sqrt[3]{2})^2}$$
$$= \frac{\sqrt[3]{2} \cdot \sqrt[3]{2}}{(\sqrt[3]{2})^3}$$
$$= \frac{\sqrt[3]{4}}{2}.$$

Frequently, we must rationalize a denominator containing an expression such as $\sqrt{x} + \sqrt{y}$ or $\sqrt{x} - \sqrt{y}$. Since for any real numbers a and b,

$$(a - b)(a + b) = a^2 - b^2,$$

we observe that the product

$$(\sqrt{x} + \sqrt{y})(\sqrt{x} - \sqrt{y}) = (\sqrt{x})^2 - (\sqrt{y})^2$$
$$= x - y$$

is free of radicals. The use of this technique is illustrated in the following examples.

Example 13   Rationalize the denominator of the expression

$$\frac{1}{\sqrt{x} + \sqrt{y}}.$$

Solution   To eliminate radicals from the denominator without changing the value of the expression, we multiply numerator and denominator by $\sqrt{x} - \sqrt{y}$:

$$\frac{1}{\sqrt{x} + \sqrt{y}} \cdot \frac{\sqrt{x} - \sqrt{y}}{\sqrt{x} - \sqrt{y}} = \frac{\sqrt{x} - \sqrt{y}}{(\sqrt{x})^2 - (\sqrt{y})^2}$$

$$= \frac{\sqrt{x} - \sqrt{y}}{x - y}.$$

$\blacksquare$

Example 14   Rationalize the denominator of the expression

$$\frac{\sqrt{3} + \sqrt{2}}{\sqrt{3} - \sqrt{2}}.$$

Solution

$$\frac{\sqrt{3} + \sqrt{2}}{\sqrt{3} - \sqrt{2}} \cdot \frac{\sqrt{3} + \sqrt{2}}{\sqrt{3} + \sqrt{2}} = \frac{\sqrt{3}\sqrt{3} + \sqrt{3}\sqrt{2} + \sqrt{2}\sqrt{3} + \sqrt{2}\sqrt{2}}{(\sqrt{3})^2 - (\sqrt{2})^2}$$

$$= \frac{3 + \sqrt{2}\sqrt{3} + \sqrt{2}\sqrt{3} + 2}{3 - 2}$$

$$= \frac{5 + \sqrt{6} + \sqrt{6}}{1}$$

$$= 5 + 2\sqrt{6}.$$

$\blacksquare$

The concept of the pth root of a number enables us to extend the definition of x^n from integral to rational exponents.

Rational Exponents

For any real number x and any positive integer n, we define

$$x^{1/n} = \sqrt[n]{x},$$

provided that $\sqrt[n]{x}$ is a real number. And for any integer m such that m/n is reduced to lowest terms

$$x^{m/n} = (x^{1/n})^m.$$

This definition is illustrated in the following examples.

Example 15 **a.** $(25)^{1/2} = \sqrt{25} = 5.$

              b. $(64)^{1/3} = \sqrt[3]{64} = 4.$ ■

Example 16 $(0.09)^{5/2} = [(0.09)^{1/2}]^5 = (\sqrt{0.09})^5 = (0.3)^5 = 0.00243.$ ■

Example 17 $(-27)^{-5/3} = [(-27)^{1/3}]^{-5} = [\sqrt[3]{-27}]^{-5} = (-3)^{-5} = -\dfrac{1}{243}.$ ■

Example 18 $5^{7/2} = (5^{1/2})^7 = (\sqrt{5})^7 = (\sqrt{5})^6\sqrt{5} = 125\sqrt{5}.$ ■

For $x > 0$ it can be shown that

$$(x^m)^{1/n} = (x^{1/n})^m = x^{m/n}.$$

However, for $x < 0$ and certain choices of m and n, $(x^{1/n})^m$ is not defined, while the expression $(x^m)^{1/n}$ is defined. For example, if $x = -9$, $m = 2$, and $n = 2$, we have

$$(x^m)^{1/n} = [(-9)^2]^{1/2} = 81^{1/2} = 9.$$

But

$$(x^{1/n})^m = [(-9)^{1/2}]^2 = [\sqrt{-9}]^2$$

involves the square root of a *negative* number and hence is not real. Note also that

$$x^{m/n} = (-9)^{2/2} = (-9)^1 = -9.$$

It is true, however, that when all three expressions $x^{m/n}$, $(x^{1/n})^m$, and $(x^m)^{1/n}$ are real numbers, then they are equal. For the remainder of this section we shall assume that all variable bases represent positive numbers so that all rational powers are defined.

The laws for integer exponents (1)–(5) also hold true for rational exponents. For any rational numbers r and s and any real numbers x and y,

$$x^r x^s = x^{r+s} \tag{12}$$

$$(x^r)^s = x^{rs} \tag{13}$$

$$(xy)^r = x^r y^r \tag{14}$$

$$\left(\frac{x}{y}\right)^r = \frac{x^r}{y^r} \tag{15}$$

$$\frac{x^r}{x^s} = x^{r-s}, \tag{16}$$

provided that each expression represents a real number.

As shown in the following examples, these laws enable us to simplify algebraic expressions. Recall that we are assuming that the variables represent positive numbers.

Example 19

$$(3x^{1/2})(2x^{1/5}) = 3(2)x^{1/2}x^{1/5}$$
$$= 6x^{1/2 + 1/5} \qquad [\text{by (12)}]$$
$$= 6x^{(5 + 2)/10}$$
$$= 6x^{7/10}.$$

Example 20

$$(a^2b^{-8})^{1/4} = (a^2)^{1/4}(b^{-8})^{1/4} \qquad [\text{by (14)}]$$
$$= a^{2/4}b^{-8/4} \qquad [\text{by (13)}]$$
$$= a^{1/2}b^{-2}$$
$$= \frac{a^{1/2}}{b^2}.$$

Example 21

$$\frac{x^{2/3}y^{1/2}}{x^{1/4}y^{3/2}} = x^{2/3 - 1/4}y^{1/2 - 3/2} \qquad [\text{by (16)}]$$
$$= x^{(8 - 3)/12}y^{-1}$$
$$= \frac{x^{5/12}}{y}.$$

Example 22

$$\left(\frac{3x^{3/4}}{y^{1/3}}\right)^3 = \frac{(3x^{3/4})^3}{(y^{1/3})^3} \qquad [\text{by (15)}]$$
$$= \frac{3^3(x^{3/4})^3}{(y^{1/3})^3} \qquad [\text{by (14)}]$$
$$= \frac{27x^{9/4}}{y}. \qquad [\text{by (13)}]$$

When simplifying an expression such as $(3x^{3/4})^3$ in the previous example, it is a common mistake to *incorrectly* apply (14) and obtain

$$3(x^{3/4})^3 = 3x^{9/4} \quad \text{instead of} \quad 3^3(x^{3/4})^3 = 27x^{9/4}.$$

As shown in the next two examples, it is usually easier to work with rational exponents than radicals.

Example 23   Write $\sqrt[3]{16}/\sqrt{2}$ as a single radical.

Solution   We rewrite the expression using rational exponents:

$$\frac{\sqrt[3]{16}}{\sqrt{2}} = \frac{16^{1/3}}{2^{1/2}}.$$

Now we find a common base and use the properties of rational exponents to simplify the resulting expression:

$$\frac{16^{1/3}}{2^{1/2}} = \frac{(2^4)^{1/3}}{2^{1/2}}$$

$$= \frac{2^{4/3}}{2^{1/2}}$$

$$= 2^{4/3 - 1/2}$$

$$= 2^{5/6}$$

$$= (2^5)^{1/6}$$

$$= \sqrt[6]{2^5}$$

$$= \sqrt[6]{32}.$$ ∎

Example 24   Write $\sqrt{x\sqrt[4]{x}}$ as a single radical.

Solution   We rewrite $\sqrt{x\sqrt[4]{x}}$ using rational exponents and then simplify.

$$\sqrt{x\sqrt[4]{x}} = (x\sqrt[4]{x})^{1/2}$$

$$= (x \cdot x^{1/4})^{1/2}$$

$$= (x^{5/4})^{1/2}$$

$$= x^{5/8}$$

$$= \sqrt[8]{x^5}.$$ ∎

Exercise 0.2

Throughout this exercise set, assume that all variables represent positive numbers. In Problems 1–4, write the given expression with a positive exponent.

1. $\dfrac{1}{8 \cdot 8 \cdot 8}$

2. $-3 \cdot 3 \cdot 3$

3. $2y \cdot 2y \cdot 2y \cdot 2y$

4. $(-5) \cdot (-5) \cdot (-5)$

In Problems 5–8, write the given expression with a negative exponent.

5. $\dfrac{x^2}{y^2}$

6. $\dfrac{1}{4^5}$

7. $\dfrac{1}{x^3}$

8. x^6

In Problems 9–18, find the indicated numbers.

9. a. 3^4 b. 3^{-4} 10. a. $\left(\frac{1}{3}\right)^3$ b. $\left(-\frac{1}{3}\right)^{-3}$

11. a. $(-7)^2$ b. $(-7)^{-2}$ 12. a. $\left(-\frac{2}{3}\right)^5$ b. $\left(-\frac{2}{3}\right)^{-5}$

13. a. $(5)^0$ b. $(-5)^0$ 14. a. $\dfrac{0^1}{1^0}$ b. $\dfrac{1^0}{2^0}$

15. $2^{-1} - 2^1$

16. $\dfrac{2^{-2}}{3^{-3}}$

17. $\dfrac{2^{-1} - 3^{-1}}{2^{-1} + 3^{-1}}$

18. $\dfrac{(-1)^5 - (2)^6}{(-1)^{-1}}$

In Problems 19–42, eliminate negative exponents and simplify.

19. $x^6 x^{-2}$

20. $2^{10} 2^{12}$

21. $(7x^4)(-3x^2)$

22. $(-5x^2y^3)(3xy^{-2})$

23. $\dfrac{2^8}{2^3}$

24. $\dfrac{3^4}{3^{-2}}$

25. $\dfrac{10^{-7}}{10^4}$

26. $\dfrac{35y^8x^5}{-21y^{-1}x^9}$

27. $(5x)^2$

28. $(-4x)^3$

29. $(5^2)^3$

30. $(x^4)^{-5}$

31. $(4x^2y^{-1})^3$

32. $(3x^2y^4)^{-2}$

33. $x^2x^3x^{-4}$

34. $\dfrac{-x^5(y^2)^3}{(xy)^2}$

35. $\dfrac{(7a^2b^3)^2}{a^3b^5}$

36. $\dfrac{(-4x^5y^{-2})^3}{x^7y^{-3}}$

37. $(-3xy^5)^2(x^3y)^{-1}$

38. $\left(\dfrac{a^4b^{-5}}{b^2}\right)^{-1}$

39. $\left(\dfrac{a^3b^3}{b^{-2}}\right)^2$

40. $(-x^2y^4)^3(x^3y^{-1})^2$

41. $\dfrac{-xy^2z^3}{(xy^2z^3)^{-1}}$

42. $\dfrac{(3abc)^3}{(2a^{-1}b^{-2}c)^2}$

In Problems 43–48, write the given number in scientific notation.

43. 2371

44. 0.00359

45. 2,453,000

46. 0.0424242...

47. 675×10^3

48. $2890 \times 10,000$

49. Newton's universal gravitational constant is 0.0000000000667 N · m²/kg². Express this number in scientific notation.

50. The closest known star to our solar system is called Proxima Centauri. Its distance is (approximately) 1.3 parsecs. If one parsec is equivalent to 206,000 astronomical units and one astronomical unit is equivalent to 150 million kilometers, express the distance to Proxima Centauri in terms of kilometers.

In Problems 51–58, evaluate the given radical.

51. $\sqrt[3]{-125}$

52. $\sqrt[5]{100,000}$

53. $\sqrt[4]{0.0001}$

54. $\sqrt{\frac{1}{16}}$

55. $\sqrt[4]{\frac{1}{4}}\sqrt[4]{\frac{1}{4}}$

56. $\sqrt[3]{16}$

57. $\sqrt[4]{x^8y^4}$

58. $\sqrt[3]{8x^3y^5}$

In Problems 59–66, rationalize the denominator of the given expression.

59. $\dfrac{2}{\sqrt{3}}$

60. $\dfrac{1}{\sqrt{7}}$

61. $\dfrac{1}{\sqrt[3]{10}}$

62. $\dfrac{2}{\sqrt[4]{4}}$

63. $\dfrac{\sqrt{2}-\sqrt{5}}{\sqrt{2}+\sqrt{5}}$

64. $\dfrac{\sqrt{3}+\sqrt{5}}{\sqrt{3}-\sqrt{5}}$

65. $\dfrac{\sqrt{x}-\sqrt{y}}{\sqrt{x}+\sqrt{y}}$

66. $\dfrac{1}{\sqrt{a}-\sqrt{b}}$

In Problems 67–70, rationalize the numerator of the given expression.

67. $\dfrac{2+\sqrt{10}}{6}$

68. $\dfrac{\sqrt{5}-\sqrt{8}}{3}$

69. $\dfrac{\sqrt{x+h}-\sqrt{x}}{h}$

70. $\dfrac{\sqrt{3(x+h)+1}-\sqrt{3x+1}}{h}$

In Problems 71–76, find the indicated numbers.

71. a. $(49)^{1/2}$ **b.** $(49)^{-1/2}$

72. a. $(-8)^{1/3}$ **b.** $(-8)^{-1/3}$

73. a. $(0.04)^{7/2}$ **b.** $(0.04)^{-7/2}$

74. a. $(0.125)^{5/3}$ **b.** $(0.0001)^{-1/2}$

75. a. $(27)^{-1/3}$ **b.** $\left(-\frac{8}{27}\right)^{2/3}$

76. a. $\left(\frac{1}{64}\right)^{2/3}$ **b.** $(64)^{-3/2}$

In Problems 77–88, eliminate negative exponents and simplify.

77. $x^{1/2}x^{1/4}x^{1/8}x^{-3/8}$

78. $(100x^4)^{-3/2}$

79. $(a^2b^4)^{1/4}$

80. $(4x^4y^{-6})^{1/2}$

81. $(25x^{1/3}y)^{3/2}$

82. $(4x^{1/2})(3x^{1/3})$

83. $\left(\dfrac{2x^{1/2}}{z^{-1/6}y^{2/3}}\right)^6$

84. $\left(\dfrac{-y^{1/2}}{y^{-1/2}}\right)^{-1}$

85. $[(-27a^3b^{-6})^{1/3}]^2$

86. $a^{1/3}(a^{2/3}+a^{5/3}-a^{-1/3})$ **87.** $[2(w^{-2/3})^6]^{-1}$

88. $\dfrac{4x^{1/2}}{(8x)^{1/3}}$

In Problems 89–94, rewrite the expression as a single radical.

89. $\sqrt{5}\sqrt[3]{2}$

90. $\sqrt[3]{4}\sqrt{2}$

91. $\dfrac{\sqrt[3]{16}}{\sqrt[6]{4}}$

92. $\dfrac{\sqrt[3]{81}}{\sqrt[3]{3}}$

93. $\sqrt{x\sqrt{x}}$

94. $\sqrt{x\sqrt[3]{x}}$

Review Problems

95. Write $-0.7454545\ldots$ as a quotient of integers.

96. Write the expression $|x+5|$ without absolute value symbols if $x<-5$.

97. True or false: $|-3x|=3|x|$ for any real number x.

98. True or false: $-\frac{1}{3}$ is greater than $-\frac{1}{2}$.

99. Find the midpoint m of the line segment joining the points x and $x + h$, $h > 0$.

100. Let $m = 20$ be the midpoint of the line segment joining the points a and b, and let $d(m, b) = 8$. Find a and b.

0.3 The Binomial Theorem

Binomials

In the next two sections we examine certain consequences of the distributive law,*

$$A(B + C) = AB + AC. \tag{1}$$

By using the distributive law and the laws of exponents, we can multiply out, or expand, positive integer powers of the **binomial** $a + b$. For example,

$$(a + b)^2 = (a + b)(a + b)$$
$$= (a + b)a + (a + b)b$$
$$= a^2 + ba + ab + b^2$$
$$= a^2 + 2ab + b^2 \tag{2}$$

and

$$(a + b)^3 = (a + b)(a + b)^2$$
$$= (a + b)(a^2 + 2ab + b^2)$$
$$= (a + b)a^2 + (a + b)2ab + (a + b)b^2$$
$$= a^3 + ba^2 + a2ab + b2ab + ab^2 + b^3$$
$$= a^3 + a^2b + 2a^2b + 2ab^2 + ab^2 + b^3$$
$$= a^3 + 3a^2b + 3ab^2 + b^3. \tag{3}$$

Example 1   Expand: **a.** $(2x + 5)^2$ **b.** $(2x + 5)^3$

Solutions   **a.** From (2) with $a = 2x$ and $b = 5$, we have

$$(2x + 5)^2 = (2x)^2 + 2(2x)(5) + (5)^2$$
$$= 4x^2 + 20x + 25.$$

b. From (3) with $a = 2x$ and $b = 5$, we have

$$(2x + 5)^3 = (2x)^3 + 3(2x)^25 + 3(2x)(5)^2 + (5)^3$$
$$= 8x^3 + 60x^2 + 150x + 125. \qquad \blacksquare$$

* For a discussion of the properties of real numbers, see Appendix A-1.

Example 2   Expand $(\sqrt{2x} + \sqrt{6y})^2$.

Solution   If we identify $a = \sqrt{2x}$ and $b = \sqrt{6y}$, then

$$(\sqrt{2x} + \sqrt{6y})^2 = (\sqrt{2x})^2 + 2\sqrt{2x}\sqrt{6y} + (\sqrt{6y})^2$$
$$= 2x + 2\sqrt{12xy} + 6y$$
$$= 2x + 2\sqrt{4 \cdot 3xy} + 6y$$
$$= 2x + 4\sqrt{3xy} + 6y. \qquad \blacksquare$$

For convenience, we list three important expansions:

$$\boxed{\begin{aligned} (a + b)^2 &= a^2 + 2ab + b^2 \\ (a + b)^3 &= a^3 + 3a^2b + 3ab^2 + b^3 \\ (a + b)^4 &= a^4 + 4a^3b + 6a^2b^2 + 4ab^3 + b^4. \end{aligned}} \qquad (4)$$

We note that the products $(a - b)^2$, $(a - b)^3$, and $(a - b)^4$ are included in (4) since $a - b = a + (-b)$.

Example 3   Expand $(y^{-2} - 1)^4$

Solution   From (4) and the laws of exponents, it follows that

$$(y^{-2} - 1)^4 = [y^{-2} + (-1)]^4$$
$$= (y^{-2})^4 + 4(y^{-2})^3(-1) + 6(y^{-2})^2(-1)^2 + 4(y^{-2})(-1)^3 + (-1)^4$$
$$= y^{-8} - 4y^{-6} + 6y^{-4} - 4y^{-2} + 1. \qquad \blacksquare$$

The Pattern of the Exponents

In the expansion of $(a + b)^n$ the exponents of a and b follow a definite pattern. For example, in $(a + b)^4$, we observe that

decreasing by 1

$$a^4 + 4a^3b^1 + 6a^2b^2 + 4a^1b^3 + b^4. \qquad (5)$$

increasing by 1

Thus, the exponents of *a decrease* by one from the first through the fourth term and the exponents of *b increase* by one from the second through the fifth term. To extend this pattern we may consider the first and last terms to be multiplied by b^0 and a^0, respectively; that is,

$$a^4b^0 + 4a^3b + 6a^2b^2 + 4ab^3 + a^0b^4.$$

We also note that the sum of the exponents in each term of the expansion $(a + b)^4$ is 4.

The Coefficients

The coefficients in the expansion of $(a + b)^n$ are displayed below for a few values of n.

$$
\begin{array}{cccccccccc}
n = 0 & & & & & 1 & & & & \\
n = 1 & & & & 1 & & 1 & & & \\
n = 2 & & & 1 & & 2 & & 1 & & \\
n = 3 & & 1 & & 3 & & 3 & & 1 & \\
n = 4 & 1 & & 4 & & 6 & & 4 & & 1
\end{array}
$$

Note that each number in the interior of the triangular array is the *sum* of the two numbers directly above it. Thus, the next line in the array can be obtained as follows:

$$
\begin{array}{ccccccccccc}
1 & & 4 & & 6 & & 4 & & 1 & \\
& 1 & & 5 & & 10 & & 10 & & 5 & & 1.
\end{array}
$$

As you might expect, these numbers are the coefficients of the powers of a and b in the expansion of $(a + b)^5$. This array of numbers is known as **Pascal's triangle**.

Example 4 Expand $(3 - x)^5$.

Solution From the discussion above we can write

$$(3 - x)^5 = [3 + (-x)]^5$$

$$= \mathbf{1}(3)^5 + \mathbf{5}(3)^4(-x) + \mathbf{10}(3)^3(-x)^2 + \mathbf{10}(3)^2(-x)^3 + \mathbf{5}(3)(-x)^4 + \mathbf{1}(-x)^5$$

$$= 243 - 405x + 270x^2 - 90x^3 + 15x^4 - x^5. \qquad \blacksquare$$

Example 5 Expand $(x + y + z)^3$.

Solution The trinomial $x + y + z$ can be treated as a binomial by grouping the terms as $(x + y) + z$. Thus,

$$(x + y + z)^3 = [(x + y) + z]^3$$

$$= \mathbf{1}(x + y)^3 + \mathbf{3}(x + y)^2 z + \mathbf{3}(x + y)z^2 + \mathbf{1}z^3$$

$$= (x^3 + 3x^2 y + 3xy^2 + y^3) + 3(x^2 + 2xy + y^2)z + 3(x + y)z^2 + z^3$$

$$= x^3 + 3x^2 y + 3xy^2 + y^3 + 3x^2 z + 6xyz + 3y^2 z + 3xz^2 + 3yz^2 + z^3. \qquad \blacksquare$$

Factorial Notation

Before giving a general formula for the expansion of $(a + b)^n$, it will be helpful to introduce **factorial notation**. The symbol $r!$ is defined for any positive integer r as the product

$$r! = r(r - 1)(r - 2) \cdots 3 \cdot 2 \cdot 1,$$

and is read "*r* factorial." For example,

$$4! = 4 \cdot 3 \cdot 2 \cdot 1 = 24.$$

Also it is convenient to define $0! = 1.$

The Binomial Theorem

The formula for the expansion of $(a + b)^n$ for any positive integer n is given by the **binomial theorem**:

$$(a + b)^n = a^n + \frac{n}{1!}a^{n-1}b + \frac{n(n-1)}{2!}a^{n-2}b^2 + \cdots$$

$$+ \frac{n(n-1)\cdots(n-k+1)}{k!}a^{n-k}b^k + \cdots + b^n. \qquad (6)$$

There are $n + 1$ terms in the expansion of $(a + b)^n$. The $(k + 1)$st term is

$$\frac{n(n-1)\cdots(n-k+1)}{k!}a^{n-k}b^k. \qquad (7)$$

For $k = 0, 1, \ldots, n$, the binomial coefficients of $a^{n-k}b^k$ are the same as those obtained from Pascal's triangle. For a proof of the binomial theorem using mathematical induction, see Appendix A-2.

Example 6 Using the binomial theorem (6) to expand $(a + b)^4$, we have

$$(a + b)^4 = a^4 + \frac{4}{1!}a^{4-1}b + \frac{4(3)}{2!}a^{4-2}b^2 + \frac{4(3)(2)}{3!}a^{4-3}b^3 + \frac{4(3)(2)(1)}{4!}b^4$$

$$= a^4 + 4a^3b + \frac{12}{2}a^2b^2 + \frac{24}{6}ab^3 + \frac{24}{24}b^4$$

$$= a^4 + 4a^3b + 6a^2b^2 + 4ab^3 + b^4. \qquad \blacksquare$$

Example 7 Find the third term in the expansion of $(r + s)^5$.

Solution From (7) we see that the third term corresponds to $k = 2$. Thus, with $n = 5, a = r$, and $b = s$, we have

$$\frac{5 \cdot 4}{2!}r^{5-2}s^2 = 10r^3s^2. \qquad \blacksquare$$

Example 8 Find the sixth term in the expansion of $(x^2 - 2y)^7$.

Solution From (7) with $k = 5$, $n = 7$, $a = x^2$, and $b = -2y$, we have that

$$\frac{7 \cdot 6 \cdot 5 \cdot 4 \cdot 3}{5!}(x^2)^{7-5}(-2y)^5 = 21x^4(-32y^5)$$

$$= -672x^4 y^5.$$

Exercise 0.3

In Problems 1–16, expand the given expression.

1. $(x + y)^2$ **2.** $(x - y)^2$ **3.** $(2x + 1)^2$ **4.** $(3x - 5)^2$

5. $(2a^2 - 3)^2$ **6.** $(3x + 4y)^3$ **7.** $(r - s)^3$ **8.** $(x^{-1} + y^{-1})^3$

9. $(x^2 - y^2)^3$ **10.** $(\sqrt{2} - \sqrt{3})^4$ **11.** $(x^{-1} + 1)^4$ **12.** $(2 - y^2)^4$

13. $(x^2 + y^2)^5$ **14.** $\left(2x + \dfrac{1}{x}\right)^5$ **15.** $(x + 2y + 1)^2$ **16.** $(a - b - c)^3$

In Problems 17–24, evaluate the given expression.

17. $3!$ **18.** $5!$ **19.** $\dfrac{3!}{5!}$ **20.** $\dfrac{5!}{4!}$

21. $2!3!$ **22.** $\dfrac{n!}{(n-1)!}$ **23.** $\dfrac{5!}{2!(5-2)!}$ **24.** $\dfrac{2 \cdot 3!}{(2 \cdot 3)!}$

In Problems 25–34, find the indicated term in the expansion of the given expression.

25. Sixth term, $(a + b)^6$ **26.** Second term, $(x - y)^5$

27. Fourth term, $(x^2 - y^2)^6$ **28.** Third term, $(x - 5)^5$

29. Fifth term, $(4 + x)^7$ **30.** Seventh term, $(a - b)^7$

31. Tenth term, $(x + y)^{14}$ **32.** Fifth term, $(r + 1)^4$

33. Eighth term, $(2 - y)^9$ **34.** Ninth term, $(3 - z)^{10}$

35. Find the term in the expansion of $(\sqrt{x} + \sqrt{y})^6$ that involves y^2.

36. Find the term in the expansion of $(2a^3 + a^{-4}b^2)^7$ that does not involve the symbol a.

In Problems 37 and 38, use Pascal's triangle to find all binomial coefficients in the expansion of $(a + b)^n$ for the given value of n.

37. $n = 8$ **38.** $n = 10$

In Problems 39–42, simplify the given expression. Assume $h \neq 0$.

39. $\dfrac{(x + h)^2 - x^2}{h}$ **40.** $\dfrac{(2x + 2h - 5)^2 - (2x - 5)^2}{h}$

41. $\dfrac{(x + h)^3 - x^3}{h}$ **42.** $\dfrac{(2 + h)^4 - 16}{h}$

In Problems 43 and 44, use the binomial theorem as an aid in calculating the given number.

43. $(0.99)^3$ [*Hint:* $0.99 = 1 - 0.01$.]   **44.** $(1.1)^4$

Review Problems

45. Write $\sqrt{\sqrt[3]{\sqrt[4]{x}}}$ as a single radical.

46. True or false: $\dfrac{1}{\sqrt{8}} = \dfrac{\sqrt{2}}{4}$.

47. Write $\sqrt{x^6 + x^4 y^2}$ in terms of rational exponents.

48. Write the expression $|x - 1| + |6x| + |5 - 2x|$ without absolute value symbols if $x < 0$.

49. True or false: $\dfrac{5}{0}$ is a rational number.

50. True or false: A real number that has a nonrepeating and nonterminating decimal representation must be irrational.

0.4 Products

To multiply two algebraic expressions each containing two terms, we may use the distributive law, as follows:

$$(AX + B)(CX + D) = (AX + B)CX + (AX + B)D$$
$$= ACX^2 + BCX + ADX + BD$$
$$= ACX^2 + (AD + BC)X + BD. \qquad (1)$$

Example 1

$$(3x + 4)(2x + 5) = (3)(2)x^2 + [(3)(5) + (4)(2)]x + (4)(5)$$
$$= 6x^2 + (15 + 8)x + 20$$
$$= 6x^2 + 23x + 20. \qquad \blacksquare$$

Example 2

$$(6x - 1)(3x + 2) = 6(3)x^2 + [(6)(2) + (-1)(3)]x + (-1)(2)$$
$$= 18x^2 + (12 - 3)x - 2$$
$$= 18x^2 + 9x - 2. \qquad \blacksquare$$

Formula (1), illustrated in Figure 0.10, is sometimes called the **FOIL** method after the first letter in each of the boldface words.

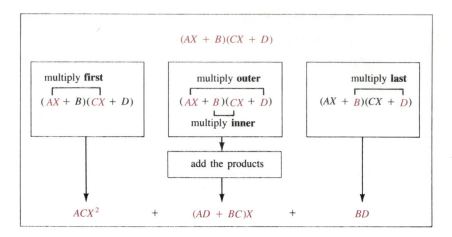

Figure 0.10

Example 3

$$(7x - 1)(8x + 2) = 7(8)x^2 + [(7)(2) + (-1)(8)]x + (-1)(2)$$
$$= 56x^2 + (14 - 8)x - 2$$
$$= 56x^2 + 6x - 2.$$ ∎

Example 4

$$(\tfrac{2}{3}x - 2)(x - \tfrac{1}{3}) = \tfrac{2}{3}(1)x^2 + [(\tfrac{2}{3})(-\tfrac{1}{3}) + (-2)(1)]x + (-2)(-\tfrac{1}{3})$$
$$= \tfrac{2}{3}x^2 + (-\tfrac{2}{9} - 2)x + \tfrac{2}{3}$$
$$= \tfrac{2}{3}x^2 - \tfrac{20}{9}x + \tfrac{2}{3}.$$ ∎

At first glance, some products may not appear to be in the form

$$(AX + B)(CX + D), \tag{2}$$

when in fact they are. With practice the student can become adept at recognizing this form.

Example 5

$$(5x^2 + 2)(x^2 - 4) = 5(1)(x^2)^2 + [5(-4) + 2(1)]x^2 + 2(-4)$$
$$= 5x^4 + (-20 + 2)x^2 - 8$$
$$= 5x^4 - 18x^2 - 8.$$ ∎

Example 6

$$(2\sqrt{x} + 1)(4\sqrt{x} - 3) = 2(4)(\sqrt{x})^2 + [2(-3) + 1(4)]\sqrt{x} + 1(-3)$$
$$= 8x - 2\sqrt{x} - 3.$$ ∎

Example 7   Note in the product

$$(2y - \sqrt{3})(5 - y)$$

that we must factor -1 from the second factor to obtain form (2). We have

$$(2y - \sqrt{3})(5 - y) = -(2y - \sqrt{3})(y - 5)$$
$$= -\{2y^2 + [2(-5) + (-\sqrt{3})(1)]y + (-\sqrt{3})(-5)\}$$
$$= -2y^2 + (10 + \sqrt{3})y - 5\sqrt{3}.$$ ∎

Important Product Formulas

Other commonly used product formulas are given below.

$$(X + Y)(X + Y) = X^2 + 2XY + Y^2 \tag{3}$$
$$(X - Y)(X - Y) = X^2 - 2XY + Y^2 \tag{4}$$
$$(X + Y)(X - Y) = X^2 - Y^2 \tag{5}$$
$$(X + Y)(X^2 - XY + Y^2) = X^3 + Y^3 \tag{6}$$
$$(X - Y)(X^2 + XY + Y^2) = X^3 - Y^3. \tag{7}$$

As illustrated in the next example, these formulas may be verified by using the distributive law. (See Problems 41 and 42.) Although equations (3) and (4) were discussed in the previous section, we include them here because they are frequently encountered.

Example 8   Verify $(X + Y)(X - Y) = X^2 - Y^2$.

Solution   By the distributive law, we have

$$(X + Y)(X - Y) = (X + Y)X + (X + Y)(-Y)$$
$$= X^2 + YX + X(-Y) - Y^2$$
$$= X^2 + XY - XY - Y^2$$
$$= X^2 - Y^2.$$ ∎

If two algebraic expressions have the appropriate form, we can find their product by using either (1) or equations (3)–(7). Otherwise, use the distributive law.

Example 9   By equation (3) we find that

$$(2x + 3)(2x + 3) = (2x)^2 + 2(2x)(3) + (3)^2$$
$$= 4x^2 + 12x + 9.$$ ∎

Example 10 Observe that the product

$$(x^2 - y)(x^4 + x^2y + y^2)$$

has form (7). Hence, we can write

$$(x^2 - y)(x^4 + x^2y + y^2) = [(x^2) - y][(x^2)^2 + (x^2)y + y^2]$$
$$= (x^2)^3 - y^3$$
$$= x^6 - y^3.$$

▪

Example 11 Since the product

$$(x^3 - 1)(x^2 - 3x + 7)$$

is not one of the appropriate forms, we apply the distributive law:

$$(x^3 - 1)(x^2 - 3x + 7) = x^3(x^2 - 3x + 7) - 1(x^2 - 3x + 7)$$
$$= x^5 - 3x^4 + 7x^3 - x^2 + 3x - 7.$$

▪

Example 12 By the distributive law we have

$$(3y^2 + y^{-1} + y^{-2})(y^3 - y) = 3y^2(y^3 - y) + y^{-1}(y^3 - y) + y^{-2}(y^3 - y)$$
$$= 3y^2y^3 - 3y^2y + y^{-1}y^3 - y^{-1}y + y^{-2}y^3 - y^{-2}y$$
$$= 3y^5 - 3y^3 + y^2 - 1 + y - y^{-1}$$
$$= 3y^5 - 3y^3 + y^2 + y - y^{-1} - 1.$$

▪

Example 13

$$(2x + y)(2x - y)(4x^2 + y^2) = [(2x + y)(2x - y)](4x^2 + y^2)$$
$$= (4x^2 - y^2)(4x^2 + y^2) \quad [\text{by (5)}]$$
$$= 16x^4 - y^4. \quad [\text{by (5)}]$$

▪

Exercise 0.4

In Problems 1–40, find the given product.

1. $(x - 1)(x + 2)$
2. $(x^2 + 3)(x^2 - 5)$
3. $(2x^3 + 1)(x^3 - 7)$
4. $(x + 3)(4x - 5)$
5. $(5x - 7)(2x + 8)$
6. $(3x - 5)(7x + 1)$
7. $(4\sqrt{x} + 1)(6\sqrt{x} - 2)$
8. $(9x - 5)(8x + 3)$
9. $(0.3x + 0.7)(10.0x + 2.1)$
10. $(1.2x + 1.0)(2.0x - 1.3)$
11. $(\frac{1}{2}x - \frac{1}{3})(2x + \frac{1}{4})$
12. $(\frac{2}{5}x + 5)(\frac{1}{5}x + 1)$

13. $(1 + 5x)(7 + 2x)$

14. $(4 - 3x)(4 + 2x)$

15. $(a + b)(a - b)$

16. $(2 + \sqrt{3})(2 - \sqrt{3})$

17. $[4(x + 1) + 3][5(x + 1) - 1]$

18. $[5(2 - z) + 5][7(2 - z) + 2]$

19. $(y^{-1} + 2x)(y^{-1} + 3x)$

20. $(z^2 + z)(z^2 - z)$

21. $(1 + x)(1 - x)$

22. $(5 - \sqrt{7}x)(5 + \sqrt{7}x)$

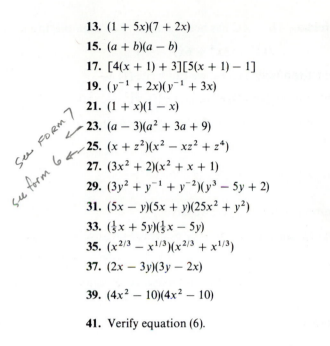

see form 7

see form 6

23. $(a - 3)(a^2 + 3a + 9)$

24. $(2 - y)(4 + 2y + y^2)$

25. $(x + z^2)(x^2 - xz^2 + z^4)$

26. $(9 + y)(81 - 9y + y^2)$

27. $(3x^2 + 2)(x^2 + x + 1)$

28. $(1 - y^2)(y^{-2} + y^{-1} + 1)$

29. $(3y^2 + y^{-1} + y^{-2})(y^3 - 5y + 2)$

30. $(x^2 - 4x)(x^4 + 5x + 6)$

31. $(5x - y)(5x + y)(25x^2 + y^2)$

32. $(3a + 2b)(3a - 2b)(9a^2 + 4b^2)$

33. $(\frac{1}{5}x + 5y)(\frac{1}{5}x - 5y)$

34. $(\sqrt{x} + \sqrt{y})(\sqrt{x} - \sqrt{y})$

35. $(x^{2/3} - x^{1/3})(x^{2/3} + x^{1/3})$

36. $(y - x)(x^2 + xy + y^2)$

37. $(2x - 3y)(3y - 2x)$

38. $(3x + 5)(5 + 3x)$

39. $(4x^2 - 10)(4x^2 - 10)$

40. $\left(\dfrac{1}{y^2} - \dfrac{1}{x^2}\right)\left(\dfrac{1}{y^4} + \dfrac{1}{y^2x^2} + \dfrac{1}{x^4}\right)$

41. Verify equation (6).

42. Verify equation (7).

Review Problems

43. Find the midpoint of the line segment joining $-\frac{3}{2}$ and $\frac{9}{2}$. Find the distance between the points.

44. Evaluate: $(-1000)^{-2/3}$

45. True or false: $\sqrt{(-13)^2} = 13$.

46. Expand: $(4x - y^2)^3$

47. Find the term in the expansion of $(t + s^3)^{10}$ that involves s^{21}.

48. True or false: $10 \cdot 9! = 10!$

0.5 Factoring

When the distributive law

$$A(B + C) = AB + AC$$

is read from the right to left,

$$AB + AC = A(B + C),$$

we say that the expression $AB + AC$ has been **factored**. For example, in each term of

$$3x^4 + 6x^2 \tag{1}$$

there is a common factor of $3x^2$. Hence, we can write (1) as

$$3x^4 + 6x^2 = 3x^2(x^2 + 2).$$

In general, the first step in factoring any algebraic expression is to determine whether each term has a common factor.

Example 1 Factor $6x^4y^4 - 4x^2y^2 + 10\sqrt{2}xy^3 - 2xy^2$.

Solution Since there is a common factor of $2xy^2$, we have

$$6x^4y^4 - 4x^2y^2 + 10\sqrt{2}xy^3 - 2xy^2 = 2xy^2(3x^3y^2 - 2x + 5\sqrt{2}y - 1). \quad \blacksquare$$

When the terms of an expression do not have a common factor, it may still be possible to factor by *grouping* the terms in an appropriate manner.

Example 2 Factor $x^2 + 2yx - x - 2y$.

Solution Grouping the first two terms and the last two terms, we have

$$x^2 + 2yx - x - 2y = (x^2 + 2yx) + (-x - 2y)$$
$$= x(x + 2y) + (-1)(x + 2y).$$

We observe the common factor $x + 2y$ and complete the factorization as

$$x^2 + 2yx - x - 2y = (x - 1)(x + 2y). \quad \blacksquare$$

Some Important Factorizations

By reading the product formulas in Section 1.4 from right to left, we have the following important factorizations:

Quadratic expression:	$ACX^2 + (AD + BC)X + BD$ $\qquad$ (2)
	$= (AX + B)(CX + D)$
Square of a sum:	$X^2 + 2XY + Y^2$ $\qquad$ (3)
	$= (X + Y)(X + Y)$
Square of a difference:	$X^2 - 2XY + Y^2$ $\qquad$ (4)
	$= (X - Y)(X - Y)$
Difference of two squares:	$X^2 - Y^2$ $\qquad$ (5)
	$= (X - Y)(X + Y)$

$$\text{Sum of two cubes:} \qquad X^3 + Y^3 \qquad\qquad (6)$$
$$= (X + Y)(X^2 - XY + Y^2)$$
$$\text{Difference of two cubes:} \qquad X^3 - Y^3 \qquad\qquad (7)$$
$$= (X - Y)(X^2 + XY + Y^2).$$

Observe that the summary above indicates that the difference of two squares, the difference of two cubes, and the sum of two cubes *always* factor.

Example 3   Factor $16x^4y^2 - 25$.

Solution   This is the difference of two squares. From (5) with $X = 4x^2y$ and $Y = 5$, we have
$$16x^4y^2 - 25 = (4x^2y)^2 - (5)^2$$
$$= (4x^2y - 5)(4x^2y + 5). \qquad \blacksquare$$

Example 4   Factor $8a^3 + 27b^6$.

Solution   Since $8a^3 + 27b^6$ is a sum of two cubes, it may be factored using (6). If we identify $X = 2a$ and $Y = 3b^2$, then
$$8a^3 + 27b^6 = (2a)^3 + (3b^2)^3$$
$$= (2a + 3b^2)[(2a)^2 - (2a)(3b^2) + (3b^2)^2]$$
$$= (2a + 3b^2)(4a^2 - 6ab^2 + 9b^4). \qquad \blacksquare$$

Example 5   Factor $x^8 - 256$.

Solution   We apply the formula (5) for the difference of two squares three times.
$$x^8 - 256 = (x^4)^2 - (16)^2$$
$$= (x^4 - 16)(x^4 + 16)$$
$$= [(x^2)^2 - (4)^2](x^4 + 16)$$
$$= (x^2 - 4)(x^2 + 4)(x^4 + 16)$$
$$= (x - 2)(x + 2)(x^2 + 4)(x^4 + 16). \qquad \blacksquare$$

Example 6   Factor $(x^2 + 1)^3 - (y^2 + 1)^3$.

Solution   This is a difference of two cubes. From (7) we have
$$(x^2 + 1)^3 - (y^2 + 1)^3 = [(x^2 + 1) - (y^2 + 1)]$$
$$\cdot [(x^2 + 1)^2 + (x^2 + 1)(y^2 + 1) + (y^2 + 1)^2]$$
$$= (x^2 - y^2)(x^4 + 3x^2 + y^4 + 3y^2 + x^2y^2 + 3).$$

Factoring the difference of two squares, we have

$$(x^2 + 1)^3 - (y^2 + 1)^3 = (x - y)(x + y)(x^4 + 3x^2 + x^2y^2 + 3y^2 + y^4 + 3). \quad \blacksquare$$

Example 7 Factor $x^6 - y^6$.

Solution We can view the expression $x^6 - y^6$ in two ways: as a difference of two squares or as a difference of two cubes. Using the difference of two cubes, we write

$$x^6 - y^6 = (x^2)^3 - (y^2)^3$$
$$= (x^2 - y^2)(x^4 + x^2y^2 + y^4)$$
$$= (x - y)(x + y)(x^4 + x^2y^2 + y^4).$$

From this we might conclude that the factorization is complete. However, treating the expression $x^6 - y^6$ as a difference of two squares is more revealing, since

$$x^6 - y^6 = (x^3)^2 - (y^3)^2$$
$$= (x^3 - y^3)(x^3 + y^3)$$
$$= (x - y)(x^2 + xy + y^2)(x + y)(x^2 - xy + y^2)$$
$$= (x - y)(x + y)(x^2 + xy + y^2)(x^2 - xy + y^2).$$

Thus, we have discovered the additional factorization

$$x^4 + x^2y^2 + y^4 = (x^2 + xy + y^2)(x^2 - xy + y^2) \quad \blacksquare$$

For real numbers a, b, and c, an algebraic expression of the form $ax^2 + bx + c$ is said to be **quadratic in x**. When a, b, and c are integers, it is sometimes possible to factor $ax^2 + bx + c$ as

$$(Ax + B)(Cx + D),$$

where A, B, C, and D are also integers.

First we assume the quadratic expression has lead coefficient $a = 1$. If $x^2 + bx + c$ has a factorization using integer coefficients, then it will be of the form

$$(x + B)(x + D),$$

where B and D are integers. By examining the product

$$(x + B)(x + D) = x^2 + (B + D)x + BD = x^2 + bx + c,$$

we see that

$$B + D = b \quad \text{and} \quad BD = c.$$

Thus to factor $x^2 + bx + c$ with integer coefficients, we list all possible factorizations of c as a product of two integers B and D. We then check which, if any, of the sums $B + D$ equals b.

Example 8   Factor $x^2 - 9x + 18$.

Solution   Since $a = 1$ the factors will be $(x + \quad)(x + \quad)$. Now with $b = -9$ and $c = 18$, we seek integers B and D such that

$$B + D = -9 \quad \text{and} \quad BD = 18.$$

We can write 18 as a product BD in the following ways:

$$1(18), \quad 2(9), \quad 3(6), \quad (-1)(-18), \quad (-2)(-9), \quad \text{or} \quad (-3)(-6).$$

Since -9 is the sum of -3 and -6, the factorization is

$$x^2 - 9x + 18 = (x - 3)(x - 6).$$ ∎

Example 9   Factor $x^2 + 3x - 1$.

Solution   The number -1 can be written as a product of two integers BD in only one way, $(-1)(1)$. Since the sum

$$B + D = -1 + 1 \neq 3,$$

$x^2 + 3x - 1$ cannot be factored using *integer* coefficients.   ∎

It is more complicated to factor the general quadratic polynomial $ax^2 + bx + c$, with $a \neq 1$, since we must also consider factors of a as well as c. When we examine the product

$$(Ax + B)(Cx + D) = ACx^2 + (AD + BC)x + BD = ax^2 + bx + c,$$

we see that $ax^2 + bx + c$ factors as $(Ax + B)(Cx + D)$ if

$$AC = a, \quad AD + BC = b, \quad \text{and} \quad BD = c.$$

Example 10   Factor $2x^2 + 11x - 6$.

Solution   The factors will be

$$(2x + \quad)(1x + \quad)$$

where the blanks are to be filled with a pair of integers whose product BD is -6. Possible pairs are

$$1 \text{ and } -6, \quad -1 \text{ and } 6, \quad 3 \text{ and } -2, \quad -3 \text{ and } 2.$$

Now we must check to see if one of the pairs gives 11 as the value of $AD + BC$ (the coefficient of the middle term), where $A = 2$ and $C = 1$. We find that

$$2(6) + (-1)(1) = 11,$$

and therefore

$$2x^2 + 11x - 6 = (2x - 1)(x + 6).$$ ■

Example 11 Factor $15x^2 + 17x + 4$.

Solution The factors could have the form

$$(5x + \quad)(3x + \quad) \quad \text{or} \quad (15x + \quad)(1x + \quad). \qquad (8)$$

There is no need to consider the cases

$$(-5x + \quad)(-3x + \quad) \quad \text{and} \quad (-15x + \quad)(-x + \quad).$$

(Why?)

The blanks in (8) must be filled with a pair of integers whose product is 4. Possible pairs are

$$1 \text{ and } 4, \quad -1 \text{ and } -4, \quad 2 \text{ and } 2, \quad -2 \text{ and } -2.$$

We check each pair with the possible forms in (8) to see which combination, if any, gives a coefficient of 17 for the middle term. We find

$$15x^2 + 17x + 4 = (5x + 4)(3x + 1).$$ ■

Example 12 Factor $2x^4 + 11x^2 + 12$.

Solution Letting $X = x^2$, we may view this expression as a quadratic in X,

$$2X^2 + 11X + 12.$$

We then factor the expression using formula (2). The factors will be

$$(X + \quad)(2X + \quad), \qquad (9)$$

where the blanks are to be filled with a pair of integers whose product is 12. Possible pairs are

$$1 \text{ and } 12, \quad -1 \text{ and } -12, \quad 2 \text{ and } 6, \quad -2 \text{ and } -6, \quad 3 \text{ and } 4, \quad -3 \text{ and } -4.$$

We check each pair with (9) to see which combination, if any, gives a coefficient of 11 for the middle term. We find that

$$2X^2 + 11X + 12 = (X + 4)(2X + 3).$$

Substituting x^2 for X, we have

$$2x^4 + 11x^2 + 12 = (x^2 + 4)(2x^2 + 3).$$ ■

As we have done, it is customary to seek factorizations of quadratic expressions with integer coefficients. However, the reader should not think that quadratic expressions such as

$$x^2 + x - 1, \quad x^2 - \tfrac{1}{4}x - \tfrac{1}{8}, \quad \text{and} \quad x^2 + 1$$

do not factor. In fact, it can be shown that *any* expression of the form

$$ax^2 + bx + c$$

can be factored if one is willing to use rational, irrational, or even complex numbers as coefficients.

Exercise 0.5

In Problems 1–46, factor the given expressions if possible.

1. $12x^3 + 2x^2 + 6x$
2. $6x^3y^4 - 3\sqrt{3}x^2y^2 - 3x^2y + 3xy$
3. $2y^2 - yz + 6y - 3z$
4. $6x^5y^5 + \sqrt{2}x^2y^3 + 14xy^3$
5. $36x^2 - 25$
6. $a^2 - 4b^2$
7. $4x^2y^2 - 1$
8. $49x^2 - 64y^2$
9. $x^4 + y^4$
10. $x^2 + 1$
11. $x^8 - y^8$
12. $a^3 - 64b^3$
13. $8x^3y^3 + 27$
14. $y^3 + 125$
15. $y^6 - 1$
16. $1 - x^3$
17. $x^2 - 5x + 6$
18. $x^2 - 10x + 24$
19. $x^2 + 7x + 10$
20. $x^4 + 10x^2 + 21$
21. $x^4 - 3x^2 - 4$
22. $x^2 + 4x - 12$
23. $x^2 + 2x - 1$
24. $x^2 + 5x - 2$
25. $x^2 - xy - 2y^2$
26. $x^2 - 4xy + 3y^2$
27. $x^2 + 10x + 25$
28. $4x^2 + 12x + 9$
29. $x^2 - 8xy + 16y^2$
30. $9x^2 - 6xy + y^2$
31. $2x^2 + 7x + 5$
32. $8x^2 + 2x - 3$
33. $6x^4 + 13x^2 - 15$
34. $10x^4 - 23x^2 + 12$
35. $2x^2 - 7xy + 3y^2$
36. $-3x^2 - 5xy + 12y^2$
37. $(x^2 + 1)^3 + (y^2 - 1)^3$
38. $(4 - x^2)^3 - (4 - y^2)^3$
39. $x(x - y) + y(y - x)$
40. $x(x - y) - y(y - x)$
41. $(1 - x^2)^3 - (1 - y^2)^3$
42. $(x^2 - 4)^3 + (4 - y^2)^3$
43. $1 - 256x^8$
44. $x^8 - 6561$
45. $x^6 + 7x^3 - 8$
46. $16x^8 + 15x^4 - 2$

Review Problems

47. Find the products **a.** $(5z^3 - 2y^2)(5z^3 + 2y^2)$ **b.** $(a + b + 1)(a - b - 1)$

48. Rationalize the denominator of $\dfrac{1}{\sqrt{x} + \sqrt{x + 2}}$.

49. Write $|6^{-1} - 3^{-1}|$ without absolute value symbols.

50. Simplify $\dfrac{(4r^3s^2t^{-5})^3}{(2r^2st^{-1})^5}$ and write without negative exponents.

51. Find the indicated number:

    a. $\sqrt[5]{(\frac{1}{32})^{-3}}$ **b.** $(0.008)^{-2/3}$

52. Simplify $\sqrt{\dfrac{25x^{12}y^{16}}{z^{-20}}}$.

0.6 Rational Expressions

Polynomial

An algebraic expression of the form

$$a_nx^n + a_{n-1}x^{n-1} + \cdots + a_2x^2 + a_1x + a_0, \qquad a_n \neq 0, \qquad (1)$$

where n is a nonnegative integer and a_i, $i = 0, 1, \ldots, n$ are real constants, is called a **polynomial** of **degree n** in the variable x. The number a_n, the coefficient of the highest power of x, is called the **leading coefficient** of the polynomial. Each a_ix^i is called a **term** of the polynomial.

Example 1 The expression

$$8x^5 + (-4)x^3 + 10x^2 + (-3)$$

is a polynomial of degree 5. The leading coefficient is 8 and the terms of the polynomial are $8x^5$, $(-4)x^3$, $10x^2$, and (-3). For simplicity, we may disregard the parentheses around the negative coefficients and write the polynomial as

$$8x^5 - 4x^3 + 10x^2 - 3.$$

∎

Bear in mind that the essential characteristic of a polynomial is that it is a sum of terms in which the exponent of the variable is a *nonnegative integer*. Thus the algebraic expressions

$$3x^2 - 4x^{1/2} + 10 \quad \text{and} \quad 15x^{-1} + x^3$$

are *not* polynomials. A polynomial is usually written so that, reading left to right, the powers of the variable decrease. Thus we rewrite

$$20x^3 + 0.5x^6 - 7x^{10} + 2x^5 \quad \text{as} \quad -7x^{10} + 0.5x^6 + 2x^5 + 20x^3.$$

We also note that there is nothing special about using the symbol x as a variable. The expressions

$$z^4 + 6z^3 + 11z - 30 \quad \text{and} \quad t^2 + 4t + 9$$

are polynomials in the variables z and t, respectively.

Rational Expression

A quotient of two polynomials is called a **rational expression**. For example,

$$\frac{2x^2 + 5}{x + 1} \quad \text{and} \quad \frac{3}{2x^3 - x + 8}$$

are rational expressions.

Rules for Fractions

A rational expression represents a real number for any value of the variable such that the denominator is not zero. Therefore we may apply properties of the real number system to combine and simplify rational expressions. The following **rules for fractions** are particularly helpful.

For any real numbers a, b, c, and d

$$\frac{ac}{bc} = \frac{a}{b}, \qquad c \neq 0 \tag{1}$$

$$\frac{a}{b} \pm \frac{c}{b} = \frac{a \pm c}{b} \tag{2}$$

$$\frac{a}{b} \cdot \frac{c}{d} = \frac{ac}{bd} \tag{3}$$

$$\frac{a}{b} \div \frac{c}{d} = \frac{a}{b} \cdot \frac{d}{c}, \tag{4}$$

provided that each denominator is not zero.

Example 2 Simplify $\dfrac{2x^2 - x - 1}{x^2 - 1}$.

Solution We factor the numerator and the denominator and cancel common factors.

$$\frac{2x^2 - x - 1}{x^2 - 1} = \frac{(2x + 1)(x - 1)}{(x + 1)(x - 1)}$$

$$= \frac{2x + 1}{x + 1}. \qquad [\text{by (1)}] \qquad \blacksquare$$

Note that the cancellation of the factor $x - 1$ is valid only for those values of x such that $x - 1$ is nonzero; that is, $x \neq 1$. For the remainder of this section, however, we shall assume without further comment that all denominators are nonzero.

Example 3 Simplify $\dfrac{4x^2 + 11x - 3}{2 - 5x - 12x^2}$.

Solution

$$\frac{4x^2 + 11x - 3}{2 - 5x - 12x^2} = \frac{(4x - 1)(x + 3)}{(1 - 4x)(2 + 3x)} = \frac{(4x - 1)(x + 3)}{-(4x - 1)(2 + 3x)}$$

$$= -\frac{x + 3}{2 + 3x}. \qquad\qquad [\text{by (1)}]$$

Least Common Denominator

In order to add or subtract rational expressions, we first find a common denominator and then apply (2). Although any common denominator will do, less work is involved if we use the **least common denominator** (LCD). This is found by factoring each denominator completely and forming a product of the distinct factors, using each factor with the largest exponent with which it occurs in any single denominator.

Example 4   Find the LCD of $\dfrac{1}{x^4 - x^2}, \dfrac{x + 2}{x^2 + 2x + 1}$, and $\dfrac{1}{x}$.

Solution   Factoring the denominators in the rational expressions, we obtain

$$\frac{1}{x^2(x - 1)(x + 1)}, \quad \frac{x + 2}{(x + 1)^2}, \quad \text{and} \quad \frac{1}{x}.$$

The distinct factors of the denominators are x, $x - 1$, and $x + 1$. We use each factor with the largest exponent with which it occurs in any single denominator. Thus the LCD is

$$x^2(x - 1)(x + 1)^2.$$

Example 5   Combine and simplify

$$\frac{x}{x^2 - 4} + \frac{1}{x^2 + 4x + 4}.$$

Solution   In factored form the denominators are $(x - 2)(x + 2)$ and $(x + 2)^2$. Thus the LCD is $(x - 2)(x + 2)^2$. We use (1) to rewrite each rational expression with the LCD as denominator, and then we use (2) to combine the expressions:

$$\frac{x}{(x - 2)(x + 2)} + \frac{1}{(x + 2)^2} = \frac{x(x + 2)}{(x - 2)(x + 2)(x + 2)} + \frac{1(x - 2)}{(x + 2)^2(x - 2)} \qquad [\text{by (1)}]$$

$$= \frac{x(x + 2) + x - 2}{(x - 2)(x + 2)^2} \qquad\qquad [\text{by (2)}]$$

$$= \frac{x^2 + 2x + x - 2}{(x - 2)(x + 2)^2}$$

$$= \frac{x^2 + 3x - 2}{(x - 2)(x + 2)^2}.$$

To multiply or divide rational expressions we apply (3) or (4) as illustrated in the following examples.

Example 6 Combine and simplify

$$\frac{x}{5x^2 + 21x + 4} \cdot \frac{25x^2 + 10x + 1}{3x^2 + x}.$$

Solution

$$\frac{x}{5x^2 + 21x + 4} \cdot \frac{25x^2 + 10x + 1}{3x^2 + x} = \frac{x(25x^2 + 10x + 1)}{(5x^2 + 21x + 4)(3x^2 + x)} \qquad [\text{by (3)}]$$

$$= \frac{x(5x + 1)(5x + 1)}{(5x + 1)(x + 4)x(3x + 1)}$$

$$= \frac{5x + 1}{(x + 4)(3x + 1)}. \qquad [\text{by (1)}] \quad \blacksquare$$

Example 7 Combine and simplify

$$\frac{2x^2 + 9x + 10}{x^2 + 4x + 3} \div \frac{2x + 5}{x + 3}.$$

Solution

$$\frac{2x^2 + 9x + 10}{x^2 + 4x + 3} \div \frac{2x + 5}{x + 3} = \frac{2x^2 + 9x + 10}{x^2 + 4x + 3} \cdot \frac{x + 3}{2x + 5} \qquad [\text{by (4)}]$$

$$= \frac{(2x^2 + 9x + 10)(x + 3)}{(x^2 + 4x + 3)(2x + 5)} \qquad [\text{by (3)}]$$

$$= \frac{(2x + 5)(x + 2)(x + 3)}{(x + 3)(x + 1)(2x + 5)}$$

$$= \frac{x + 2}{x + 1}. \qquad [\text{by (1)}] \quad \blacksquare$$

As shown in the following example, the techniques illustrated above enable us to simplify more complicated quotients.

Example 8 Simplify $\dfrac{\dfrac{1}{t} - \dfrac{t}{t + 1}}{1 + \dfrac{1}{t}}.$

Solution First we obtain single rational expressions for the numerator and the denominator.

$$\frac{\dfrac{1}{t} - \dfrac{t}{t+1}}{1 + \dfrac{1}{t}} = \frac{\dfrac{1(t+1)}{t(t+1)} - \dfrac{tt}{(t+1)t}}{\dfrac{t}{t} + \dfrac{1}{t}} \qquad \text{[by (1)]}$$

$$= \frac{\dfrac{t+1-t^2}{t(t+1)}}{\dfrac{t+1}{t}} \qquad \text{[by (2)]}$$

Now we apply (4) to this quotient to obtain

$$\frac{\dfrac{-t^2+t+1}{t(t+1)}}{\dfrac{t+1}{t}} = \frac{-t^2+t+1}{t(t+1)} \cdot \frac{t}{t+1} \qquad \text{[by (4)]}$$

$$= \frac{-t^2+t+1}{(t+1)^2}. \qquad \text{[by (1)]} \qquad \blacksquare$$

The techniques discussed in this section are often applicable to expressions containing negative exponents.

Example 9 Simplify $(a^{-1} + b^{-1})^{-1}$.

Solution Frequently, there is more than one way to simplify such an expression. One solution is

$$(a^{-1} + b^{-1})^{-1} = \left(\frac{1}{a} + \frac{1}{b}\right)^{-1} = \left(\frac{b+a}{ab}\right)^{-1}$$

$$= \frac{1}{\dfrac{b+a}{ab}} = \frac{ab}{b+a}.$$

Alternatively,

$$(a^{-1} + b^{-1})^{-1} = \frac{1}{a^{-1} + b^{-1}} = \frac{1}{\dfrac{1}{a} + \dfrac{1}{b}}$$

$$= \frac{1}{\dfrac{b+a}{ab}} = \frac{ab}{b+a}. \qquad \blacksquare$$

Exercise 0.6

In Problems 1–4, determine the values of x for which the given rational expression represents a real number.

1. $\dfrac{1}{x^2 - 4}$
2. $\dfrac{x + 1}{x^2 + 6x + 5}$
3. $\dfrac{x}{x^2 + x}$
4. $\dfrac{x}{x^2 + 1}$

In Problems 5–8, simplify the given rational expression.

5. $\dfrac{x^2 + 3x + 2}{x^2 + 6x + 8}$
6. $\dfrac{t^4 + 4t^2 + 4}{4 - t^4}$
7. $\dfrac{x^2 - 9}{x^3 + 27}$
8. $\dfrac{x^2 - 2xy - 3y^2}{x^2 - 4xy + 3y^2}$

In Problems 9–12, find the LCD of the given expressions.

9. $\dfrac{1}{x^2 + x - 2}, \dfrac{1}{x + 2}$

10. $\dfrac{1}{x^2 + 2x + 1}, \dfrac{1}{x^2 - 3x - 4}$

11. $\dfrac{1}{x^3 + x^2 - 6x}, \dfrac{1}{x^3 - 6x^2}, \dfrac{1}{x - 2}$

12. $\dfrac{1}{z^2 - 10z + 25}, \dfrac{1}{z^2 - 25}, \dfrac{1}{z^2 + 10z + 25}$

In Problems 13–38, combine and simplify the given expressions.

13. $\dfrac{4x}{4x + 5} + \dfrac{5}{4x + 5}$

14. $\dfrac{3}{x - 2} + \dfrac{4}{2 - x}$

15. $\dfrac{7x}{7x - 1} - \dfrac{1}{1 - 7x}$

16. $\dfrac{3}{x - 2} - \dfrac{6}{x^2 + 4}$

17. $\dfrac{2x}{x + 1} + \dfrac{5}{x^2 - 1}$

18. $\dfrac{x}{2x + 1} - \dfrac{2x}{x - 2}$

19. $\dfrac{y}{y - x} - \dfrac{x}{y + x}$

20. $\dfrac{x}{x - y} + \dfrac{x}{y - x}$

21. $\dfrac{2}{x^2 - x - 12} + \dfrac{x}{x + 3}$

22. $\dfrac{1}{x + 3} + \dfrac{x}{x + 1} + \dfrac{x^2 + 1}{x^2 + 4x + 3}$

23. $\dfrac{x}{2x^2 + 3x - 2} - \dfrac{1}{2x - 1} - \dfrac{4}{x + 2}$

24. $\dfrac{x}{2x + 3} - \dfrac{3}{4x^2 - 3x - 1} + \dfrac{4x + 1}{2x^2 + x - 3}$

25. $\dfrac{x - 4}{x + 3} \cdot \dfrac{x + 5}{x - 2}$

26. $\dfrac{x^2 + x}{x^2 - 1} \cdot \dfrac{x + 1}{x^2}$

27. $(x^2 - 2x + 1) \cdot \dfrac{x + 1}{x^3 - 1}$

28. $\dfrac{2x + 8}{x - 1} \cdot \dfrac{x + 4}{2x}$

29. $\dfrac{6x + 5}{3x + 3} \cdot \dfrac{x + 1}{6x^2 - 7x - 10}$

30. $\dfrac{1 + x}{2 + x} \cdot \dfrac{x^2 + x - 12}{3 + 2x - x^2}$

31. $\dfrac{x + 1}{x + 2} \div \dfrac{x + 1}{x + 7}$

32. $\dfrac{3x + 1}{x - 4} \div \dfrac{2x + 1}{x}$

33. $\dfrac{x}{x + 4} \div \dfrac{x + 5}{x}$

34. $\dfrac{x - 3}{x + 1} \div \dfrac{x + 1}{2x + 1}$

35. $\dfrac{x^2 - 1}{x^2 + 2x - 3} \div \dfrac{x - 4}{x + 3}$

36. $\dfrac{x^2 - 3x + 2}{x^2 - 7x + 12} \div \dfrac{x - 2}{x - 3}$

37. $\dfrac{x^2 - 5x + 6}{x^2 + 7x + 10} \div \dfrac{2 - x}{x + 2}$

38. $\dfrac{x}{x + y} \div \dfrac{y}{x + y}$

In Problems 39–54, simplify the given expression.

39. $\dfrac{\dfrac{1}{x^2} - x}{\dfrac{1}{x^2} + x}$

40. $\dfrac{\dfrac{1}{x} + \dfrac{1}{y}}{\dfrac{1}{x} - \dfrac{1}{y}}$

41. $\dfrac{x + \dfrac{1}{2}}{2 + \dfrac{1}{x}}$

42. $\dfrac{\dfrac{1 + x}{x} + \dfrac{x}{1 - x}}{\dfrac{1 - x}{x} + \dfrac{x}{1 + x}}$

43. $\dfrac{x^2 + xy + y^2}{\dfrac{x^2}{y} - \dfrac{y^2}{x}}$

44. $\dfrac{\dfrac{x}{x - 1} - \dfrac{x + 1}{x}}{1 - \dfrac{x}{x - 1}}$

45. $(a^{-1} - b^{-1})^{-1}$

46. $\dfrac{a + b}{a^{-1} + b^{-1}}$

47. $(1 - a^{-1})^{-1}$

48. $\left(\dfrac{1}{s^2} + \dfrac{1}{t^2}\right)^{-2}$

49. $(s + t^{-2})^{-1}$

50. $\dfrac{y^{-3} + x^{-3}}{y + x}$

51. $\dfrac{x^2 - a^2}{x - a}$

52. $\dfrac{x^3 - a^3}{x - a}$

53. $\dfrac{\dfrac{x}{3x + 4} - \dfrac{a}{3a + 4}}{x - a}$

54. $\dfrac{\dfrac{1}{x^2 + 1} - \dfrac{1}{a^2 + 1}}{x - a}$

Review Problems

55. Factor **a.** $rs - rt - 5s + 5t$ **b.** $r^4u^2 - s^4u^2 - r^4v^2 + s^4v^2$

56. Find the products **a.** $(2\sqrt{x} + 8\sqrt{z})(2\sqrt{x} - 8\sqrt{z})$ **b.** $(w - t)(w^2 + wt + t^2)$

57. Find the fifth term in the expansion $\left(\dfrac{2}{x^2} + \dfrac{y^3}{2}\right)^7$.

58. True or false: 0 is a nonnegative number.

59. The midpoint of the line segment joining the points a and $25/3$ is $7/2$. Find a.

60. Write $5\sqrt[4]{5}/\sqrt{125}$ as a single radical.

0.7 Polynomial Equations

Polynomial Equations

Recall from Section 0.6 that an algebraic expression of the form

$$a_n x^n + a_{n-1} x^{n-1} + \cdots + a_2 x^2 + a_1 x + a_0,$$

where the coefficients a_i, $i = 0, 1, \ldots, n$, are real constants, is called a **polynomial**. The statement

$$a_n x^n + a_{n-1} x^{n-1} + \cdots + a_2 x^2 + a_1 x + a_0 = 0, \qquad a_n \neq 0, \tag{1}$$

is said to be a **polynomial equation of degree** n. A number r is said to be a **solution** or **root** of the equation if a true statement results when r is substituted for x. The set of all roots of an equation is called its **solution set**. For example,

$$x^2 + 2x - 3 = 0$$

is a polynomial equation of degree 2 with roots $r_1 = 1$ and $r_2 = -3$, since

$$(1)^2 + 2(1) - 3 = 0 \quad \text{and} \quad (-3)^2 + 2(-3) - 3 = 0.$$

The solution set of $x^2 + 2x - 3 = 0$ is then $\{1, -3\}$.

Although we do not prove it in this text, any polynomial equation of degree n has *at most n real* roots. Moreover, it can be shown that such an equation has *exactly n roots*, where each root is counted according to its multiplicity.* Also, some of the roots may be complex numbers. As we shall see, the problem of finding the roots of a polynomial equation of degree $n = 1$ or $n = 2$ is easily solved. However, finding roots for $n \geq 3$ is significantly more difficult. In this case we shall rely on our ability to factor the given expression.

Linear Equation

A first-degree equation

$$ax + b = 0, \qquad a \neq 0, \tag{2}$$

is also called a **linear equation**. Since

$$ax + b = c \tag{3}$$

can be written as

$$ax + (b - c) = 0,$$

we see that equation (3) is equivalent to (2). It is a simple matter to solve (2) by elementary algebra:

$$ax + b = 0$$

---

* For example, the root $r = 3$ of the polynomial equation $x^2 - 6x + 9 = (x - 3)(x - 3) = 0$ is said to have multiplicity 2 and the root $r = 1$ of $x^5 - 3x^4 + 4x^3 - 4x^2 + 3x - 1 = (x^2 + 1)(x - 1)(x - 1)(x - 1) = 0$ has multiplicity 3.

$$ax + b + (-b) = -b$$

$$ax + 0 = -b$$

$$ax = -b$$

$$\frac{1}{a}(ax) = \frac{1}{a}(-b)$$

$$1x = -\frac{b}{a}$$

$$x = -\frac{b}{a}.$$

Example 1   Solve $-3x + 4 = -8$.

Solution

$$-3x + 4 + (-4) = -8 + (-4)$$

$$-3x = -12$$

$$x = \frac{-12}{-3}$$

$$x = 4.$$

The solution set is $\{4\}$.   ■

Quadratic Equations

A second-degree equation

$$ax^2 + bx + c = 0, \qquad a \neq 0, \tag{4}$$

is called a **quadratic equation**. From our previous remarks it follows that a quadratic equation always has two roots.

Example 2   Solve $x^2 - 9 = 0$.

Solution   The expression $x^2 - 9$ is a difference of two squares and can be readily factored. We have

$$(x - 3)(x + 3) = 0. \tag{5}$$

Since a product of real numbers $A \cdot B = 0$ if and only if $A = 0$ or $B = 0$ (or, of course, both $A = 0$ and $B = 0$), we conclude from (5) that

$$x - 3 = 0 \quad \text{or} \quad x + 3 = 0.$$

Thus, $x = 3 \quad \text{or} \quad x = -3,$

and so the solution set is $\{-3, 3\}$.   ■

The Quadratic Formula

It is not necessary to factor the expression $ax^2 + bx + c$ in order to solve the equation

$$ax^2 + bx + c = 0.$$

We will show that the solutions of this general quadratic equation can always be expressed in terms of the coefficients a, b, and c. The technique we use is called **completing the square.** First we write the equation as

$$ax^2 + bx = -c \quad \text{or} \quad x^2 + \frac{b}{a}x = -\frac{c}{a}. \tag{6}$$

Now we wish to write the left-hand side of (6) as a perfect square,

$$x^2 + 2Bx + B^2 = (x + B)^2. \tag{7}$$

Comparing (6) and (7), we see that we must have

$$2B = \frac{b}{a} \quad \text{or} \quad B = \frac{b}{2a}.$$

Thus, if we add

$$B^2 = \left(\frac{b}{2a}\right)^2$$

to the left-hand side of (6), it will be the perfect square

$$x^2 + 2\left(\frac{b}{2a}\right)x + \left(\frac{b}{2a}\right)^2 = \left(x + \frac{b}{2a}\right)^2.$$

Of course to preserve equality we also add $(b/2a)^2$ to the right-hand side of (6). We then have

$$x^2 + \left(\frac{b}{a}\right)x + \left(\frac{b}{2a}\right)^2 = \frac{-c}{a} + \left(\frac{b}{2a}\right)^2$$

$$\left(x + \frac{b}{2a}\right)^2 = \frac{b^2}{4a^2} - \frac{c}{a}$$

$$\left(x + \frac{b}{2a}\right)^2 = \frac{b^2 - 4ac}{4a^2}$$

$$x + \frac{b}{2a} = \pm\sqrt{\frac{b^2 - 4ac}{4a^2}}$$

$$x = \frac{-b}{2a} \pm \frac{\sqrt{b^2 - 4ac}}{2a},$$

so that

$$\boxed{x = \frac{-b \pm \sqrt{b^2 - 4ac}}{2a}.} \tag{8}$$

This result is known as the **quadratic formula.**

The Discriminant   From formula (8) we see that equation (4) has two roots,

$$r_1 = \frac{-b + \sqrt{b^2 - 4ac}}{2a} \quad \text{and} \quad r_2 = \frac{-b - \sqrt{b^2 - 4ac}}{2a}.$$

The nature of these roots is determined by the radicand $b^2 - 4ac$, which is called the **discriminant**. The three possible cases are summarized in the table.

Discriminant	Roots
$b^2 - 4ac > 0$	r_1 and r_2 real and distinct
$b^2 - 4ac = 0$	r_1 and r_2 real but $r_1 = r_2$
$b^2 - 4ac < 0$	No real roots (r_1 and r_2 are complex numbers)

Example 3   Solve $6x^2 + x - 2 = 0$.

Solution   From the quadratic formula with $a = 6$, $b = 1$, and $c = -2$, we have

$$x = \frac{-1 \pm \sqrt{(1)^2 - 4(6)(-2)}}{2(6)}$$

$$= \frac{-1 \pm \sqrt{49}}{12}$$

$$= \frac{-1 \pm 7}{12}.$$

It follows that the two roots are

$$r_1 = \frac{-1 + 7}{12} = \frac{1}{2} \quad \text{and} \quad r_2 = \frac{-1 - 7}{12} = -\frac{2}{3}.$$

The solution set is $\left\{ \dfrac{1}{2}, -\dfrac{2}{3} \right\}$. ∎

Example 4   Solve $3x^2 - 2x - 4 = 0$.

Solution   From the quadratic formula we can write

$$x = \frac{-(-2) \pm \sqrt{(-2)^2 - 4(3)(-4)}}{2(3)}$$

$$= \frac{2 \pm \sqrt{52}}{6} = \frac{2 \pm 2\sqrt{13}}{6} = \frac{1 \pm \sqrt{13}}{3}$$

so that

$$r_1 = \frac{1 + \sqrt{13}}{3} \quad \text{and} \quad r_2 = \frac{1 - \sqrt{13}}{3}.$$

The solution set is $\left\{ \dfrac{1 + \sqrt{13}}{3}, \dfrac{1 - \sqrt{13}}{3} \right\}$.

Example 5 Solve $9x^2 - 24x + 16 = 0$.

Solution From the quadratic formula we have

$$x = \frac{-(-24) \pm \sqrt{(-24)^2 - 4(9)(16)}}{2(9)}$$

$$= \frac{24 \pm \sqrt{576 - 576}}{18}$$

$$= \frac{24}{18} = \frac{4}{3}.$$

Since the discriminant is 0 the roots are equal and the solution set is $\left\{\frac{4}{3}\right\}$.

Example 6 Solve $x^2 + x + 3 = 0$.

Solution From the quadratic formula, we find that

$$x = \frac{-1 \pm \sqrt{(1)^2 - (4)(1)(3)}}{2(1)}$$

$$= \frac{-1 \pm \sqrt{-11}}{2}.$$

Since the discriminant is negative, the solutions are complex numbers; there are no real roots.

As we have just seen, the quadratic formula provides a means of finding all roots, real or complex, for a second-degree equation. However, we are primarily concerned with real numbers in this course, since, in general, complex numbers are not used in calculus. Consequently, in the following discussion we shall seek only real roots for polynomial equations of degree $n \geq 3$.

Example 7 Solve $x^4 - 2x^2 - 2 = 0$.

Solution This polynomial can be considered as a quadratic in the variable x^2. That is,

$$(x^2)^2 - 2(x^2) - 2 = 0.$$

Using the quadratic formula to solve for x^2, we have

$$x^2 = \frac{2 \pm \sqrt{12}}{2}$$

$$= 1 \pm \sqrt{3},$$

so either $x^2 = 1 + \sqrt{3}$ or $x^2 = 1 - \sqrt{3}$. From $x^2 = 1 + \sqrt{3}$, we have the real roots

$$x = \pm\sqrt{1 + \sqrt{3}}.$$

From $x^2 = 1 - \sqrt{3}$, we obtain no real roots, since $1 < \sqrt{3}$ and thus $1 - \sqrt{3} < 0$. ∎

Finding Rational Roots

We now discuss a special technique for finding the *possible* **rational roots** of a polynomial equation. This will simplify the problem of finding all the real roots of a polynomial equation of degree $n \geq 3$. The technique is based on the following result.

If $r = p/q$ is a rational root (reduced to lowest terms) of the polynomial equation of degree n,

$$a_n x^n + a_{n-1} x^{n-1} + \cdots + a_1 x + a_0 = 0, \qquad (9)$$

where the coefficients a_i, $i = 0, 1, \ldots, n$, are *integers*, then p divides a_0 and q divides a_n. Thus any rational roots of (9) are necessarily contained in the set of all fractions

$$\left\{ \frac{p}{q} \,\middle|\, p \text{ divides } a_0 \text{ and } q \text{ divides } a_n \right\}.$$

We illustrate an application of this result in the following example.

Example 8 Find the rational roots of $2x^3 + x^2 - 3x + 1 = 0$.

Solution The rational roots must be of the form p/q, where p divides 1 and q divides 2. Thus, possible values for p are ± 1 and for q are $\pm 2, \pm 1$. Hence, the only possible rational roots are

$$\pm 1 \quad \text{or} \quad \pm\tfrac{1}{2}.$$

We simply check to see which of these, if any, satisfy the given equation.

$$2x^3 + x^2 - 3x + 1 = 0.$$

We find

$$2(1)^3 + (1)^2 - 3(1) + 1 = 1 \neq 0$$

$$2(-1)^3 + (-1)^2 - 3(-1) + 1 = 3 \neq 0$$

$$2(\tfrac{1}{2})^3 + (\tfrac{1}{2})^2 - 3(\tfrac{1}{2}) + 1 = 0$$

$$2(-\tfrac{1}{2})^3 + (-\tfrac{1}{2})^2 - 3(-\tfrac{1}{2}) + 1 = \tfrac{5}{2} \neq 0.$$

Thus, $x = \tfrac{1}{2}$ is the only rational root. ∎

Factors and Roots   It is customary to denote a polynomial

$$a_n x^n + a_{n-1} x^{n-1} + \cdots + a_1 x + a_0$$

by $p(x)$. For convenience we shall use this notation in the following discussion. It is clear that if $(x - r)$ is a factor of a polynomial $p(x)$, that is, if

$$p(x) = (x - r)q(x),$$

where $q(x)$ is a polynomial, then $x = r$ is a root of the equation $p(x) = 0$. The converse is also true: If r is a root of a polynomial equation $p(x) = 0$, then $(x - r)$ is a factor of the polynomial $p(x)$; that is, there is a polynomial $q(x)$ such that

$$p(x) = (x - r)q(x).$$

This fact is very important in solving higher-degree polynomial equations.

Example 9   Solve $2x^3 + x^2 - 3x + 1 = 0$.

Solution   In the preceding example, we saw that $x = \frac{1}{2}$ is a rational root of the equation. It follows from the discussion above that

$$2x^3 + x^2 - 3x + 1 = (x - \tfrac{1}{2})q(x).$$

We can determine $q(x)$ by long division as follows:

$$
\begin{array}{r}
2x^2 + 2x - 2 \\
x - \tfrac{1}{2} \overline{\smash{\big)}\ 2x^3 + x^2 - 3x + 1} \\
\underline{2x^3 - x^2} \\
2x^2 - 3x \\
\underline{2x^2 - x} \\
-2x + 1 \\
\underline{-2x + 1} \\
0
\end{array}
$$

Thus, we can write

$$2x^3 + x^2 - 3x + 1 = (x - \tfrac{1}{2})(2x^2 + 2x - 2) = 0.$$

By applying the quadratic formula to the second factor, we have

$$r_2 = \frac{-1 + \sqrt{5}}{2} \quad \text{and} \quad r_3 = \frac{-1 - \sqrt{5}}{2}.$$

The solution set of the original equation is then $\left\{ \dfrac{1}{2}, \dfrac{-1 + \sqrt{5}}{2}, \dfrac{-1 - \sqrt{5}}{2} \right\}.$ ∎

Example 10   Solve $x^4 = 81$.

Solution We write the equation as

$$x^4 - 81 = 0$$

and factor $x^4 - 81$ as a difference of two squares

$$(x^2 - 9)(x^2 + 9) = 0.$$

We may then factor again as

$$(x - 3)(x + 3)(x^2 + 9) = 0.$$

Since there are no real numbers satisfying $x^2 + 9 = 0$, the only real solutions are found from

$$x - 3 = 0 \quad \text{or} \quad x + 3 = 0.$$

Hence, the real roots are $x = \pm 3$. ∎

Exercise 0.7

In Problems 1–6, solve the given equation.

1. $2x + 14 = 0$ **2.** $3x - 5 = 0$ **3.** $-5x + 1 = 2$

4. $\frac{1}{2}x - \frac{1}{4} = \frac{1}{8}$ **5.** $\frac{1}{3}x - \frac{2}{3} = \frac{1}{3}$ **6.** $2(3x - 1) = 4(1 - x)$

Solve equations 7–18, by factoring.

7. $x^2 + x - 6 = 0$ **8.** $2x^2 + x - 1 = 0$ **9.** $x^2 - 16 = 0$

10. $x^2 + 2x - 35 = 0$ **11.** $3x^2 - 13x + 4 = 0$ **12.** $64 - x^2 = 0$

13. $8x^2 - 22x + 15 = 0$ **14.** $x^3 - 9x = 0$ **15.** $16x^4 - x^2 = 0$

16. $x^4 - 18x^2 + 32 = 0$ **17.** $4x^5 - 25x^3 = 0$ **18.** $x^6 = -3x^5 + 28x^4$

Solve equations 19–32, by the quadratic formula.

19. $2x^2 + x - 1 = 0$ **20.** $x^2 + 3x - 2 = 0$ **21.** $5x^2 - 4x - 4 = 0$

22. $2x^2 + 5x + 3 = 0$ **23.** $x^2 - 2x + 5 = 0$ **24.** $4x^2 + 3x - 5 = 0$

25. $3x^2 - 7x + 2 = 0$ **26.** $1 + 2x - 6x^2 = 0$ **27.** $2 + 5x - 10x^2 = 0$

28. $9x^2 + 30x + 25 = 0$ **29.** $4x^2 - 12x + 9 = 0$ **30.** $8x^2 + 10x + 5 = 0$

31. $x^4 - 6x^2 + 7 = 0$ **32.** $x^4 - 2x^2 - 4 = 0$

In Problems 33–40, find the real roots of the given equation.

33. $2x^3 + x^2 - 2x - 1 = 0$ **34.** $3x^3 + 4x^2 - 12x - 16 = 0$

35. $x^3 + 2x^2 + 2x + 1 = 0$ **36.** $3x^3 - 5x^2 + 7x + 3 = 0$

37. $4x^3 - 11x + 3 = 0$ **38.** $12x^3 + 13x^2 - 12x + 2 = 0$

39. $24x^3 + 14x^2 - 29x + 6 = 0$ **40.** $36x^4 + 36x^3 - 115x^2 + 48x - 5 = 0$

41. Determine d so that $(d + 3)x + 2d = 0$ has $x = 4$ as a solution.

42. Determine d so that $(2d - 1)x + 3 = d$ has $x = -2$ as a solution.

43. Determine all values of d so that $x^2 + (d + 6)x + 8d = 0$ has two equal real roots.

44. Determine all values of d so that $3dx^2 - 4dx + d + 1 = 0$ has two equal real roots.

45. Determine the other root of $(k - 2)x^2 - x - 4k = 0$ given that one root is -3.

46. Find the error in the following "solution."

$$x^2 - 5x + 6 = x^2 - 9$$
$$(x - 3)(x - 2) = (x - 3)(x + 3)$$
$$x - 2 = x + 3$$
$$-2 = 3.$$

47. If T_f represents temperature measured in degrees Fahrenheit and T_c represents temperature measured in degrees Celsius, the equation $T_c = \frac{5}{9}(T_f - 32)$ represents the conversion from the Fahrenheit to the Celsius scale. Find the degrees Fahrenheit corresponding to $120°C$.

48. The height s, measured from ground level, of an arrow shot upward with a velocity of 96 ft/sec from an initial height of 22 ft above the ground, is given by the formula $s = -16t^2 + 96t + 22$. Determine the times when the arrow is 102 ft above the ground.

49. A farmer encloses a rectangular pasture with 1000 ft of fence. Find the dimensions of the pasture if the fence encloses 62,500 ft² of land.

50. A box with no top is made from a square piece of cardboard by cutting square pieces from each corner and then folding up the sides. See Figure 0.11. The length of one side of the cardboard is 10 inches. Find the length of one side of the squares that were cut from the corners if the volume of the box is 48 in³.

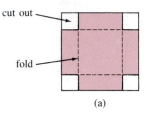
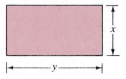

(a) (b)

Figure 0.11

51. The lengths of the sides of the rectangle shown in Figure 0.12 are said to satisfy the **golden ratio** if

$$\frac{x}{y} = \frac{y}{x + y}.$$

Find the dimensions of a piece of paper with area 100 in² if the lengths of its sides satisfy the golden ratio.

Figure 0.12

Review Problems

52. Find b so that the midpoint of $(3, b)$ is $17/4$.

53. Rationalize the denominator of $\dfrac{2}{x + \sqrt{a}}$.

54. Eliminate negative exponents and simplify:
 a. $(16x^4 y^{-2})^{-3/4}$ **b.** $(2x^{2/3})^2(3x^{-5/3})$

55. Simplify $\dfrac{\dfrac{(x + h)^2}{x + h + 1} - \dfrac{x^2}{x + 1}}{h}$.

56. Simplify $\left(\dfrac{1}{x^{-2}} + \dfrac{1}{y^{-2}}\right)^{-1}$.

57. Factor and simplify: $(x - 2)^4(2x) + 4(x^2 + 1)(x - 2)^3$.

0.8 Miscellaneous Equations

Many equations that are not polynomial equations can be put into that form by suitable algebraic manipulations. For example, in order to solve

$$\frac{1}{x - 2} = \frac{5}{3x} \tag{1}$$

we multiply both sides by $(x - 2)(3x)$ and simplify:

$$(x - 2)(3x)\left(\frac{1}{x - 2}\right) = (x - 2)(3x)\left(\frac{5}{3x}\right)$$

$$3x = 5(x - 2)$$

$$-2x + 10 = 0. \tag{2}$$

We then solve the linear equation (2).

$$-2x = -10$$

$$x = 5.$$

By substituting $x = 5$ into (1), we see that

$$\frac{1}{5 - 2} = \frac{5}{3(5)} \quad \text{or} \quad \frac{1}{3} = \frac{1}{3},$$

and hence $x = 5$ is a solution of (1).

Extraneous Solutions

When solving an equation such as the above one, care must be taken that **extraneous solutions** are not introduced. Consider the following examples.

Example 1   Solve $2 - \dfrac{1}{x} = \dfrac{x-1}{x}$.   (3)

Solution   If we multiply both sides of the given equation by x, we obtain

$$2x - 1 = x - 1$$

or $x = 0$.   (4)

Then $x = 0$ is a solution of (4) but not a solution of (3), since division by 0 is not defined. Thus, 0 is an extraneous solution of (3). In fact, the given equation has no solutions and so we say its solution set is the empty set $\varnothing$.   ∎

Example 2   Solve $x - 5 = \sqrt{x + 7}$.   (5)

Solution   Squaring both sides of the given equation, we have

$$x^2 - 10x + 25 = x + 7 \quad \text{or} \quad x^2 - 11x + 18 = 0.$$

This equation factors as

$$(x - 9)(x - 2) = 0.$$

The solutions of this latter equation are $x = 9$ and $x = 2$. If we check $x = 9$ in the original equation, we find that

$$9 - 5 = 4 = \sqrt{9 + 7},$$

so 9 is a solution of (5). But if we substitute $x = 2$ into equation (5), we find that

$$2 - 5 = -3 \neq \sqrt{2 + 7} = 3.$$

Therefore, 2 is an extraneous solution. Thus, the solution set of (5) is $\{9\}$.   ∎

As illustrated by the previous examples, it is absolutely essential to check your solutions if you have performed either of the following operations:

- Multiplied the equation by an expression containing a variable, or
- Raised both sides of the equation to a power.

Example 3   Solve $x - \sqrt{x} = 0$.

Solution   To solve for x we must eliminate the radical involving x from the equation. We write the equation as

$$x = \sqrt{x}$$ (6)

and square both sides,

$$x^2 = x. \tag{7}$$

Rewriting the last equation as

$$x^2 - x = 0 \quad \text{or} \quad x(x - 1) = 0,$$

we observe that the solutions of (7) are $x = 0$ and $x = 1$. By checking these in the original equation (6), we find that both are solutions. The solution set is $\{0, 1\}$. ∎

A mistake frequently made in solving an equation such as $x^2 = x$ in (7) of Example 3 is to divide by the common factor x, so that

$$\frac{x^2}{x} = \frac{x}{x} \quad \text{and} \quad x = 1.$$

As a result of the division, the solution $x = 0$ is lost. Since division by an algebraic expression which may equal zero can result in the *loss* of a solution, this procedure should be avoided.

Example 4 Solve $\sqrt[3]{x^2 + 2} = 3$.

Solution Cubing both sides of the equation, we have

$$(\sqrt[3]{x^2 + 2})^3 = (3)^3$$
$$x^2 + 2 = 27$$
$$x^2 = 25$$
$$x = \pm 5.$$

Substituting $x = 5$ and $x = -5$ in the original equation, we find that

$$\sqrt[3]{(5)^2 + 2} = \sqrt[3]{27} = 3 \quad \text{and} \quad \sqrt[3]{(-5)^2 + 2} = \sqrt[3]{27} = 3.$$

Therefore, the solution set is $\{-5, 5\}$. ∎

Example 5 Solve $\sqrt{2x - 4} - \sqrt{x - 1} - 1 = 0$.

Solution We first write the equation as

$$\sqrt{2x - 4} = \sqrt{x - 1} + 1$$

before squaring both sides to avoid obtaining the product of the two radicals. We then have

$$2x - 4 = (x - 1) + 2\sqrt{x - 1} + 1 \quad \text{or} \quad x - 4 = 2\sqrt{x - 1}.$$

We square both sides again:

$$(x - 4)^2 = 4(x - 1)$$

$$x^2 - 8x + 16 = 4x - 4$$

$$x^2 - 12x + 20 = 0$$

$$(x - 10)(x - 2) = 0.$$

The solutions of this latter equation are $x = 10$ and $x = 2$.

Checking $x = 10$:

$$\sqrt{2(10) - 4} - \sqrt{10 - 1} - 1 =$$

$$\sqrt{16} - \sqrt{9} - 1 =$$

$$4 - 3 - 1 = 0$$

Checking $x = 4$:

$$\sqrt{2(2) - 4} - \sqrt{2 - 1} - 1 =$$

$$\sqrt{0} - \sqrt{1} - 1 =$$

$$-2 \neq 0.$$

Thus, the solution set of the original equation is $\{10\}$. ∎

Example 6 Solve $x^{2/3} + 4x^{1/3} - 5 = 0$.

Solution The equation can be written as

$$(x^{1/3})^2 + 4(x^{1/3})^1 - 5 = 0,$$

and hence may be considered a quadratic in the variable $z = x^{1/3}$. The quadratic expression on the left side of

$$z^2 + 4z - 5 = 0$$

factors, so we have

$$(z + 5)(z - 1) = 0.$$

The solutions of this latter equation are $z = -5$ and $z = 1$. Since $z = x^{1/3}$, we have

$$x^{1/3} = -5 \quad \text{and} \quad x^{1/3} = 1.$$

Cubing each of these equations, we obtain

$$x = -125 \quad \text{and} \quad x = 1.$$

By substitution it can be verified that both of these satisfy the original equation. Hence, the solution set is $\{-125, 1\}$. ∎

Equations Involving Absolute Values

From the definition of the absolute value of a number in Section 1.1 it follows that for any *positive* real number a,

$$|x| = a \quad \text{if and only if} \quad x = a \quad \text{or} \quad x = -a.$$

Example 7 Solve $|5x - 3| = 8$.

Solution The given equation is equivalent to the two equations: $5x - 3 = 8$ or $5x - 3 = -8$. We solve each of these equations.

$$5x - 3 = 8 \qquad\qquad 5x - 3 = -8$$
$$5x = 11 \qquad\qquad 5x = -5$$
$$x = \tfrac{11}{5} \qquad\qquad x = -1$$

Thus, the solution set is $\{-1, \tfrac{11}{5}\}$.

■

Example 8 Since the absolute value of a real number is always nonnegative, there is no solution to the equation

$$|x - 4| = -3.$$

Hence the solution set is the empty set $\varnothing$.

■

Exercise 0.8

In Problems 1–40, solve the given equation.

1. $\dfrac{1}{x-1} + \dfrac{3}{4-x} = 0$

2. $\dfrac{1}{x-2} = \dfrac{2x+1}{x^2-4}$

3. $\dfrac{x}{x-5} = 2 + \dfrac{5}{x-5}$

4. $\dfrac{3x}{x-2} = \dfrac{6}{x-2} + 1$

5. $\dfrac{1-x}{x^2} = \dfrac{3}{x-1}$

6. $\dfrac{3-x}{x+2} = 2 + \dfrac{x}{x+2}$

7. $\dfrac{1}{(x+2)^2} + \dfrac{1}{x+2} - 6 = 0$

8. $\dfrac{6}{y^4} - \dfrac{1}{y^2} - 1 = 0$

9. $\dfrac{2}{(z+1)^2} - \dfrac{9}{z+1} - 5 = 0$

10. $\dfrac{16}{(x^2+1)^2} - \dfrac{8}{x^2+1} = -1$

11. $x^{2/5} - 7x^{1/5} - 8 = 0$

12. $6x^{1/3} - 7x^{1/6} - 3 = 0$

13. $x^{1/4} - 12x^{-1/4} - 1 = 0$

14. $(x-1)^{1/2} - (x-1)^{1/4} - 20 = 0$

15. $12(x+3)^{2/3} + (x+3)^{1/3} - 1 = 0$

16. $\sqrt{2 - 5x} - 3 = 0$

17. $\sqrt{6x^2 + 3} = 7$

18. $\sqrt{4x^2 - 2} + 5 = 0$

19. $\sqrt[3]{2x - 1} = 5$

20. $\sqrt[3]{3 - 5x} = 0$

21. $\sqrt{7 - 6x} = -1$

22. $3x + \sqrt{3x - 1} = 1$

23. $\sqrt{x+1} + x - 1 = 0$

24. $\sqrt{7 - x} = 3x$

25. $\sqrt{2x - 5} - \sqrt{x - 3} = 1$

26. $\sqrt{2x - 7} = 1 + \sqrt{x - 4}$

27. $\sqrt{2x} - \sqrt{x + 1} + 1 = 0$

28. $x + \sqrt{3x + 1} = 3$

29. $\sqrt{4x - 3} - \sqrt{2 - 2x} - 1 = 0$

30. $|4x - 3| = 8$

31. $|4x - 1| = 2$

32. $|5 - 4x| = 7$

33. $|\tfrac{1}{4} - \tfrac{3}{2}x| = 1$

34. $|x - 7| = 0$

35. $|15 - 4x| = 0$

36. $|16x - 2| = -3$

37. $|2 - 3x| = -10$

38. $|2x^2 + 1| = 3$

39. $|1 - x^2| = 5$

40. $||x| - 1| = 3$

Review Problems

41. Reduce the index of the radical $\sqrt[6]{8x^{15}y^9}$ as much as possible.

42. Simplify: $\left(\dfrac{x^4 - x^2y^2}{x + 2y}\right)^{-1} \cdot \dfrac{x(x + y)^2}{x^2 - 4y^2}$

43. Use Pascal's triangle to find the coefficient of the fourth term in the expansion of $(x - y)^7$.

44. True or false: Every polynomial equation has at least one rational root.

0.9 Inequalities

In Section 0.1 we introduced the concept of "less than" from an intuitive viewpoint. We now give a formal definition of this order relation on the set of real numbers.

> **Definition 1.3** *For any real numbers a and b we say that a is **less than** b, written $a < b$, if $b - a$ is positive. Equivalently, we also say that b is **greater than** a and write $b > a$.*

For example

$$-7 < 5 \quad \text{since} \quad 5 - (-7) = 12$$

is positive. Alternatively, we can write $5 > -7$.

Example 1   Using the greater-than relation, "$>$," compare the real numbers π and $\frac{22}{7}$.

Solution   Since $\pi = 3.14159\ldots$ and $\frac{22}{7} = 3.142857142857\ldots$, we have

$$\tfrac{22}{7} - \pi = (3.1428\ldots) - (3.1415\ldots)$$

$$= 0.001\ldots,$$

which is positive. Therefore,

$$\tfrac{22}{7} > \pi.$$

∎

Example 2   **a.** $3 > 0$ since $3 - 0 = 3$ is positive.

b. $-9 < 0$ since $0 - (-9) = 9$ is positive.

∎

Example 2 suggests the facts that:

$$a > 0 \quad \text{if and only if } a \text{ is positive;}$$
$$a < 0 \quad \text{if and only if } a \text{ is negative.}$$

Two additional order relations are: **a is less than or equal to b**, given by

$$a \leq b \quad \text{if and only if either} \quad a < b \quad \text{or} \quad a = b;$$

and **a is greater than or equal to b**, given by

$$a \geq b \quad \text{if and only if either} \quad a > b \quad \text{or} \quad a = b.$$

Example 3   Since $2 = \sqrt{4}$, we may write $2 \geq \sqrt{4}$. Also, we may write $4 \leq 9$, since $4 < 9$. ∎

Rules for the Order Relation

The following rules enable us to change the form of an inequality. We shall verify the first of these rules and leave the verification of the others to the exercises.

Let a, b, and c be real numbers.

I. If $a < b$ and c is any real number, then
$$a + c < b + c.$$
II. If $a < b$ and c is positive, then
$$a \cdot c < b \cdot c.$$
III. If $a < b$ and c is negative, then
$$a \cdot c > b \cdot c.$$

To verify rule I, assume that $a < b$; then it follows that $b - a$ is positive. If we add $c - c = 0$ to a positive number, the sum is positive. Therefore,

$$b - a + (c - c) = b + c - a - c$$
$$= (b + c) - (a + c)$$

is a positive number. Hence, we have that $a + c < b + c$.

These rules may be extended to the other order relations. For example, rule II could be more generally stated as: *If both sides of an inequality are multiplied by the same positive number, then the sense of the inequality remains unchanged.*

Solving Inequalities

Rules I–III are useful in **solving inequalities**: by this we mean finding all values of the variable for which the inequality is true.

Example 4   Solve $3x - 7 \geq 5$.

Solution   To solve an inequality we isolate the variable using rules I–III,

$$3x - 7 \geq 5$$
$$3x - 7 + 7 \geq 5 + 7 \qquad \text{(by I)}$$
$$3x \geq 12$$
$$\tfrac{1}{3}(3x) \geq \tfrac{1}{3}(12) \qquad \text{(by II)}$$
$$x \geq 4.$$

The solution set is $\{x \mid x \geq 4\}$. ■

Example 5   Solve $\tfrac{1}{2} - 3x > \tfrac{5}{2}$.

Solution

$$\tfrac{1}{2} - 3x > \tfrac{5}{2}$$
$$-\tfrac{1}{2} + \tfrac{1}{2} - 3x > -\tfrac{1}{2} + \tfrac{5}{2} \qquad \text{(by I)}$$
$$-3x > 2$$
$$-\tfrac{1}{3}(-3x) < -\tfrac{1}{3}(2) \qquad \text{(by III)}$$
$$x < -\tfrac{2}{3}.$$

The solution set is $\{x \mid x < -\tfrac{2}{3}\}$. ■

Simultaneous Inequalities

The **simultaneous inequality**

$$a < x < b$$

means that both $a < x$ *and* $x < b$. For example, the set of real numbers that satisfy

$$2 < x < 5$$

is the intersection of the sets $\{x \mid 2 < x\}$ and $\{x \mid x < 5\}$. This is illustrated in Figure 0.13. It is customary to write simultaneous inequalities with the smaller number on the left; thus $5 \geq x > -3$, while technically correct, should be written $-3 < x \leq 5$. Also, the symbols "$<$" and "$>$" are never combined in a simultaneous inequality.

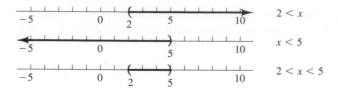

2 < x

x < 5

2 < x < 5

Figure 0.13

Solving Simultaneous Inequalities

As shown in the following examples a simultaneous inequality may be solved by isolating the variable in the middle. Rules I–III are applied to both parts of the inequality at the same time.

Example 6 Solve $-7 \leq 2x + 1 < 19$.

Solution

$$-7 \leq 2x + 1 < 19$$

$$-7 - 1 \leq 2x + 1 - 1 < 19 - 1 \quad \text{(by I)}$$

$$-8 \leq 2x < 18$$

$$\tfrac{1}{2}(-8) \leq \tfrac{1}{2}(2x) < \tfrac{1}{2}(18) \quad \text{(by II)}$$

$$-4 \leq x < 9.$$

The solution set is $\{x \mid -4 \leq x < 9\}$. ∎

Interval Notation

Interval notation is convenient for describing certain sets of real numbers. The basic intervals and their graphs are:

Set	Interval Notation	Name	Graph
$\{x \mid a < x < b\}$	(a, b)	Open interval	
$\{x \mid a \leq x \leq b\}$	$[a, b]$	Closed interval	

It is a useful convention to denote half-lines as follows.

Set	Interval Notation	Name	Graph
$\{x \mid a < x\}$	(a, ∞)	Open half-line	
$\{x \mid x \leq b\}$	$(-\infty, b]$	Closed half-line	

The symbols $[a, b)$ and $(a, b]$ have the obvious meaning and are called half-open intervals. For example, $[2, 4)$ is equivalent to $2 \leq x < 4$.

Example 7 Write the solutions of the three preceding examples in interval notation and sketch their graphs.

Solution
a. $\{x \mid x \geq 4\} = [4, \infty)$
b. $\{x \mid x < -\frac{2}{3}\} = (-\infty, -\frac{2}{3})$
c. $\{x \mid -4 \leq x < 9\} = [-4, 9)$
See Figure 0.14 for the graphs.

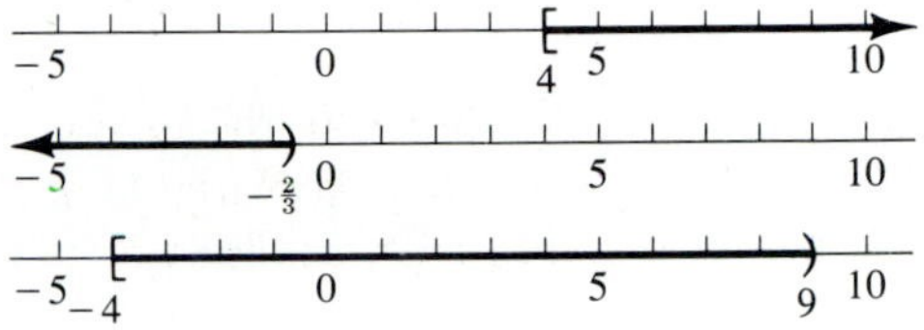

Figure 0.14

Example 8 Solve $0 < 1 - 2x < 3$ and give the solution in interval notation.

Solution
$$0 < 1 - 2x < 3$$
$$-1 + 0 < -1 + 1 - 2x < -1 + 3 \quad \text{(by I)}$$
$$-1 < -2x < 2$$
$$-\tfrac{1}{2}(-1) > -\tfrac{1}{2}(-2x) > -\tfrac{1}{2}(2) \quad \text{(by III)}$$
$$\tfrac{1}{2} > x > -1$$
or, equivalently, $-1 < x < \frac{1}{2}$. Thus, the solution is the interval $(-1, \frac{1}{2})$. ∎

Exercise 0.9

In Problems 1–4, compare the given pairs of numbers using the order relation "<."

1. $15, -3$ **2.** $-9, 0$ **3.** $\frac{4}{3}, 1.33$ **4.** $-\frac{7}{15}, -\frac{5}{11}$

In Problems 5–8, compare the given pairs of numbers using the order relation "≥."

5. $2.5, \frac{5}{2}$ **6.** $-\frac{1}{7}, -0.143$ **7.** $7\pi, 22$ **8.** $\sqrt{2}, 1.414$

9. Express the rule corresponding to rule I for the order relation "≥."
10. Express the rule corresponding to rule III for the order relation ">."

In Problems 11–22, solve the given inequalities and indicate where rules I, II, and III are used.

11. $x + 3 > -2$ **12.** $3x - 9 < 6$ **13.** $\frac{3}{2}x + 4 \leq 10$
14. $5 - \frac{5}{4}x \geq -4$ **15.** $-7 < x - 2 < 1$ **16.** $3 \leq x + 4 \leq 10$
17. $7 < 3 - \frac{1}{2}x \leq 8$ **18.** $-3 < 4x < 0$ **19.** $-x < x + 1 < 3 - x$
20. $x - 1 \leq 4x - 3$ **21.** $4 < -x < 8$ **22.** $3x + 2 \geq 5x + 10$

In Problems 23–40, solve the given inequality. Express the solution in interval form and graph the solution set.

23. $2 + 3x < 0$

24. $\frac{1}{2} + 5x > 2$

25. $2 + x \geq 5$

26. $-7x + 3 \leq 4$

27. $\pi + 6 \geq 3x - 2$

28. $\sqrt{2} - 4 \leq \frac{3}{2} - \frac{1}{2}x$

29. $5x - 8 < 2x - 1$

30. $-10 < \frac{2}{3}x < 4$

31. $-3 \leq -x < 2$

32. $0 < 3x - 7 \leq 1$

33. $-\frac{1}{2} < 2 - 4x < 0$

34. $2 \leq \frac{1}{2}x - 6 < 8$

35. $\sqrt{2} + 1 < 5x + 1 < 8$

36. $100 < 41 - 6x < 121$

37. $0 < \dfrac{1}{x} < 4$

38. $-2 < \dfrac{1}{x} < 0$

39. $x^2 < (x - 4)^2$

40. $x(x - 2) \geq (x - 1)(x + 3)$

In Problems 41–44, describe the given interval using (a) inequality notation, (b) interval notation.

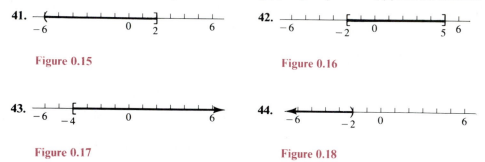

41.

Figure 0.15

42.

Figure 0.16

43.

Figure 0.17

44.

Figure 0.18

45. The relationship between degrees Fahrenheit and degrees Celsius is $T_f = (9/5)T_c + 32$. Find the interval on the Celsius scale corresponding to $32 \leq T_f \leq 212$.

46. Use the relationship between T_f and T_c in Problem 45 to find the interval on the Fahrenheit scale corresponding to $-20 \leq T_c \leq 30$.

47. The value V of a computer, which cost $50,000 initially and is undergoing linear depreciation over 20 years, is given by $V = 50{,}000(1 - x/20)$ where x represents years. Determine the values of x so that $0 < V < 20{,}000$.

48. The International Whaling Commission has specified that the weight of a mature blue whale be defined by $W = (3.51)L - 192$, where L is its length in feet and W is measured in long tons. Determine the interval of weights corresponding to $L \geq 70^*$.

49. The pulse rate of a healthy person while jogging can vary widely. To obtain the maximum beneficial effect from jogging, the pulse rate should be maintained in a certain interval. The endpoints of that interval are determined as follows. The jogger's age is subtracted from 220. The result is multiplied by 0.70 and 0.85, respectively, to obtain the two endpoints. Write the desired interval for the pulse rate of a 40-year-old jogger as a simultaneous inequality.

50. Prove rule II using the fact that the product of two positive numbers is positive.

51. Prove rule III using the fact that the product of a positive and a negative number is negative.

_____

* For $0 < L < 70$ the whale is considered immature, and the formula is not used.

Review Problems

52. Solve: $2|x| + x = 5$.

53. Solve: $\dfrac{6}{z} - 4 = \dfrac{3}{2z}$.

54. Solve: $x^{2/3} + \frac{2}{3}(x - 1)x^{-1/3} = 0$.

55. If x and y are nonzero real numbers, simplify the expression $4x^0 + (8y)^0$.

56. Express $|h|$ without the absolute value symbol if $h \le -3$.

57. Rationalize the numerator of $\dfrac{\sqrt{2 + h} - \sqrt{2}}{h}$ and simplify.

58. Combine and simplify $\dfrac{x^2 - 1}{x^2 - 2x - 8} - \dfrac{x^2 - 9}{x^2 - 7x + 12}$.

0.10 Inequalities Involving Absolute Values

In calculus one encounters expressions of the form

$$|x - a| < b.$$

In order to solve an inequality involving an absolute value, we require two additional rules.

> Let b be a positive real number.
>
> **IV.** $|X| < b$ if and only if $-b < X < b$.
>
> **V.** $|X| > b$ if and only if $X < -b$ or $X > b$.

Rules IV and V also hold with "$\le$" in place of "$<$" and "$\ge$" in place of "$>$."

Recall from Section 0.1 that $|x|$ represents the distance along the number line from x to the origin. Thus, $|x| < b$ means that the distance from x to the origin is less than b. From Figure 0.19 we see that this is the set of real numbers x such that $-b < x < b$. On the other hand, $|x| > b$ means that the distance from x to the origin is greater than b. Hence, as seen in Figure 0.20 either $x > b$ or $x < -b$.

$|x| < b$ $|x| > b$

Figure 0.19   **Figure 0.20**

Example 1   Solve $|3x - 7| < 1$. Graph the solution.

Solution   Using rule IV, we make the identification $X = 3x - 7$ and replace $|3x - 7| < 1$ by the equivalent simultaneous inequality and solve.

$$-1 < 3x - 7 < 1$$

$$-1 + 7 < 3x < 1 + 7 \qquad \text{(by I)}$$

$$6 < 3x < 8$$

$$\tfrac{1}{3}(6) < \tfrac{1}{3}(3x) < \tfrac{1}{3}(8) \qquad \text{(by II)}$$

$$2 < x < \tfrac{8}{3}.$$

In interval notation, the solution is $(2, \tfrac{8}{3})$. The graph of this solution is given in Figure 0.21. ∎

Figure 0.21

Example 2   Solve $|4 - \tfrac{1}{2}x| \geq 7$. Graph the solution.

Solution   From rule V, for $\geq$, we make the identification $X = 4 - \tfrac{1}{2}x$, and conclude that the given inequality is equivalent to

$$4 - \tfrac{1}{2}x \leq -7 \quad \text{or} \quad 4 - \tfrac{1}{2}x \geq 7.$$

We solve each of these inequalities separately. First we have

$$4 - \tfrac{1}{2}x \leq -7 \qquad \text{(by I)}$$

$$-4 + 4 - \tfrac{1}{2}x \leq -4 - 7$$

$$-\tfrac{1}{2}x \leq -11$$

$$-2(-\tfrac{1}{2}x) \geq -2(-11) \qquad \text{(by III)}$$

$$x \geq 22.$$

In interval notation this is $[22, \infty)$. And then

$$4 - \tfrac{1}{2}x \geq 7$$

$$-4 + 4 - \tfrac{1}{2}x \geq -4 + 7$$

$$-\tfrac{1}{2}x \geq 3$$

$$-2(-\tfrac{1}{2}x) \leq -2(3)$$

$$x \leq -6.$$

In interval notation this is $(-\infty, -6]$.

Since any real number that satisfies either $4 - \frac{1}{2}x \leq -7$ or $4 - \frac{1}{2}x \geq 7$ satisfies $|4 - \frac{1}{2}x| \geq 7$, the solution set is the union of the two disjoint intervals

$$(-\infty, -6] \cup [22, \infty).$$

The graph of this solution is given in Figure 0.22.

■

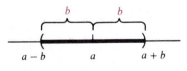

Figure 0.22

Example 3   Solve $|x + 1| > -1$.

Solution   Recall from Section 0.1 that for any real number, r, we have $|r| \geq 0$. Thus, regardless of the value of x, it is always true that $|x + 1| > -1$. Hence, the solution set is the set R of real numbers.

■

Example 4   Solve $|3x - 4| \leq 0$.

Solution   Since the absolute value of an expression is never negative, the only values satisfying the given inequality are those for which

$$|3x - 4| = 0 \quad \text{or} \quad 3x - 4 = 0.$$

Hence, the solution set is $\{\frac{4}{3}\}$.

■

Recall from Section 0.1 that $|x - a|$ represents the distance along the number line from x to a. Thus, the inequality

$$|x - a| < b$$

will be satisfied by all real numbers x whose distance from a is less than b. That is, the solution set is the open interval of length $2b$ centered at a. This interval is shown in Figure 0.23.

Figure 0.23

Example 5   Use an inequality to describe the set of real numbers which are less than 7 units from 2. Express the solution of this inequality as an interval.

Solution From the discussion above with $a = 2$ and $b = 7$, we see that the inequality is

$$|x - 2| < 7.$$

We could solve this inequality using rule IV, or we can simply note that the interval described has midpoint 2 and length $2(7) = 14$. It can thus be written as $(2 - 7, 2 + 7)$ or $(-5, 9)$. ∎

In some applications it is useful to express an interval (c, d) in terms of an inequality whose solution set is this interval.

Example 6 Find an inequality whose solution set is $(3, 8)$.

Solution We first find the midpoint a of the interval $(3, 8)$. This will be

$$a = \frac{3 + 8}{2} = 5.5.$$

Next, we find the distance b from the midpoint to an endpoint of the interval. This will be

$$b = 8 - 5.5 = 2.5.$$

The desired inequality is then

$$|x - 5.5| < 2.5. \qquad ∎$$

Exercise 0.10

In Problems 1–18, solve the given inequality. Express the solution in interval form and graph the solution set.

1. $|x| < 5$

2. $|-2x| < 9$

3. $|x - 2| < 7$

4. $|x| > 6$

5. $|\frac{1}{2}x| > 2$

6. $|x - 4| > 4$

7. $|x - 4| \leq 9$

8. $|3 + x| > 7$

9. $|2x - 7| \geq 1$

10. $|5 - \frac{1}{3}x| < \frac{1}{2}$

11. $|x + \sqrt{2}| \geq 1$

12. $|7x + 2| > 0$

13. $|17x - 3| \leq -3$

14. $|-12 - 3x| < 6$

15. $|\frac{5}{2}x - 3| \leq 0$

16. $|\sqrt{3}x - 1| > 2$

17. $|3 - 5x| \geq 8$

18. $1 < |x| < 3$

In Problems 19–24, find an inequality involving an absolute value whose solution is the set of real numbers satisfying the given condition. Express each set in interval form.

19. Less than 5 units from 12.

20. Greater than or equal to 2 units from -3.

21. Less than $\frac{1}{2}$ unit from 3.5.

22. Greater than 7 units from $\frac{5}{3}$.

23. Less than $\sqrt{3}$ units from π.

24. Less than or equal to 5 units from -1.

In Problems 25–32, find an inequality involving an absolute value whose solution set is the given interval.

25. $[3, 17]$ **26.** $(-2, 0)$ **27.** $(-5, 3)$

28. $[-3, 3]$ **29.** $(7, 21)$ **30.** $[-\frac{2}{3}, \frac{5}{3}]$

31.   **32.**

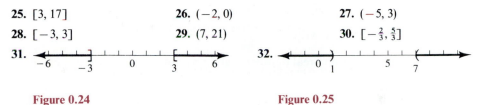

Figure 0.24   Figure 0.25

Review Problems

33. True or false: $\frac{4}{3}$ is a rational root of $3x^3 - 4x^2 - 21x + 28 = 0$.

34. Solve: $x^3 + x^2 = x$.

35. Simplify: $\dfrac{\dfrac{1}{(x-1)^3} - 1}{x - 2}$.

36. Simplify: $(x^2 + 1)^{-2}(3x^2) + x^3(-2)(x^2 + 1)^{-3}(2x)$.

37. Simplify: $\dfrac{\sqrt{2x+1} - x(2x+1)^{-1/2}}{2x + 1}$

38. True or false: $\pi/4$ is a rational number.

39. Solve: $\dfrac{4}{t^2} - \dfrac{16}{t} + 7 = 0$.

40. True or false: The solution set of the equation $\sqrt{x} + 4 = 0$ is empty.

Chapter Review

For any real number a, the **absolute value** of a, denoted by $|a|$, is

$$|a| = \begin{cases} a & \text{if } a \geq 0 \\ -a & \text{if } a < 0. \end{cases}$$

If a and b are two points on the real number line, the **distance between a and b** is

$$d(a, b) = |b - a|.$$

For any real number x and any positive integer n,

$$\overbrace{x^n = x \cdot x \cdot \cdots \cdot x}^{n \text{ factors}}$$

and

$$x^{-n} = \frac{1}{x^n}, \quad x \neq 0.$$

For any positive integer p and any real number r, we call r the **pth root** of x if

$$r^p = x$$

and we write

$$r = \sqrt[p]{x}.$$

For any real number x and any positive integer n we define

$$x^{1/n} = \sqrt[n]{x}$$

provided that $\sqrt[n]{x}$ is a real number. And for any integer m,

$$x^{m/n} = (x^{1/n})^m$$

provided that $x^{1/n}$ is defined and m/n is reduced to lowest terms.

The **laws of exponents** for any *rational* numbers r and s and any real numbers x and y are

$$x^r x^s = x^{r+s}$$

$$(x^r)^s = x^{rs}$$

$$(xy)^r = x^r y^r$$

$$\left(\frac{x}{y}\right)^r = \frac{x^r}{y^r}$$

$$\frac{x^r}{x^s} = x^{r-s}$$

provided that each expression represents a real number.

From the distributive law of real numbers,

$$A(B + C) = AB + AC,$$

we obtain the **product formulas**,

$$(X + Y)(X + Y) = X^2 + 2XY + Y^2$$

$$(X - Y)(X - Y) = X^2 - 2XY + Y^2$$

$$(X + Y)(X - Y) = X^2 - Y^2$$

$$(X + Y)(X^2 - XY + Y^2) = X^3 + Y^3$$

$$(X - Y)(X^2 + XY + Y^2) = X^3 - Y^3$$

$$(AX + B)(CX + D) = ACX^2 + (AD + BC)X + BD.$$

Reading these product formulas from right to left, we obtain **factorization formulas**.

The **binomial theorem** enables us to expand $(a + b)^n$ for any positive integer n; for example,

$$(a + b)^3 = a^3 + 3a^2b + 3ab^2 + b^3.$$

A quotient of two polynomials is called a **rational expression**. The **rules for fractions** and the properties of the real number system are used to combine and simplify rational expressions.

The solutions of the general quadratic equation

$$ax^2 + bx + c = 0$$

are given by the **quadratic formula**,

$$x = \frac{-b \pm \sqrt{b^2 - 4ac}}{2a}.$$

Any rational root of the polynomial equation of degree n,

$$a_nx^n + a_{n-1}x^{n-1} + \cdots + a_1x + a_0 = 0$$

is necessarily contained in the set

$$\left\{ \frac{p}{q} \middle| p \text{ divides } a_0 \text{ and } q \text{ divides } a_n \text{ where } p \text{ and } q \text{ are integers} \right\}.$$

When solving equations, extraneous solutions may possibly be introduced. Thus, it is important to verify all "solutions" by substituting back into the original equation.

If a, b, and c are real numbers, we have the following rules for the order relation "less than":

I. If $a < b$ and c is any real number, then $a + c < b + c$.

II. If $a < b$ and c is positive, then $ac < bc$.

III. If $a < b$ and c is negative, then $ac > bc$.

These rules extend to the other order relations. For dealing with inequalities involving absolute values, we have

IV. $|X| < b$ if and only if $-b < X < b$;

V. $|X| > b$ if and only if $X < -b$ or $X > b$;

where b is any positive real number.

Chapter 0 Review Exercises

In Problems 1–16, answer true or false.

1. The number 0 is neither positive nor negative. _____

2. The repeating decimal numbers 0.999... and 1.000... are equal. _____

3. Every terminating decimal is rational. _____

4. The solution sets of the equations $x^2 = 2x$ and $(x - 2)^2 = 0$ are both $\{2\}$. _____

5. For any real number a, $|a| > 0$. _____

6. -3 is not greater than -1. _____

7. $\sqrt{9} = \pm 3$ _____

8. $|\pi - 3.14| = 3.14 - \pi$ _____

9. $(4^3)^5 = 4^8$ _____

10. $\frac{1}{16}(2^{-4} \cdot 8^6 \cdot 4^{-5}) = 1$ _____

11. $(a^2 + b^2)^2 = a^4 + b^4$ _____

12. $0.1 \geq \frac{1}{10}$ _____

13. $\sqrt{x\sqrt{x}} = x^{3/4}$, $x > 0$ _____

14. $[(-27)^{-1}]^{2/3} = 9$ _____

15. A polynomial equation of degree n has at least n real roots. _____

16. The union of the sets $\{x \mid x < 3\}$ and $\{x \mid x > 5\}$ can be written as a simultaneous inequality. _____

In Problems 17 and 18, rationalize the denominator of the given expression. (Assume that all variables represent positive numbers.)

17. $\dfrac{\sqrt{3} - \sqrt{2}}{\sqrt{3} + \sqrt{2}}$

18. $\dfrac{1}{2\sqrt{x} + 3\sqrt{y}}$

In Problems 19 and 20, expand the given expression.

19. $(2x + 3y)^3$

20. $(4x - 3y)^3$

In Problems 21 and 22, find the indicated term in the expansion of the given expression.

21. Fourth term, $(x - y)^6$

22. Third term, $(3a - 2)^4$

In Problems 23–26, find the given product.

23. $(x + y)(y - x)$

24. $(x + 2)(3x - 1)$

25. $(x + 1)(x^2 + 2x + 1)$

26. $(3x^2 + 5)(4x^2 - 7)$

In Problems 27–32, factor the given expression.

27. $x^4 - 16y^2$

28. $3x^2y + 6xy^2 - 18x^3y^3$

29. $x^2 - 9x + 14$

30. $27a^3 - 8b^3$

31. $4x^2 - 8x + 3$

32. $y^6 + 1$

In Problems 33 and 34, perform the indicated operations and simplify.

33. $\dfrac{1}{x - 2} + \dfrac{2}{x + 2} - \dfrac{1}{x^2 - 4}$

34. $\dfrac{x + \dfrac{1}{x^2}}{x^2 + \dfrac{1}{x}}$

In Problems 35–44, solve the given equation.

35. $\frac{3}{2} - \frac{1}{2}x = \frac{5}{2}$

36. $2x - 7 = 15$

37. $4x^2 - 3x - 1 = 0$

38. $4 - x^2 = 0$

39. $2x^2 + x - 3 = 0$

40. $x^2 + 2x - 4 = 0$

41. $2x^4 + x^2 - 2 = 0$

42. $16x^2 - 8x + 1 = 0$

43. $2x^3 + 3x^2 - 6x + 2 = 0$

44. $x^4 + 2x^3 - 2x - 1 = 0$

In Problems 45–52, solve the given equation.

45. $\dfrac{1}{(x-3)^2} - \dfrac{2}{x-3} + 1 = 0$

46. $\dfrac{x}{x-4} - 2 = \dfrac{4}{x-4}$

47. $x + \sqrt{10 - 3x} = 0$

48. $3x^{2/3} + 11x^{1/3} - 4 = 0$

49. $\sqrt{3x + 1} = \sqrt{2x} - 1$

50. $\sqrt{x + 7} - \sqrt{x - 1} = 0$

51. $|3 - 5x| = 4$

52. $|2x + 7| = 4$

In Problems 53–56, solve the given inequality and express the solution in interval notation.

53. $9 - 4x > 2$

54. $2 + 3x \le 5$

55. $-6 \le 4x - 1 \le 2$

56. $0 < 3 - x \le 3$

In Problems 57 and 58, solve the inequality and express the solution in interval notation.

57. $|2 + x| \le 3$

58. $|3 - 7x| > 4$

In Problems 59–62, find an inequality whose solution set is the given interval.

59. Greater than 6 units from $\frac{1}{2}$

60. Less than $\sqrt{2}$ units from -5

61. $(\frac{1}{2}, \frac{17}{2})$

62. $(-3, 7)$

1 Coordinate Geometry

1.1 The Cartesian Coordinate System and Graphing

In Chapter 0 we saw how each real number may be associated with exactly one point on the number line. We now examine a correspondence between points in the plane and ordered pairs of real numbers.

The Cartesian Coordinate System

The **Cartesian coordinate system*** is formed in a plane by two perpendicular number lines which intersect at the point corresponding to the number 0 on each line. This point of intersection is called the **origin** and is denoted by *O*. The horizontal line is called the *x* **axis** and the vertical line the *y* **axis**. The axes divide the plane into four regions, called **quadrants**, which are numbered as in Figure 1.1.

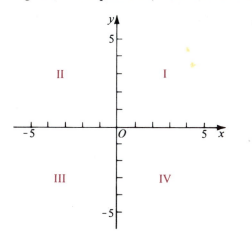

Figure 1.1

* Named after René Descartes (1596–1650), a French mathematician and philosopher. He is credited with the development of analytic geometry.

Abscissa and Ordinate

Let P be a point in the plane. We associate an ordered pair of real numbers with P by drawing a vertical line from P to the x axis and a horizontal line from P to the y axis. If the vertical line intersects the x axis at a and the horizontal line intersects the y axis at b, we associate the ordered pair (a, b) with the point P. Conversely, to each ordered pair (a, b) of real numbers, there corresponds a point P in the plane. This point is at the intersection of the vertical line passing through a on the x axis and the horizontal line passing through b on the y axis. We call a the **x coordinate** or **abscissa** of P and b the **y coordinate** or **ordinate** of P. See Figure 1.2.

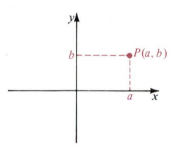

Figure 1.2

Example 1 Plot the points whose coordinates are $A(1, 2)$, $B(-4, 3)$, $C(-\frac{3}{2}, -2)$, $D(0, 4)$, and $E(3.5, 0)$. Specify in which quadrant each point lies.

Solution The points are plotted in Figure 1.3. Point A is in quadrant I, B in quadrant II, and C in quadrant III. Points on either of the axes are not considered to be in any quadrant.

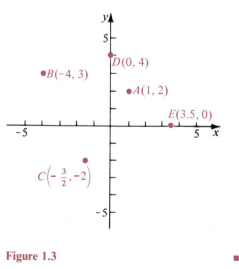

Figure 1.3

The algebraic signs of the x coordinate and y coordinate of any point (x, y) in each of the four quadrants are indicated in Figure 1.4.

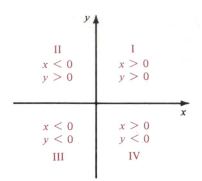

Figure 1.4

Example 2   Describe the set of points (x, y) in the plane which satisfy each of the specified conditions.

a. $x = 1$ **b.** $xy = 0$ **c.** $xy < 0$ **d.** $|y| \geq 2$

Solution   **a.** $x = 1$. The points that satisfy this condition must have x coordinate equal to 1, whereas the y coordinate may be any real number. As shown in Figure 1.5, the set of all points such that $x = 1$ is a vertical line 1 unit to the right of the y axis.

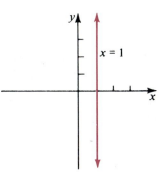

Figure 1.5

b. $xy = 0$. Since the product of two numbers is 0 exactly when one or the other or both of the numbers is 0, we have $xy = 0$ when $x = 0$ or $y = 0$. The points for which $x = 0$ constitute the y axis and the points for which $y = 0$ constitute the x axis. Thus, the points for which $xy = 0$ are the coordinate axes. See Figure 1.6.

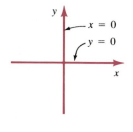

Figure 1.6

c. $xy < 0$. A product of two numbers will be negative when one of the numbers is positive and the other is negative. Thus, $xy < 0$ when $x > 0$ and $y < 0$ or when $x < 0$ and $y > 0$. We see from Figure 1.4 that $xy < 0$ for all points (x, y) in quadrants II and IV. Thus, we can represent the set of points (x, y) for which $xy < 0$ by the shaded region in Figure 1.7. The coordinate axes are shown as dashed lines to indicate that the points on these lines are not included in the solution.

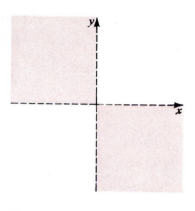

Figure 1.7

d. $|y| \geq 2$. Recall from Section 0.10 that $|y| \geq 2$ may be expressed as $y \leq -2$ or $y \geq 2$. Since x is not restricted, the points (x, y) for which $y \leq -2$ or $y \geq 2$ can be represented by the shaded region in Figure 1.8. We use solid lines on the boundaries of the region to indicate that these points are included in the solution.

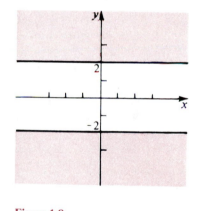

Figure 1.8

Relations and Graphs

In the preceding example we saw how equations or inequalities could be used to describe a particular set of points in the plane. In general, any set of ordered pairs of real numbers is called a **relation**, and the corresponding set of points in the plane is called the **graph of the relation**.

Example 3 Graph the relation $S = \{(-1, 2), (0, 4), (3, -\frac{5}{2}), (-3, -1)\}$.

Solution In Figure 1.9 we plot the points that correspond to the ordered pairs in the relation S.

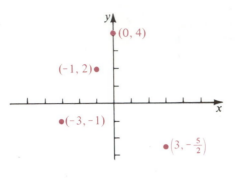

Figure 1.9

Example 4   Graph the relation $T = \{(x, y) \mid 0 \leq x \leq 2, |y| = 1\}$.

Solution   To graph T we plot the points whose abscissas are numbers in the interval $[0, 2]$ and whose ordinates are either 1 or -1. See Figure 1.10.

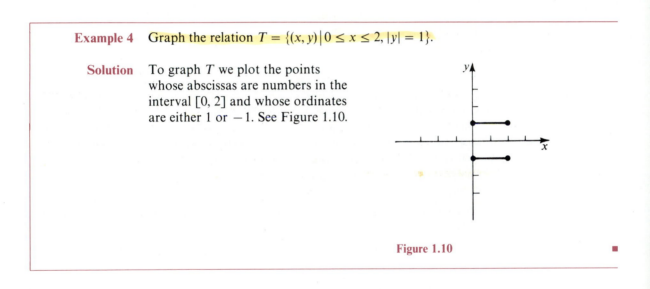

Figure 1.10

Graph of an Equation

If a relation is defined by an equation, we call the set of points in the plane corresponding to the ordered pairs in the relation the **graph of the equation**. One elementary method for sketching the graph of an equation is to plot points and then connect these points with a smooth curve. Of course we must plot enough points so that the shape of the graph is evident.

Example 5   Graph the equation $y = x^2$.

Solution   By assigning values of x we find the values of y given in the accompanying table. Plotting these points and joining them with a smooth curve, we obtain the graph in Figure 1.11.

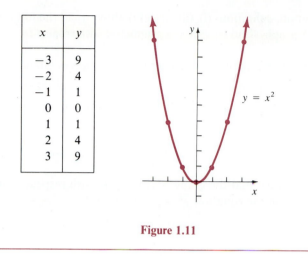

x	y
−3	9
−2	4
−1	1
0	0
1	1
2	4
3	9

$y = x^2$

Figure 1.11

Symmetry

Note that the portion of the graph of $y = x^2$ in Figure 1.11 in the second quadrant is the mirror image of the graph in the first quadrant. This is due to the fact that the y coordinate corresponding to a given $x > 0$ is the same as the y coordinate corresponding to $-x$. This is an example of **symmetry**, which can be quite an aid in graphing. Before plotting points, you can determine whether the graph of an equation has symmetry. As shown in Figure 1.12, a graph is

(i) **symmetric with respect to the y axis** if whenever (x, y) is a point on the graph, $(-x, y)$ is also a point on the graph;

(ii) **symmetric with respect to the x axis** if whenever (x, y) is a point on the graph, $(x, -y)$ is also a point on the graph;

(iii) **symmetric with respect to the origin** if whenever (x, y) is a point on the graph, $(-x, -y)$ is also a point on the graph.

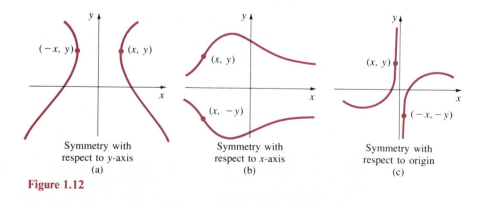

Symmetry with
respect to y-axis
(a)

Symmetry with
respect to x-axis
(b)

Symmetry with
respect to origin
(c)

Figure 1.12

**Tests for
Symmetry**

For an equation, definitions (i), (ii), and (iii) above yield the following three **tests for symmetry**: A graph of an equation is **symmetric with respect to**

    (i) **the y axis**, if replacing x by $-x$ results in an equivalent equation;

    (ii) **the x axis**, if replacing y by $-y$ results in an equivalent equation;

    (iii) **the origin**, if replacing x by $-x$ and y by $-y$ results in an equivalent equation.

Example 6   Verify that $y^2 = x^3$ is symmetric with respect to the x axis and graph the equation.

Solution   By Test (ii), to show that the graph is symmetric with respect to the x axis, we substitute $-y$ for y in the equation $y^2 = x^3$. This gives

$$(-y)^2 = x^3,$$

which is equivalent to

$$y^2 = x^3.$$

Thus, whenever the point (x, y) is on the graph of $y^2 = x^3$, so is the point $(x, -y)$. We tabulate values of x and y that satisfy the equation and graph the curve in Figure 1.13. Note that x cannot be negative, since then y would be the square root of a negative number.

x	y
0	0
$\frac{1}{2}$	$\pm\sqrt{1/8} \approx \pm 0.35$
$\frac{3}{4}$	$\pm\sqrt{27/64} \approx \pm 0.65$
1	± 1
2	$\pm\sqrt{8} \approx \pm 2.82$
3	$\pm\sqrt{27} \approx \pm 5.20$

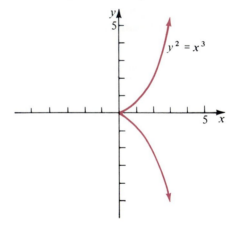

Figure 1.13

Example 7   Determine whether the graph of $y = |x| + 1$ is symmetric with respect to

    a. the y axis   **b.** the x axis   **c.** the origin

Solution   **a.** Applying Test (i), we have $y = |-x| + 1$, which is equivalent to the original equation $y = |x| + 1$, since $|-x| = |x|$. Thus the graph of $y = |x| + 1$ is symmetric with respect to the y axis.

b. Applying Test (ii), we find $-y = |x| + 1$ or $y = -|x| - 1$, which is not equivalent to the original equation.

c. Applying Test (iii), we obtain $-y = |-x| + 1$ or $y = -|x| - 1$, which is not equivalent to the original equation. ∎

Exercise 1.1

In Problems 1–4, plot the given points.

1. $(2, 3), (4, 5), (0, 2), (-1, -3)$

2. $(1, 4), (-3, 0), (-4, 2), (-1, -1)$

3. $(-\frac{1}{2}, -2), (0, 0), (-1, \frac{4}{3}), (3, 3)$

4. $(0, 0.8), (-2, 0), (1.2, -1.2), (-2, 2)$

In Problems 5–16, determine in which quadrant the given point lies if (a, b) is in quadrant I.

5. $(-a, b)$ **6.** $(a, -b)$ **7.** $(-a, -b)$ **8.** (b, a)

9. $(-b, a)$ **10.** $(-b, -a)$ **11.** (a, a) **12.** $(b, -b)$

13. $(-a, -a)$ **14.** $(-a, a)$ **15.** $(b, -a)$ **16.** $(-b, b)$

17. Plot the points given in Problems 5–16 if (a, b) is the point shown in Figure 1.14.

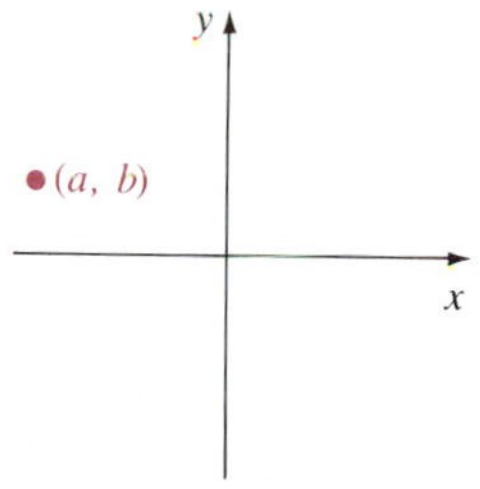

Figure 1.14

In Problems 18–25, graph the given relation.

18. $x = -3$ **19.** $y = 4$ **20.** $xy \leq 0$ **21.** $xy > 0$

22. $|x| > 4$ **23.** $|y| \leq 1$ **24.** $y = x$ **25.** $y = -x$

In Problems 26–46, determine whether the graph of the given equation is symmetric with respect to the y axis, the x axis, or the origin. Graph.

26. $y = x^2 + 1$ **27.** $y = -x^2$ **28.** $y^2 = x$

29. $y = x^2 - 3$ **30.** $y = (x - 1)^2$ **31.** $y = (x + 1)^2$

32. $x + y = 1$ **33.** $x - y = 1$ **34.** $y = x^3$

35. $y^3 = x$ **36.** $y = -x^3$ **37.** $y^3 = x^2$

38. $y = x^4$ **39.** $y = |x|$ **40.** $|y| = x$

41. $|x + y| = 0$ **42.** $|x - y| = 2$ **43.** $|x - y| = 0$

44. $y^2 = x^2$ **45.** $y = |x|^3$ **46.** $y^2 = 9$

In Problems 47–54, given a formal definition of the given statement.

47. Symmetry with respect to the line $y = 1$

48. Symmetry with respect to the line $y = -2$

49. Symmetry with respect to the line $x = -3$

50. Symmetry with respect to the line $x = h$

51. Symmetry with respect to the line $y = k$

52. Symmetry with respect to the point $(4, 1)$

53. Symmetry with respect to the point $(1, -2)$

54. Symmetry with respect to the point (h, k)

55. Show that the graph of $y = (x - 4)^2 + 6$ is symmetric with respect to the line $x = 4$.

56. Show that the graph of $y = x$ is symmetric with respect to the point (a, a) for any real number a. [*Hint:* See Problem 54.]

1.2 The Distance Formula and the Circle

In this section we develop a formula for finding the distance between two points in the plane whose coordinates are known. We then apply this formula to the problem of finding an equation of a circle.

Distance Formula
Suppose $P_1(x_1, y_1)$ and $P_2(x_2, y_2)$ are two distinct points not on a vertical or a horizontal line. Then, as shown in Figure 1.15, P_1, P_2, and $P_3(x_1, y_2)$ are vertices of a right triangle. Also, as shown in Figure 1.15, the length of the side $P_3 P_2$ is $|x_2 - x_1|$ and the length of the side $P_1 P_3$ is $|y_2 - y_1|$. If we denote the length of $P_1 P_2$ by d, we have

$$d^2 = |x_2 - x_1|^2 + |y_2 - y_1|^2, \tag{1}$$

from the Pythagorean theorem. Since the square of any real number is equal to the square of its absolute value, we may drop the absolute value signs in (1). Taking square roots on both sides of (1), we have

$$d = \sqrt{(x_2 - x_1)^2 + (y_2 - y_1)^2}.$$

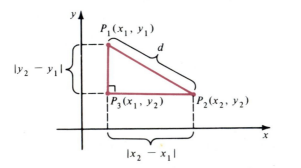

Figure 1.15

Although we derived this equation for two points not on a vertical or horizontal line, it holds in these cases as well.

Distance Formula   The distance between any two points $P_1(x_1, y_1)$ and $P_2(x_2, y_2)$, denoted by $d(P_1, P_2)$, is given by

$$d(P_1, P_2) = \sqrt{(x_2 - x_1)^2 + (y_2 - y_1)^2}.$$

(2)

Since $(x_2 - x_1)^2 = (x_1 - x_2)^2$, it makes no difference which point is taken as P_1 in the distance formula.

Example 1   Find the distance between $A(8, -5)$ and $B(3, 7)$.

Solution
$$d(A, B) = \sqrt{(3 - 8)^2 + [7 - (-5)]^2}$$
$$= \sqrt{(-5)^2 + (12)^2}$$
$$= \sqrt{25 + 144} = \sqrt{169} = 13.$$

Note that if we compute $d(B, A)$ we also have
$$d(B, A) = \sqrt{(8 - 3)^2 + (-5 - 7)^2}$$
$$= \sqrt{5^2 + (-12)^2}$$
$$= \sqrt{25 + 144} = \sqrt{169} = 13. \qquad \blacksquare$$

Example 2   Use the distance formula and the Pythagorean theorem to determine whether or not the points $P_1(7, 1)$, $P_2(-4, -1)$, and $P_3(4, 5)$ are the vertices of a right triangle.

Solution   The lengths of the sides of the triangle $P_1 P_2 P_3$ are

$$d(P_1, P_2) = \sqrt{(-4 - 7)^2 + (-1 - 1)^2}$$
$$= \sqrt{121 + 4} = \sqrt{125},$$
$$d(P_2, P_3) = \sqrt{[4 - (-4)]^2 + [5 - (-1)]^2}$$
$$= \sqrt{64 + 36} = \sqrt{100} = 10,$$

and

$$d(P_3, P_1) = \sqrt{(7 - 4)^2 + (1 - 5)^2}$$
$$= \sqrt{9 + 16} = \sqrt{25} = 5.$$

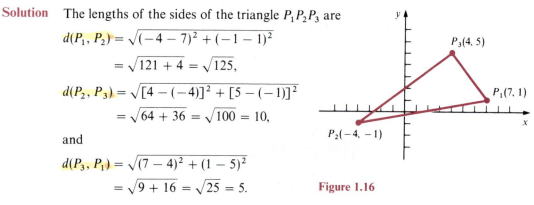

Figure 1.16

Since

$$[d(P_3, P_1)]^2 + [d(P_2, P_3)]^2 = 25 + 100 = 125 = [d(P_1, P_2)]^2,$$

it follows from the Pythagorean theorem that P_1, P_2, and P_3 are vertices of a right triangle. See Figure 1.16. $\qquad \blacksquare$

Example 3   Verify that the points $A(2, 4)$ and $B(4, -6)$ are equidistant from the point $C(3, -1)$.

Solution
$$d(A, C) = \sqrt{(3-2)^2 + (-1-4)^2} = \sqrt{1+25} = \sqrt{26}$$

and

$$d(B, C) = \sqrt{(3-4)^2 + [-1-(-6)]^2} = \sqrt{1+25} = \sqrt{26}. \qquad ■$$

Circles

In Section 1.1 we saw how to obtain the graph of a given equation. Let us consider the converse problem: Given a geometric figure in the plane, find the equation whose graph is this figure. Although in general this problem has no solution, there are a number of important cases in which it can be solved. In particular, in this chapter we will find equations of circles, straight lines, parabolas, ellipses, and hyperbolas. We now use the distance formula to find the equation of a circle.

> **Definition 1.1**   A **circle** is the set of all points P in the plane which are a given fixed distance r, called the **radius**, from a given fixed point C, called the **center**.

Equation of a Circle

In Figure 1.17 we have sketched a circle of radius r centered at the point $C(h, k)$.

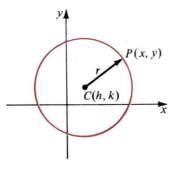

Figure 1.17

From Definition 1.1 a point $P(x, y)$ is on this circle if and only if
$$d(P, C) = r \quad \text{or} \quad \sqrt{(x-h)^2 + (y-k)^2} = r.$$
Since $(x-h)^2 + (y-k)^2$ is always nonnegative, we obtain an equivalent equation when both sides are squared:
$$(x-h)^2 + (y-k)^2 = r^2.$$
Therefore we have the following result.

Equation of a Circle   A circle of radius r centered at $C(h, k)$ has the equation

$$(x-h)^2 + (y-k)^2 = r^2. \qquad \text{signs reversed} \qquad (3)$$

Equation (3) is called the **standard form** of the equation of a circle.

Example 4 The graph of the equation

$$(x - 2)^2 + (y - 1)^2 = 16$$

is a circle of radius 4 centered at the point (2, 1). The graph is given in Figure 1.18.

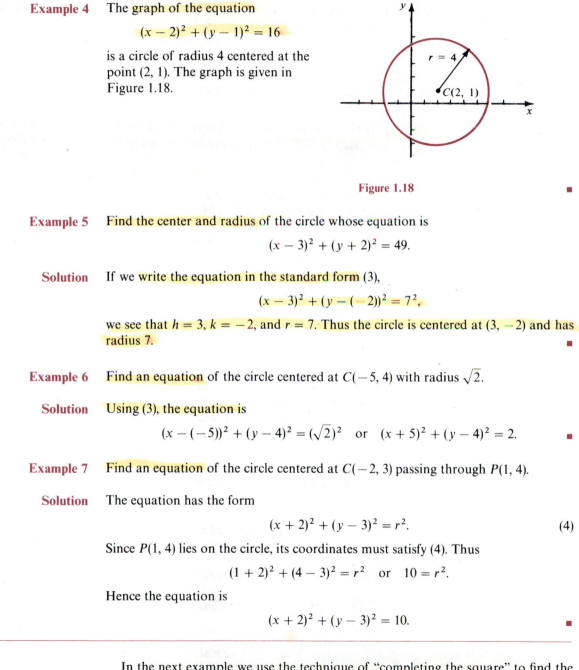

Figure 1.18

Example 5 Find the center and radius of the circle whose equation is

$$(x - 3)^2 + (y + 2)^2 = 49.$$

Solution If we write the equation in the standard form (3),

$$(x - 3)^2 + (y - (-2))^2 = 7^2,$$

we see that $h = 3$, $k = -2$, and $r = 7$. Thus the circle is centered at $(3, -2)$ and has radius 7.

Example 6 Find an equation of the circle centered at $C(-5, 4)$ with radius $\sqrt{2}$.

Solution Using (3), the equation is

$$(x - (-5))^2 + (y - 4)^2 = (\sqrt{2})^2 \quad \text{or} \quad (x + 5)^2 + (y - 4)^2 = 2.$$

Example 7 Find an equation of the circle centered at $C(-2, 3)$ passing through $P(1, 4)$.

Solution The equation has the form

$$(x + 2)^2 + (y - 3)^2 = r^2. \tag{4}$$

Since $P(1, 4)$ lies on the circle, its coordinates must satisfy (4). Thus

$$(1 + 2)^2 + (4 - 3)^2 = r^2 \quad \text{or} \quad 10 = r^2.$$

Hence the equation is

$$(x + 2)^2 + (y - 3)^2 = 10.$$

In the next example we use the technique of "completing the square" to find the center and radius of a circle. Recall from Section 0.7 that adding $(B/2)^2$ to the expression $x^2 + Bx$ yields $x^2 + Bx + (B/2)^2$, which is the perfect square $(x + B/2)^2$.

Example 8   Find the center and radius of the circle with equation

$$x^2 + y^2 + 10x - 2y + 22 = 0.$$

Solution   We want to rewrite the equation in the standard form

$$(x - h)^2 + (y - k)^2 = r^2.$$

Rearranging, we have

$$(x^2 + 10x \qquad) + (y^2 - 2y \qquad) = -22.$$

Now we complete the square of each quantity in parentheses by adding $(10/2)^2$ in the first and $(-2/2)^2$ in the second. Note that we must be careful to add these numbers to *both* sides of the equation.

$$[x^2 + 10x + (\tfrac{10}{2})^2] + [y^2 - 2y + (-\tfrac{2}{2})^2] = -22 + (\tfrac{10}{2})^2 + (-\tfrac{2}{2})^2$$

$$(x^2 + 10x + 25) + (y^2 - 2y + 1) = 4$$

$$(x + 5)^2 + (y - 1)^2 = 2^2. \tag{5}$$

We see from equation (5) that the circle is centered at $(-5, 1)$ and has radius 2.   ∎

Example 9   Find the center and radius of the circle with equation

$$3x^2 + 3y^2 - 18x + 6y + 2 = 0.$$

Solution   Rearranging, we have

$$(3x^2 - 18x \qquad) + (3y^2 + 6y \qquad) = -2.$$

We now divide both sides of this equation by 3 so that the coefficients of x^2 and y^2 are each 1.

$$(x^2 - 6x \qquad) + (y^2 + 2y \qquad) = -\tfrac{2}{3}.$$

Completing the square gives

$$(x^2 - 6x + [\tfrac{1}{2}(6)]^2) + (y^2 + 2y + [\tfrac{1}{2}(2)]^2) = -\tfrac{2}{3} + [\tfrac{1}{2}(6)]^2 + [\tfrac{1}{2}(2)]^2$$

$$(x - 3)^2 + (y + 1)^2 = -\tfrac{2}{3} + 9 + 1$$

$$(x - 3)^2 + (y + 1)^2 = \tfrac{28}{3}.$$

This is the equation of the circle centered at $(3, -1)$ with radius $\sqrt{\tfrac{28}{3}}$.   ∎

As the following examples show, not every equation of the form $Ax^2 + Ay^2 + Bx + Cy + D = 0$ necessarily represents a circle.

Example 10   When put into standard form, the equation

$$x^2 + y^2 + 10x - 2y + 26 = 0 \quad \text{becomes} \quad (x + 5)^2 + (y - 1)^2 = 0.$$

Since $(x + 5)^2$ and $(y - 1)^2$ are both nonnegative, the only values of x and y that satisfy this equation are $x = -5$ and $y = 1$. Thus the graph of this equation is the single point $(-5, 1)$. ∎

Example 11 When $x^2 + y^2 + 2x + 4y + 6 = 0$ is put into standard form, we have

$$(x + 1)^2 + (y + 2)^2 = -1.$$

Since both $(x + 1)^2$ and $(y + 2)^2$ are nonnegative, there are no real numbers x and y that satisfy this equation. Thus there is no graph. ∎

Midpoint Formula To solve the next example we will need to find the midpoint of a diameter of a circle. In general, given $P_1(x_1, y_1)$ and $P_2(x_2, y_2)$, we can find the coordinates of the midpoint $M(x, y)$ of the line segment P_1P_2 by constructing two congruent triangles.

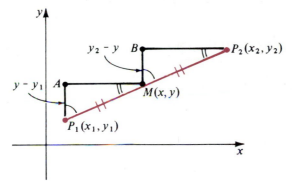

Figure 1.19

In Figure 1.19 triangles P_1AM and MBP_2 are congruent since corresponding angles are equal and $d(P_1, M) = d(M, P_2)$. Hence,

$$d(P_1, A) = d(M, B) \quad \text{and} \quad y - y_1 = y_2 - y.$$

Solving the last equation for y gives

$$y = \frac{y_1 + y_2}{2}.$$

Similarly,

$$d(A, M) = d(B, P_2), \quad \text{so} \quad x - x_1 = x_2 - x$$

and hence

$$x = \frac{x_1 + x_2}{2}.$$

Although this argument depends on the positioning of the points as shown in

Figure 1.19, the result holds in general.

> **Midpoint Formula** The coordinates of the midpoint of the line segment joining the points $P_1(x_1, y_1)$ and $P_2(x_2, y_2)$ are
>
> $$\left(\frac{x_1 + x_2}{2}, \frac{y_1 + y_2}{2}\right).$$

Example 12   Find the equation of the circle with the points $A(2, -3)$ and $B(6, 1)$ at the ends of a diameter.

Solution   The center of the circle is the midpoint of the diameter joining A and B. Thus, by the midpoint formula, the coordinates of the center are

$$\left(\frac{2 + 6}{2}, \frac{-3 + 1}{2}\right) = (4, -1).$$

The radius is the distance from the center $(4, -1)$ to the point on the circle $(2, -3)$:

$$r = \sqrt{(4 - 2)^2 + [-1 - (-3)]^2}$$
$$= \sqrt{4 + 4} = \sqrt{8}.$$

Therefore the equation of the circle is

$$(x - 4)^2 + (y + 1)^2 = 8.$$

See Figure 1.20.

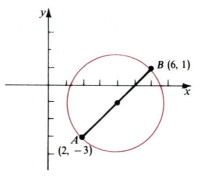

Figure 1.20

Exercise 1.2

In Problems 1–6, find the distance between the given points.

1. $A(1, 2)$, $B(-3, 4)$

2. $A(-1, 3)$, $B(5, 0)$

3. $A(2, 4)$, $B(-4, -4)$

4. $A(-12, -3)$, $B(-5, -7)$

5. $A(-\frac{3}{2}, 1)$, $B(\frac{5}{2}, -2)$

6. $A(-\frac{5}{3}, 4)$, $B(-\frac{2}{3}, -1)$

In Problems 7–10, use the distance formula and the Pythagorean theorem to determine whether the points A, B, and C are vertices of a right triangle.

7. $A(8, 1)$, $B(-3, -1)$, $C(10, 5)$

8. $A(-2, -1)$, $B(8, 2)$, $C(1, -11)$

9. $A(2, 8)$, $B(0, -3)$, $C(6, 5)$ 10. $A(4, 0)$, $B(1, 1)$, $C(2, 3)$

In Problems 11 and 12, verify that the points A and B are equidistant from the point C.

11. $A(2, -3)$, $B(-5, 6)$, $C(3, 5)$ 12. $A(1, -1)$, $B(5, -5)$, $C(2, -4)$

In Problems 13–16, find the midpoint of the line segment joining A and B.

13. $A(4, 1)$, $B(-2, 4)$ 14. $A(-1, 0)$, $B(-8, 5)$

15. $A(\frac{2}{3}, 1)$, $B(\frac{7}{3}, -3)$ 16. $A(\frac{1}{2}, -\frac{3}{2})$, $B(-\frac{5}{2}, 1)$

In Problems 17–20, find B if M is the midpoint of the line segment joining A and B.

17. $A(-2, 1)$, $M(\frac{3}{2}, 0)$ 18. $A(4, \frac{1}{2})$, $M(7, -\frac{5}{2})$

19. $A(-\frac{2}{3}, \frac{5}{2})$, $M(\frac{4}{7}, -\frac{3}{8})$ 20. $A(1.4, -2.3)$, $M(12.5, 4.7)$

In Problems 21–32, find the center and radius of the circle. Graph the circle.

21. $(x - 1)^2 + (y - 3)^2 = 49$ 22. $(x + 3)^2 + (y - 5)^2 = 25$

23. $(x - \frac{1}{2})^2 + (y - \frac{3}{2})^2 = 5$ 24. $(x + 5)^2 + (y + 8)^2 = \frac{1}{4}$

25. $x^2 + y^2 + 8y = 0$ 26. $x^2 + y^2 + 2x - 4y - 4 = 0$

27. $x^2 + y^2 - 18x - 6y - 10 = 0$ 28. $x^2 + y^2 - 16y + 3x + 63 = 0$

29. $8x^2 + 8y^2 + 16x + 64y - 40 = 0$ 30. $5x^2 + 5y^2 + 25x + 100y + 50 = 0$

31. $3x^2 + 3y^2 - 5x + 7y - 20 = 0$ 32. $4x^2 + 4y^2 + x - 2y - 3 = 0$

In Problems 33 and 34, show that the given equation does not describe a circle.

33. $x^2 + y^2 + 2y + 9 = 0$ 34. $2x^2 + 2y^2 - 2x + 6y + 7 = 0$

In Problems 35–44, find an equation of the circle that satisfies the given conditions.

35. Center $(0, 0)$, radius 1   36. Center $(1, -3)$, radius 5

37. Center $(0, 2)$, radius $\sqrt{2}$ 38. Center $(-9, -4)$, radius $\frac{3}{2}$

39. Endpoints of a diameter at $(-1, 4)$ and $(3, 8)$

40. Endpoints of a diameter at $(4, 2)$ and $(-3, 5)$

41. Center $(0, 0)$, passing through $(-1, -2)$

42. Center $(4, -5)$, passing through $(7, -3)$

43. Center $(5, 6)$, tangent to the x axis

44. Center $(-4, 3)$, tangent to the y axis

45. At what point(s) does the circle of radius 7 centered at $(3, -6)$ intersect the x axis? The y axis?

46. At what point(s) does the circle $x^2 + y^2 + 5x - 6y + 2 = 0$ intersect the x axis? The y axis?

47. Does $(1/4, \sqrt{15}/4)$ lie on the circle of radius 1 centered at the origin?

48. Does $(5, -1)$ lie on the circle of radius 4 centered at $(2, 3)$?

49. A garden archway is constructed by mounting a semicircle of radius 3 feet atop two walls of height 4 feet. See Figure 1.21. Find the height of the archway at a point 1 foot from one of the walls.

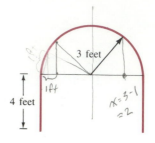

4 feet

Figure 1.21

50. A cylindrical gas tank 5 feet in diameter has a circular hole with a radius of 6 inches drilled in one end of the tank. The center of the hole is 12 inches below the center of the end of the tank. If the origin is placed at the center of the end of the tank, find an equation of the circular edge of the tank and an equation of the circular edge of the hole.

51. Solve Problem 50 if the origin is placed at the center of the small circular hole.

52. Use the distance formula to verify that

$$\left(\frac{x_1 + x_2}{2}, \frac{y_1 + y_2}{2} \right)$$

is equidistant from (x_1, y_1) and (x_2, y_2). Does this *prove* the midpoint formula?

Review Problems

53. If (a, b) lies in quadrant III, determine the quadrant in which each of the following points lies:
    a. (b, a) **b.** $(a, -b)$ **c.** $(-a, b)$ **d.** $(-b, -a)$

54. Graph $y = x^5$

In Problems 55 and 56 determine whether the given equation is symmetric with respect to (a) the y axis; (b) the x axis; (c) the origin.

55. $x^2 + y^2 = 1$ **56.** $y = -x^4$

57. The graph of $(x - 1)^2 + (y - 2)^2 = 4$ is symmetric with respect to what vertical line? What horizontal line? What point?

1.3 Equations of Straight Lines

Slope

Any pair of distinct points in the plane determines a unique *straight line*. If $P_1(x_1, y_1)$ and $P_2(x_2, y_2)$ are two points such that $x_1 \neq x_2$, then the ratio

$$m = \frac{y_2 - y_1}{x_2 - x_1} \qquad (1)$$

is called the **slope** of the line determined by these two points. It is customary to call $y_2 - y_1$ the **rise** and $x_2 - x_1$ the **run**. In calculus the rise is often denoted by Δy and the run by Δx, where the capital Greek letter delta, Δ, is used to designate a *change* in a variable. Thus, in the expressions Δy and Δx, y and x are not multiplied by Δ, but rather

$$\Delta y = y_2 - y_1 \quad \text{and} \quad \Delta x = x_2 - x_1.$$

The slope of a line is then

$$m = \frac{\text{rise}}{\text{run}} = \frac{\Delta y}{\Delta x}. \qquad (2)$$

Example 1 Find the slope of the line through $(-2, 6)$ and $(3, -4)$.

Solution Let $(-2, 6)$ be the point $P_1(x_1, y_1)$ and $(3, -4)$ be the point $P_2(x_2, y_2)$. From equation (1) the slope of the line is

$$m = \frac{y_2 - y_1}{x_2 - x_1} = \frac{-4 - 6}{3 - (-2)}$$

$$= \frac{-10}{5} = -2.$$

The straight line through the points P_1 and P_2 is shown in Figure 1.22.

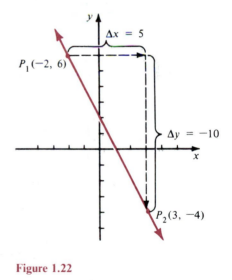

Figure 1.22 ∎

Notice that in Example 1 if we had let $P_1(x_1, y_1)$ be the point $(3, -4)$ and $P_2(x_2, y_2)$ be the point $(-2, 6)$, then equation (1) would have given

$$m = \frac{y_2 - y_1}{x_2 - x_1} = \frac{6 - (-4)}{-2 - 3} = \frac{10}{-5} = -2.$$

In general since

$$\frac{y_2 - y_1}{x_2 - x_1} = \frac{-(y_1 - y_2)}{-(x_1 - x_2)} = \frac{y_1 - y_2}{x_1 - x_2},$$

it does not matter which of the two points is called $P_1(x_1, y_1)$ and which is called $P_2(x_2, y_2)$.

In Figure 1.23 we compare the graphs of lines with positive, negative, and zero slopes.

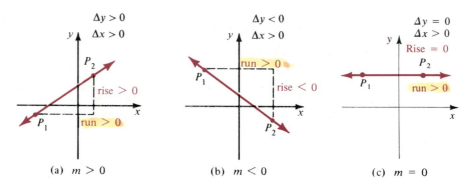

(a) $m > 0$ (b) $m < 0$ (c) $m = 0$

Figure 1.23

If $P_1(x_1, y_1)$ and $P_2(x_2, y_2)$ are two points on a vertical line, then $x_1 = x_2$, so $x_2 - x_1 = 0$. Thus the slope of this line is undefined.

Any pair of distinct points on a line will determine the same slope. To prove this, consider the similar triangles $P_1 Q_1 P_2$ and $P_3 Q_2 P_4$ shown in Figure 1.24. Since the ratios of corresponding sides are equal, we have

$$\frac{y_2 - y_1}{x_2 - x_1} = \frac{y_4 - y_3}{x_4 - x_3}.$$

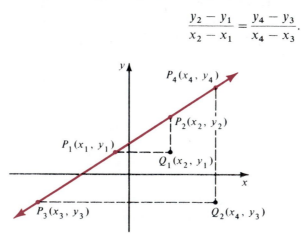

Figure 1.24

Thus the slope of the line is independent of the choice of points on the line. Although this argument was based on the positioning of P_1, P_2, P_3, and P_4 on the line, the discussion remains valid for any placement of these four points.

Example 2   Draw the lines through the given pairs of points and determine their slopes.

      a. $(-4, -1)$ and $(5, 2)$ **b.** $(-3, 3)$ and $(4, -4)$ **c.** $(-5, 2)$ and $(-5, -4)$

Solution   The points are plotted and the lines drawn in Figure 1.25. The slopes are computed using equation (1).

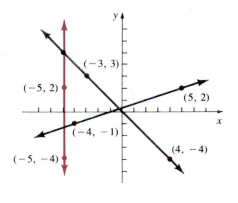

a. $m = \dfrac{2 - (-1)}{5 - (-4)} = \dfrac{3}{9} = \dfrac{1}{3}$

b. $m = \dfrac{-4 - 3}{4 - (-3)} = \dfrac{-7}{7} = -1$

c. Since $(-5, 2)$ and $(-5, -4)$ determine a vertical line, the slope is undefined.

Figure 1.25

Example 3   Graph the line with slope $-\frac{5}{3}$ that passes through the point $P(-2, 3)$.

Solution   Using the point $P(-2, 3)$ as a vertex, we construct a right triangle by moving 3 units to the *right* (since the run is $+3$) to the point $Q(1, 3)$. From Q we move *down* 5 units (since the rise is -5) to the point $R(1, -2)$. Then R is another point on the line which is graphed in Figure 1.26.

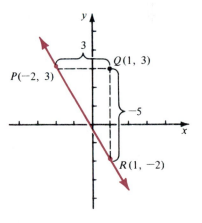

Figure 1.26

Alternatively, we could have considered the slope to be $5/(-3)$ and
constructed the triangle by moving
to the *left* 3 units (since the run is
-3) to the point $S(-5, 3)$. From S
we move *up* 5 units (since the rise is
$+5$) to the point $T(-5, 8)$. The line
is now drawn through P and T as
shown in Figure 1.27.

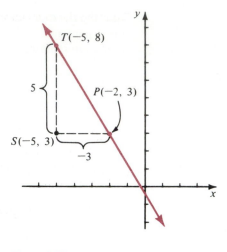

Figure 1.27

■

Equation of a Line   We turn now to the problem of finding an **equation of a line** through two points
$P_1(x_1, y_1)$ and $P_2(x_2, y_2)$ which do not both lie on the same vertical line, so that
$x_1 \neq x_2$. First, let

$$m = \frac{y_2 - y_1}{x_2 - x_1} \tag{3}$$

be the slope of the line. If $P(x, y)$ is any point on the line distinct from P_1, as in Figure
1.28, we can then compute the slope using P_1 and P:

$$m = \frac{y - y_1}{x - x_1}. \tag{4}$$

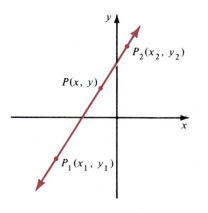

Figure 1.28

Equating the expressions in (3) and (4) and multiplying by $x - x_1$ gives the

Two-point form for the equation of a line:

$$y - y_1 = \frac{y_2 - y_1}{x_2 - x_1}(x - x_1). \tag{5}$$

Since in (5) the ratio $(y_1 - y_2)/(x_1 - x_2)$ is the slope of the line, we may express the equation in terms of its slope and one point. The resulting equation is called the

Point-slope form for the equation of a line:

$$y - y_1 = m(x - x_1). \tag{6}$$

Example 4 Find an equation of the line passing through the points $(4, 3)$ and $(-2, 5)$.

Solution The slope of the line through the two points is

$$m = \frac{5 - 3}{-2 - 4} = \frac{2}{-6} = -\frac{1}{3}.$$

From equation (6) with $x_1 = 4$ and $y_1 = 3$ we have

$$y - 3 = -\tfrac{1}{3}(x - 4)$$
$$3y - 9 = -x + 4$$
$$x + 3y - 13 = 0,$$

or solving for y,

$$y = -\tfrac{1}{3}x + \tfrac{13}{3}.$$

The reader should check that the same equation will be obtained if the point $(-2, 5)$ is used instead of $(4, 3)$. ∎

Example 5 Find an equation of the line with slope 4 passing through $(-\tfrac{1}{2}, 2)$.

Solution Letting $m = 4$, $x_1 = -\tfrac{1}{2}$, and $y_1 = 2$, we obtain from equation (6)

$$y - 2 = 4[x - (-\tfrac{1}{2})],$$
$$y - 2 = 4(x + \tfrac{1}{2}),$$
$$y = 4x + 4.$$ ∎

y Intercept

Any nonvertical line intersects the y axis. If this point of intersection is $(0, b)$, then b is called the **y intercept** of the line. The equation of the line with slope m and y intercept b can be obtained from (6). Substituting $y_1 = b$ and $x_1 = 0$ gives

$$y - b = m(x - 0).$$

Solving for y, we obtain the

Slope-intercept form for the equation of a line:

$$y = mx + b. \tag{7}$$

Example 6   Find an equation of the line with slope $\frac{2}{5}$ and y intercept -3.

Solution   Using $m = \frac{2}{5}$ and $b = -3$ in (7), we have

$$y = \tfrac{2}{5}x + (-3)$$
$$y = \tfrac{2}{5}x - 3.$$
∎

It can also be shown that the graph of any equation of the form $y = mx + b$ is a straight line with slope m and y intercept b. Thus, the slope-intercept form enables us to find the slope and y intercept of a line whose equation is known.

Example 7   Find the slope and y intercept of the line

$$3x - 7y + 5 = 0.$$

Solution   We solve the equation of the line for y:

$$3x - 7y + 5 = 0$$
$$7y = 3x + 5$$
$$y = \tfrac{3}{7}x + \tfrac{5}{7}.$$

From equation (7) we see that the slope of the line is $m = \frac{3}{7}$ and the y intercept is $b = \frac{5}{7}$.
∎

Horizontal and Vertical Lines

The horizontal line through the point (a, b) has slope $m = 0$. Therefore, from the point slope form (6), we obtain $y - b = 0(x - a)$. This simplifies to the

---

Equation of a horizontal line through (a, b):

$$y = b.$$

Even though the slope of a vertical line is undefined, it is possible to find an equation for a vertical line. If a vertical line passes through the point (a, b), then any point on the vertical line has x coordinate a and conversely. This fact gives us the

---

Equation of a vertical line through (a, b):

$$x = a.$$

Example 8 The vertical line through $(3, -1)$ has equation

$$x = 3.$$

The horizontal line through $(3, -1)$ has equation

$$y = -1.$$

Both lines are graphed in Figure 1.29.

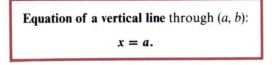

Figure 1.29

Example 9 Find the point of intersection of the line through $(-2, 3)$ and $(1, -5)$ with the line $x = 4$.

Solution The slope of the line through $(-2, 3)$ and $(1, -5)$ is

$$m = \frac{-5 - 3}{1 - (-2)} = \frac{-8}{3} = -\frac{8}{3}.$$

From equation (6) it follows that

$$y - 3 = -\frac{8}{3}(x + 2)$$

$$3(y - 3) = -8(x + 2)$$

$$3y = -8x - 7$$

$$y = -\frac{8}{3}x - \frac{7}{3}.$$

To find the intersection of the two lines, let $x = 4$.

$$y = -\frac{8}{3}(4) - \frac{7}{3} = -\frac{39}{3} = -13.$$

The point of intersection is $(4, -13)$. See Figure 1.30.

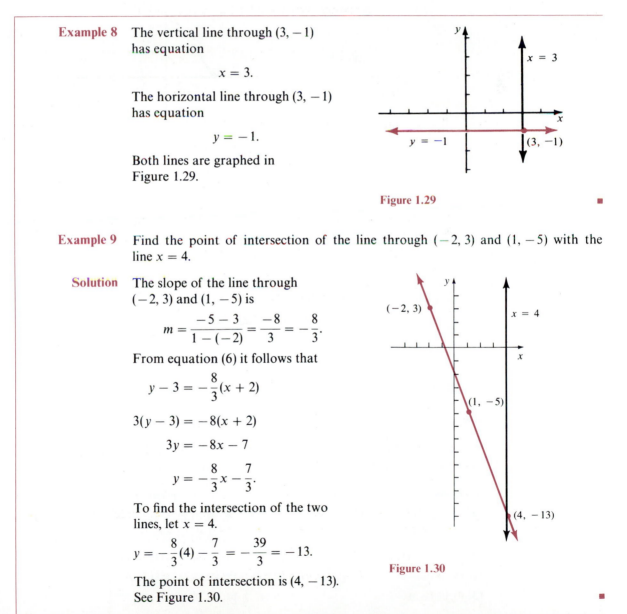

Figure 1.30

***x* Intercept**

The point at which a nonhorizontal line crosses the x axis is called the x **intercept** of the line. If the x and y intercepts are distinct, the graph of the line can be drawn through these two points.

Example 10   Graph the line $3x - 2y + 8 = 0$.

Solution   Set $x = 0$ to find the y intercept.

$$3(0) - 2y + 8 = 0$$

$$-2y + 8 = 0$$

$$2y = 8$$

$$y = 4$$

Set $y = 0$ to find the x intercept.

$$3x - 2(0) + 8 = 0$$

$$3x + 8 = 0$$

$$3x = -8$$

$$x = -\frac{8}{3}$$

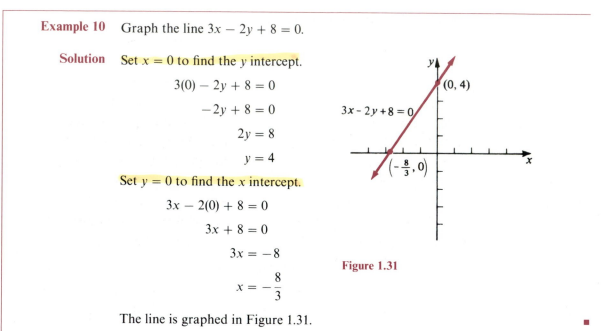

Figure 1.31

The line is graphed in Figure 1.31. ∎

The x and y intercepts of a line can be used to derive yet another form for the equation of a straight line. See Problem 65.

Exercise 1.3

In Problems 1–4, find Δx and Δy for the given pair of points $P_1(x_1, y_1)$ and $P_2(x_2, y_2)$.

1. $P_1(-2, 5), P_2(-1, 4)$ **2.** $P_1(-1, 0), P_2(0, -3)$

3. $P_1(0, -1), P_2(4, 5)$ **4.** $P_1(-3, 2), P_2(1, 2)$

In Problems 5–14, find the slope of the line through the given points, if possible.

5. $(3, -7), (1, 0)$ **6.** $(5, 2), (4, -3)$

7. $(-1, 2), (3, -2)$ **8.** $(0, 6), (-5, 6)$

9. $(-2, -4), (1, 1)$ **10.** $(-4, -1), (1, -1)$

11. $(1, 4), (6, -2)$ **12.** $(8, -\frac{1}{2}), (2, \frac{5}{2})$

13. $(0, 0), (a, b)$ **14.** $(0, 0), (4, -5)$

In Problems 15–24, graph the line through (1, 2) with the indicated slope.

15. $\frac{2}{3}$ **16.** $-\frac{2}{3}$ **17.** 0 **18.** $\frac{1}{10}$ **19.** 3

20. $-\frac{1}{2}$ **21.** -2 **22.** 1 **23.** -1 **24.** 1.2

In Problems 25–46 find an equation of the indicated line.

25. Through (5, 6) with slope 2

26. Through (2, -2) with slope -1

27. Through (0, 4) with slope $\frac{1}{4}$

28. Through (5, -6) and (4, 0)

29. Through (1, -3) with slope 4

30. Through (2, 3) and (6, -5)

31. Through (1, 8) and (1, -3)

32. Through (0, 7) and (7, -2)

33. Through (2, 2) and (-2, -2)

34. Through (4, -3) with slope 0

35. Through (-3, 1) with slope $-\frac{2}{3}$

36. Through (-2, 0) and (-2, 6)

37. Through (0, -3) and (-1, 4)

38. Through (0, -5) with slope $\frac{1}{2}$

39. Through (0, 0) with slope m

40. Through (0, 0) and (a, b)

41. Through (-4, -4) and (2, -4)

42. Through (5, 0) and (-3, -2)

43. Through (0, 0) with slope 22

44. Through (2, -4) with slope -3

45. Through (0, 2) and (5, 0)

46. Through (3, 2) with slope 1

WATCH See notes

In Problems 47–52, find the slope and y intercept of the given line.

47. $2x - 4y - 7 = 0$

48. $-3x + y = 8$

49. $\frac{1}{2}x - 3y + 2 = 0$

50. $x + y + 1 = 0$

51. $-4x - 2y = 0$

52. $ax + by + c = 0$

In Problems 53–58, graph the given line.

53. $3x - 4y + 12 = 0$

54. $x - 3y = 9$

55. $2x + 5y - 8 = 0$

56. $\frac{1}{2}x - 3y = 3$

57. $-4x - 2y + 6 = 0$

58. $y = -\frac{2}{3}x + 1$

In Problems 59–64, find the point of intersection of the given lines.

59. $2x - 5y + 1 = 0$ and $x = 2$

60. $-3x + 7y + 6 = 0$ and $y + 3 = 0$

61. $-x - 2y + 4 = 0$ and $y = -4$

62. $5x + \frac{2}{3}y + 2 = 0$ and $x = \frac{1}{3}$

63. $y = 7$ and $x - 4 = 0$

64. $y = 2x - \frac{1}{3}$ and $y = 7$

The **two-intercept form** *of the equation of a straight line is*

$$\frac{x}{a} + \frac{y}{b} = 1,$$

where a is the x intercept of the line and b is the y intercept. See Figure 1.32.

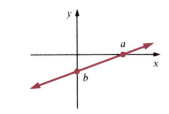

Figure 1.32

65. Derive the two-intercept form by finding an equation of the line through (a, 0) and (0, b).

66. Graph $\dfrac{x}{(-2)} + \dfrac{y}{5} = 1$.

In Problems 67–72, find an equation of the line through the given intercepts.

67. $(-3, 0), (0, 5)$

68. $(7, 0), (0, -2)$

69. $(4, 0), (0, 0)$

70. $(-\frac{1}{2}, 0), (0, -\frac{2}{3})$

71. $(0.2, 0), (0, 0.8)$

72. $(0, 0), (0, -7)$

In Problems 73–76, find the point of intersection of the two lines by solving one of the equations for y and substituting into the other equation.

73. $2x - 2y - 2 = 0$
$3x + 5y - 11 = 0$

74. $4x - y + 1 = 0$
$x + 3y + 9 = 0$

75. $x - 4y + 1 = 0$
$3x + 2y - 1 = 0$

76. $2x + y - 2 = 0$
$3x - 2y + 4 = 0$

In Problems 77–80, find the point(s) of intersection of the given line and the given circle by solving the linear equation for y and substituting into the equation of the circle.

77. $(x - 1)^2 + (y - 2)^2 = 2, \quad y - x - 1 = 0$

78. $(x + 2)^2 + (y + 4)^2 = 8, \quad y - 2x = 0$

79. $x^2 + y^2 - 2y - 8 = 0, \quad 3y + x - 3 = 0$

80. $x^2 + y^2 - 2x - 2y - 11 = 0, \quad 3y + 2x - 4 = 0$

81. The line $y = 2x + 3$ intersects the circle $x^2 + y^2 + 2x - 8y + 12 = 0$ in two points A and B. Find the midpoint of the line segment joining A and B.

82. Does $(4, 1/3)$ lie on the line $(1/3)x + 3y - 1 = 0$?

83. Does $(0.25, 0.75)$ lie on the line passing through $(1, 3)$ and $(1, -3)$?

84. Show that the slope of the line through $(1, 3)$ and $(1 + h, 3 + 2h)$ is 2.

85. Show that the slope of the line through $(2, 4)$ and $(2 + h, (2 + h)^2)$ is $4 + h$.

86. Find the slope of the diameter of the circle $x^2 + 4x + y^2 - 8y + 16 = 0$ if one endpoint of the diameter is $(5, 1)$.

87. If the slope of the line containing a diameter of the circle $x^2 - 2x + y^2 - 4y + 4 = 0$ is 10, find the endpoints of the diameter.

88. A circle with center $(8/7, 4/3)$ passes through the point $(-5, 3/2)$. Find the coordinates of the other endpoint of the diameter drawn from $(-5, 3/2)$.

89. If a triangle has vertices $A(2, -1), B(-3, 1), C(1, 4)$, find an equation of the median drawn from vertex A.

90. In Problem 89, at what point does the median drawn from vertex A intersect the median drawn from vertex B?

91. The cross section of a water trough is an equilateral triangle 60 centimeters on a side with one vertex down. If the origin is placed at the bottom vertex, find an equation of the right edge of the cross section of the trough.

92. The end of a flower box is a trapezoid with lower base 10 inches, upper base 16 inches, and height 12 inches. Find an equation of the left edge if the origin is placed at the center of the lower base.

93. Solve Problem 92 if the origin is placed at the right endpoint of the upper base.

94. A conical tank has a circular base of diameter 10 feet and a height of 8 feet. If, as shown in Figure 1.33, the origin of the xy plane lies at the center of the circular base and the vertex A lies on the y axis, find an equation of the line containing the slant side AB.

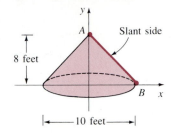

Figure 1.33

Review Problems

In Problems 95 and 96, find an equation of the circle that satisfies the given conditions.

95. Center $(-2, 1)$ passing through $(0, 5)$.

96. Center $(-4, -5)$ tangent to the x axis.

97. Determine whether the equation $x^2 - 10x + y^2 = 0$ describes a circle.

98. Graph $y = -x^4$. Is the graph symmetric with respect to the x axis? The y axis? The origin?

99. For the triangle with vertices $A(2, 1)$, $B(3, -1)$, and $C(0, -2)$, find (a) the midpoint of the side AB; (b) the length of the median from vertex C.

1.4 Parallel and Perpendicular Lines

Two distinct lines in a plane either intersect in a point or they are parallel. In this section we discuss how the slopes of two lines can be used to determine if they are parallel or perpendicular.

Parallel Lines

Suppose that two lines l_1 and l_2 are parallel to each other. If both lines are vertical, then neither has a slope; however, we shall show that if the lines are not vertical, then they have equal slopes. The case in which both are horizontal is clear, since then each line has slope zero. As indicated in Figure 1.34, we note that two non-horizontal parallel lines make equal acute angles with the y axis. Let $P_1(0, y_1)$ and $P_2(0, y_2)$ be the points of intersection of l_1 and l_2 with the y axis.

Now construct two right triangles by drawing a horizontal line through the midpoint Q of the line segment joining P_1 and P_2. Let $P_3(x_3, y_3)$ and $P_4(x_4, y_4)$ be the points of intersection of the horizontal line through Q with l_1 and l_2, respectively.

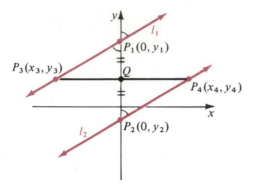

Figure 1.34

Then triangles P_1QP_3 and P_2QP_4 are congruent and so corresponding sides are equal. That is,

$$y_1 - y_3 = y_4 - y_2 \tag{1}$$

and

$$0 - x_3 = x_4 - 0. \tag{2}$$

Since the slope of l_1 is

$$m_1 = \frac{y_1 - y_3}{0 - x_3}$$

and the slope of l_2 is

$$m_2 = \frac{y_4 - y_2}{x_4 - 0},$$

it follows from (1) and (2) that $m_1 = m_2$. That is, nonvertical parallel lines have equal slopes.

Example 1 Find an equation of the line through $(-3, 1)$ which is parallel to $2x + 5y - 6 = 0$.

Solution Express the equation of the line in slope-intercept form:

$$2x + 5y - 6 = 0$$
$$5y = -2x + 6$$
$$y = -\frac{2}{5}x + \frac{6}{5}.$$

The slope of the line is $-\frac{2}{5}$, so the slope of a parallel line will also be $-\frac{2}{5}$. Using the point-slope form, we find the equation of the line.

$$y - 1 = -\frac{2}{5}(x + 3)$$

$$5y - 5 = -2x - 6$$

$$2x + 5y + 1 = 0.$$ ∎

Conversely, if two lines have equal slopes, it can be shown that they are parallel. (See Problem 47.)

Example 2   Determine which of the following lines are parallel:

$$l_1: x + 3 = 0 \qquad l_2: 3x - 5y - 2 = 0 \qquad l_3: -6x + 10y = 0$$

$$l_4: x = -8 \qquad l_5: 5x + 3y + 7 = 0$$

Solution   The equation of l_1 may be written as $x = -3$. Then, since l_1 and l_4 are vertical lines, they are parallel. We express each of the remaining lines in slope-intercept form to find their slopes.

$$l_2: 3x - 5y - 2 = 0$$

$$y = \frac{3}{5}x - \frac{2}{5}$$

$$m_2 = \frac{3}{5}$$

$$l_3: -6x + 10y = 0$$

$$y = \frac{3}{5}x$$

$$m_3 = \frac{3}{5}$$

$$l_5: 5x + 3y + 7 = 0$$

$$y = -\frac{5}{3}x - \frac{7}{3}$$

$$m_5 = -\frac{5}{3}$$

Since $m_2 = m_3$, it follows that l_2 and l_3 are parallel. ∎

Example 3   Determine if the points $P_1(-1, 3)$, $P_2(\frac{3}{5}, \frac{1}{3})$, and $P_3(2, -2)$ lie on a straight line.

Solution If we plot the three points as in Figure 1.35 we see that the three points *appear* to lie on the same straight line. However, this is not sufficient to *prove* that they do, since any graph is only an approximation. To determine if the points actually do lie on the same line, we find the slope of the line through P_1 and P_2,

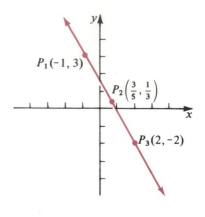

Figure 1.35

$$m_1 = \frac{\frac{1}{3} - 3}{\frac{3}{5} - (-1)} = \frac{-\frac{8}{3}}{\frac{8}{5}} = -\frac{5}{3},$$

and the slope of the line through P_3 and P_2,

$$m_3 = \frac{\frac{1}{3} - (-2)}{\frac{3}{5} - 2} = \frac{\frac{7}{3}}{-\frac{7}{5}} = -\frac{5}{3}.$$

Since these two lines have the same slopes, they are parallel. Because they also have a point in common, P_2, they must be the same line. Thus, the three points lie on the same straight line. ∎

We summarize our results on parallel lines.

> **Slopes of Parallel Lines** Two distinct nonvertical lines, with slopes m_1 and m_2, are parallel if and only if
>
> $$m_1 = m_2.$$

Perpendicular Lines

We can also determine when two lines are perpendicular by comparing their slopes. Let l_1 and l_2 be two nonvertical and nonhorizontal lines which intersect at $P_0(x_0, y_0)$. As shown in Figure 1.36, let $P_1(x_1, y_1)$ and $P_2(x_2, y_2)$ be points different from P_0 on l_1 and l_2, respectively. By the Pythagorean theorem, triangle $P_0 P_1 P_2$ will be a right triangle if and only if

$$a^2 + b^2 = c^2. \tag{3}$$

Using the distance formula, we express (3) as

$$\underbrace{(x_1 - x_0)^2 + (y_1 - y_0)^2}_{a^2} + \underbrace{(x_2 - x_0)^2 + (y_2 - y_0)^2}_{b^2} = \underbrace{(x_2 - x_1)^2 + (y_2 - y_1)^2}_{c^2} \tag{4}$$

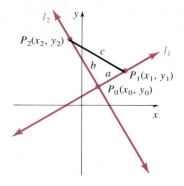

Figure 1.36

Multiplying and simplifying, (4) can be written as

$$x_0^2 + y_0^2 - x_1 x_0 - y_1 y_0 - x_2 x_0 - y_2 y_0 = -x_2 x_1 - y_2 y_1.$$

We rearrange and factor:

$$y_1 y_2 - y_1 y_0 - y_0 y_2 + y_0^2 = -x_1 x_2 + x_1 x_0 + x_0 x_2 - x_0^2$$

$$y_1(y_2 - y_0) - y_0(y_2 - y_0) = -x_1(x_2 - x_0) + x_0(x_2 - x_0)$$

$$(y_1 - y_0)(y_2 - y_0) = -(x_1 - x_0)(x_2 - x_0),$$

from which we find that

$$\frac{y_1 - y_0}{x_1 - x_0} \cdot \frac{y_2 - y_0}{x_2 - x_0} = -1. \tag{5}$$

The slopes of l_1 and l_2 are

$$m_1 = \frac{y_1 - y_0}{x_1 - x_0} \quad \text{and} \quad m_2 = \frac{y_2 - y_0}{x_2 - x_0},$$

respectively. Thus, from (5), we see that l_1 and l_2 are perpendicular if and only if

$$m_1 m_2 = -1. \tag{6}$$

Since neither l_1 nor l_2 is horizontal, neither m_1 nor m_2 is zero, and we may write equation (6) as

$$m_1 = -\frac{1}{m_2} \quad \text{or} \quad m_2 = -\frac{1}{m_1}.$$

We thus say that the slopes of perpendicular lines are *negative reciprocals* of each other.

Example 4 Find an equation of the line through $(0, -3)$ which is perpendicular to $4x - 3y + 6 = 0$.

Solution Express the given line in slope-intercept form:

$$4x - 3y + 6 = 0$$

$$-3y = -4x - 6$$

$$y = \tfrac{4}{3}x + 2.$$

This line has slope $\tfrac{4}{3}$. The slope of a line perpendicular to it will be the negative reciprocal of $\tfrac{4}{3}$, or $-\tfrac{3}{4}$. The line we are looking for, then, has slope $-\tfrac{3}{4}$ and y intercept -3. Its equation is

$$y = -\tfrac{3}{4}x - 3.$$

See Figure 1.37.

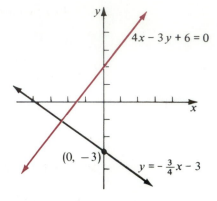

Figure 1.37

Example 5 Determine which of the following lines are perpendicular:

$$l_1: \text{the line through } (5, -7) \text{ and } (3, -4)$$

$$l_2: 6x - 4y - 9 = 0$$

$$l_3: 2x + 3y - 1 = 0$$

Solution Find the slopes of each of these lines.

$$l_1: m_1 = \frac{-4 - (-7)}{3 - 5} = -\frac{3}{2}$$

$$l_2: 6x - 4y - 9 = 0$$

$$y = \frac{6}{4}x - \frac{9}{4}$$

$$m_2 = \frac{3}{2}$$

$$l_3: 2x + 3y - 1 = 0$$

$$y = -\frac{2}{3}x + \frac{1}{3}$$

$$m_3 = -\frac{2}{3}.$$

Since $m_2 m_3 = (\tfrac{3}{2})(-\tfrac{2}{3}) = -1$, l_2 is perpendicular to l_3. Since $m_1 m_2 = (-\tfrac{3}{2})(\tfrac{3}{2}) = \tfrac{9}{4}$, and $m_1 m_3 = (-\tfrac{3}{2})(-\tfrac{2}{3}) = 1$, we conclude that l_1 is not perpendicular to either l_2 or l_3.

We summarize our results on perpendicular lines.

Slopes of Perpendicular Lines Two nonvertical lines with slopes m_1 and m_2 are perpendicular if and only if

$$m_1 m_2 = -1.$$

Exercise 1.4

In Problems 1–10, find an equation of the given line.

1. Through $(-2, 4)$, parallel to $3x + y - 2 = 0$
2. Through $(0, 0)$, parallel to $2x - 4y - 8 = 0$
3. Through $(-1, -4)$, parallel to $y = 1$
4. Through $(0, 2)$, parallel to $-3x + 3y - 1 = 0$
5. Through $(5, 3)$, parallel to $y - 2 = 0$
6. Through $(2, 7)$, parallel to $x + 4 = 0$
7. Through $(7, 0)$, parallel to $-4x - y = 0$
8. Through $(1, 6)$, parallel to $x = -5$
9. Through $(1, -3)$, parallel to $2x - 5y + 4 = 0$
10. Through $(-3, -1)$, parallel to $x - 2y + 4 = 0$

In Problems 11–20, determine whether the given points lie on the same straight line.

11. $(0, 3), (5, -2), (-7, 10)$ 12. $(3, 4), (5, -4), (6, -8)$
13. $(2, 6), (5, -1), (\frac{10}{3}, 3)$ 14. $(1, 3), (-1, \frac{7}{2}), (9, 1)$
15. $(0, 0), (7, 2), (-4, -1)$ 16. $(-8, 4), (-5, -4), (-4, -7)$
17. $(-2, 3), (1, 2), (4, 1), (6, \frac{1}{2})$ 18. $(-4, 8), (-2, 3), (0, -2), (2, -7)$
19. $(-4, 2), (0, 0), (2, -1), (6, -3)$ 20. $(-2, 7), (2, -2), (-1, 5), (3, -4)$

In Problems 21–24, determine which of the given lines are parallel to each other and which are perpendicular to each other.

21. a. $3x - 5y + 7 = 0$ b. $5x + 3y = 0$
    c. $-3x + 5y - 2 = 0$ d. $3x + 5y + 3 = 0$
    e. $-5x - 3y + 1 = 0$ f. $5x - 3y - 2 = 0$

22. a. $2x + 4y - 1 = 0$ b. $x = 3$
    c. $2x - y - 2 = 0$ d. $y - 5 = 0$
    e. $x + 7 = 0$ f. $-x - 2y - 3 = 0$

23. a. $x - 3y + 2 = 0$ b. $3x - y + 1 = 0$
    c. $3x + y = 0$ d. $x + 3y = 2$
    e. $6x - 3y + 1 = 0$ f. $x + 2y + 5 = 0$

24. a. $y + 7 = 0$ **b.** $12x - 9y + 4 = 0$
  c. $4x + 6y - 2 = 0$ **d.** $2x - 3y - 1 = 0$
  e. $x = 5$ **f.** $3x + 4y - 5 = 0$

In Problems 25–34, find an equation of the given line.

25. Through $(2, -3)$, perpendicular to $x - 3y + 1 = 0$

26. Through $(-1, 0)$, perpendicular to $2x + 5y + 2 = 0$

27. Through $(0, 0)$, perpendicular to $x - 2 = 0$

28. Through $(3, 5)$, perpendicular to $y = -4$

29. Through $(7, -1)$, perpendicular to $-3x + y = 0$

30. Through $(-2, -2)$, perpendicular to $x = 3$

31. Through $(3, 7)$, perpendicular to $3x + 7y = 0$

32. Through $(0, -2)$, perpendicular to $3x + 4y + 5 = 0$

33. Through $(1, 6)$, perpendicular to $y + 6 = 0$

34. Through $(-2, -4)$, perpendicular to $x - 4y + 2 = 0$

In Problems 35–40, determine whether the given points are vertices of a right triangle.

35. $(-1, 2), (2, 4), (5, 0)$ **36.** $(-2, 1), (1, 2), (3, -4)$

37. $(-3, 1), (-1, -3), (3, 1)$ **38.** $(-3, 4), (1, -2), \left(8, \dfrac{7}{2}\right)$

39. $(-1, 5), (1, 1), (7, 4)$ **40.** $(2, 2), (4, -3), (10, 0)$

In Problems 41–44, find an equation of the tangent line to the given circle at the indicated point. (Recall that the tangent to a circle at a point P on the circle is perpendicular to the line through the center and P, as shown in Figure 1.38.)

41. $x^2 + y^2 = 5$ at $(1, 2)$

42. $x^2 + y^2 = 4$ at $(-1, \sqrt{3})$

43. $x^2 + y^2 - 4x + 3 = 0$ at $(3, 0)$

44. $x^2 + y^2 + 2x - 6y + 1 = 0$ at $(2, 3)$

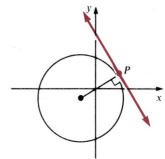

Figure 1.38

45. If $(2, 1), (6, 3)$, and $(4, t)$ are the vertices of a right triangle, find t.

46. If $(-2, 1), (3, 3)$, and $(s, -1)$ are the vertices of a right triangle, find s.

47. Prove that if two lines have equal slopes, they are parallel.

Problems 48–51 refer to the triangle with vertices at $A(3, 1), B(-2, 2),$ and $C(2, 3)$.

48. Find an equation of the altitude drawn from vertex A.

49. At what point does the altitude from vertex B intersect the line through A and C?

50. Find an equation of the line through vertex C parallel to the opposite side.

51. At what point do the altitudes of the triangle ABC intersect?

Review Problems

52. Find an equation of the circle with center $(4, 5)$ passing through $(1, 2)$.

53. Find the center and radius of the circle $x^2 - 10x + y^2 + 4y + 25 = 0$. Graph.

54. Find an equation of the line passing through $(2, 1)$ and $(-1, -5)$.

55. Find the slope, y intercept, and x intercept of $2x - 5y + 10 = 0$. Graph.

56. Find the length of the longest side of the triangle with vertices $(2, 1), (3, 2),$ and $(1, 4)$.

57. Determine b so that $y = mx + b$ is symmetric with respect to the origin.

Chapter Review

The **Cartesian coordinate system** is formed by two perpendicular number lines, called the **x axis** and **y axis**. These axes divide the plane into four **quadrants**. There is a one-to-one correspondence between points in the plane and ordered pairs (a, b) of real numbers. We call a the **x coordinate** or **abscissa** and b the **y coordinate** or **ordinate**.

Any set of ordered pairs of real numbers is called a **relation**, and the corresponding set of points in the plane is the **graph of the relation**. If a relation is defined by an equation such as $\{(x, y) \mid x^3 = y^2\}$, the set of points corresponding to the ordered pairs in the relation is called the **graph of the equation**. We say that a graph is **symmetric with respect to the y axis** if, whenever the point (x, y) is on the graph, $(-x, y)$ is also on the graph. In a similar manner, symmetry with respect to any horizontal or vertical line or any point may be defined.

If $P_1(x_1, y_1)$ and $P_2(x_2, y_2)$ are two points in the plane, the **distance** between P_1 and P_2 is given by the formula

$$d(P_1, P_2) = \sqrt{(x_2 - x_1)^2 + (y_2 - y_1)^2}.$$

The **midpoint** of the line segment joining P_1 and P_2 is

$$\left(\frac{x_1 + x_2}{2}, \frac{y_1 + y_2}{2}\right).$$

The distance formula is used to derive an equation of the **circle**,

$$(x - h)^2 + (y - k)^2 = r^2, \tag{1}$$

with **center** (h, k) and **radius** r. By *completing the square* the equation

$$x^2 + y^2 + Dx + Ey + F = 0$$

can be put in the form of equation (1) for appropriate choices of the coefficients D, E, and F.

The **slope** of the straight line through the two points (x_1, y_1) and (x_2, y_2) is

$$m = \frac{y_2 - y_1}{x_2 - x_1}, \qquad x_1 \neq x_2.$$

An equation of this line is

$$y - y_1 = m(x - x_1). \tag{2}$$

We call (2) the **point-slope** form of the equation of a line. The **slope-intercept** form of the equation of a line is

$$y = mx + b,$$

where m is the slope of the line and b is the **y intercept**.

Let two nonvertical lines have slopes m_1 and m_2. The lines are *parallel* if and only if $m_1 = m_2$; the lines are *perpendicular* if and only if $m_1 m_2 = -1$.

Chapter 1 Review Exercises

In Problems 1–10, answer true or false.

1. For all points (x, y) in the second quadrant, x is negative and y is positive. _____
2. The lines $2x - y - 2 = 0$ and $x + 2y + 4 = 0$ are perpendicular. _____
3. The graph of $x^2 + y^2 + 2x + 1 = 0$ is a circle. _____
4. The point $(0.1, -2.1)$ lies on the line $20x + 10y + 19 = 0$. _____
5. The graph of the circle $(x - 5)^2 + (y + 2)^2 = 7$ is symmetric with respect to the origin. _____
6. An equation of the horizontal line passing through $(\sqrt{2}, \sqrt{3})$ is $x = \sqrt{2}$. _____
7. There is no point on the circle $x^2 - 10x + y^2 + 22 = 0$ with x coordinate equal to 2. _____
8. If a straight line has y intercept $1/4$, the point $(\frac{1}{4}, 0)$ lies on the line. _____
9. The midpoint of the line segment joining the points $(-\sqrt{2}, \pi)$ and $(3\sqrt{2}, -\pi)$ is $(2, 0)$. _____
10. If $(-\frac{1}{2}, \frac{3}{2})$ lies on a line with slope -1, then $(\frac{1}{2}, -\frac{3}{2})$ also lies on that line. _____

In Problems 11 and 12, graph the given relation.

11. $|x| \leq 2$
12. $x + y > 0$

In Problems 13–15, graph the given equation and check for symmetry with respect to the x and y axes and the origin.

13. $x^2 = 4y$ **14.** $|y| = |x|$ **15.** $x^2 = y^2$

16. Find the distance between the points $(4, -6)$ and $(-2, 0)$ and the midpoint of the line segment joining them.

17. Find the center and radius of the circle $x^2 + y^2 - 10x + 2y + 17 = 0$.

In Problems 18–20, find an equation of the circle that satisfies the given conditions.

18. Center $(3, -5)$, radius 4

19. Center $(-3, 2)$, passing through $(5, 4)$

20. Center $(5, 1)$, tangent to the line $y = -3$

In Problems 21–24, find an equation of the indicated line.

21. Through $(2, 7)$ and $(-4, 3)$

22. Through $(6, -1)$, with slope $\frac{2}{3}$

23. Through $(-2, -4)$, parallel to $2x - 4y - 1 = 0$

24. Through $(1, -5)$, perpendicular to $5x + y - 2 = 0$

In Problems 25 and 26, find the slope, the y intercept, and the x intercept of the given line.

25. $5x - 7y + 1 = 0$ **26.** $-\frac{2}{3}x + \frac{3}{2}y - 3 = 0$

27. Find an equation of the tangent line to the circle $x^2 + y^2 - 4x - 1 = 0$ at the point $(3, -2)$.

28. Find an equation of the line passing through the origin and the point of intersection of the two lines $3x - 4y = 4$ and $2x + y = 10$.

29. Find an equation of the perpendicular bisector of the line segment from $(1, 1)$ to $(3, 7)$.

30. If (a, b) lies in quadrant IV, in which quadrant does (b, a) lie?

31. If $(a, a + \sqrt{3})$ lies on the line $y = 2x$, find a.

32. If $(1/2, w)$ lies on the line $y = 4\sqrt{3}$, find w.

In Problems 33 and 34, find the point of intersection of the given lines.

33. $x = 5, y = 7$ **34.** $5x - 2y + 1 = 0, 3x + 4y - 15 = 0$

In Problems 35 and 36, find the point(s) of intersection of the given circle and the given line.

35. $x^2 + y^2 + 8y = 0, 4y - x = 0$ **36.** $x^2 + y^2 = 1, \sqrt{3}y + x - 2 = 0$

37. Three vertices of a rectangle are $(3, 5), (-3, 7)$, and $(-6, -2)$. Find the fourth vertex.

38. Find the point of intersection of the diagonals of the rectangle in Problem 37.

39. For the line $ax + by + c = 0$, what can be said about a, b, and c if
 a. the line passes through the origin
 b. the slope of the line is zero
 c. the slope is undefined

40. Find k if the lines $4x - y = 6$ and $x + ky + 14 = 0$ are to be (a) parallel; (b) perpendicular.

2 Functions and Graphs

2.1 Definition and Basic Properties of a Function

If the question "What is the most important mathematical concept?" were posed to a group of mathematics teachers, certainly the concept of function would appear near or even at the top of the list of their responses. Functions play a very important role throughout mathematics, engineering, the physical and social sciences, and business.

Definition of a Function

Many mathematics students think of a function as a formula. Although this can be a useful characterization, it is not entirely accurate. A function is a rule of correspondence that associates members in one set with those in another set. We formally define this concept as follows.

> *Definition 2.1*   A **function from a set X to a set Y** is a rule which assigns to each element x in X one and only one element y in Y. The set X is called the **domain** of the function.

For the remainder of this chapter we shall assume that both X and Y are subsets of the real numbers.

As an example of a function we consider the table

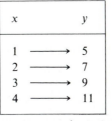

that associates the numbers in the set $\{1, 2, 3, 4\}$ with the numbers in the set $\{5, 7, 9, 11\}$ as indicated by the arrows. The domain of this function is $X = \{1, 2, 3, 4\}$.

The correspondence can be quite arbitrary provided that we do not pair more than one y with a given x. Thus, the correspondence

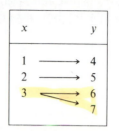

is *not* a function, since two values of y, 6 and 7, are associated with a single value of x.

Since the rule of correspondence will generate pairs of numbers, we could alternatively define a function as a set of ordered pairs of real numbers (x, y) such that no two distinct ordered pairs in the set have the same first element.

Example 1 The set of ordered pairs

$$\{(1, 3), (3, 5), (6, 7), (8, 7)\}$$

is equivalent to the correspondence shown. Notice that it is perfectly acceptable to have distinct values of x correspond to the same y. Thus, this set of ordered pairs describes a function.

Function Notation A function is usually denoted by a letter such as f or g. As previously mentioned, many functions are given by explicit formulas, for example,

$$f: y = x^2,$$

where x can be any real number. The number y that is associated with x by the function f is usually written

$$y = f(x).$$

The symbol $f(x)$ is read "f of x" or "f at x." This **function notation** is extremely important. It will be used in the remainder of this text and throughout calculus.

With this notation we may now write

$$f: y = x^2 \quad \text{as} \quad f(x) = x^2.$$

Thus, the value of the function corresponding to, say, $x = 3$ is

$$f(3) = (3)^2 = 9,$$

and the value of the function at $x = -4$ is

$$f(-4) = (-4)^2 = 16.$$

Strictly speaking, the function f is the rule given by $y = f(x)$, while $f(x)$ is simply the number associated with x. However, we shall frequently ignore this distinction and refer to "the function $f(x)$."

A function may also be compared with a computing machine. A number x is the *input* of the "machine," and the corresponding function value $f(x)$ is the *output* obtained after the "machine" has acted upon x, as illustrated in Figure 2.1.

input x function f output $f(x)$

Figure 2.1

Example 2   Find the values of $f(x) = \sqrt{x + 4}$ corresponding to the values $x = 0, 5, 8,$ and 12.

Solution   The function "machine" takes a value of x, such as $x = 0$, adds the number 4 to it, and then extracts the square root. That is,

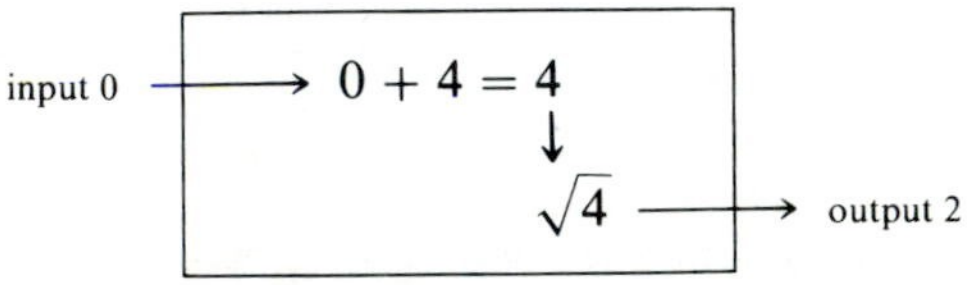

input 0 $0 + 4 = 4$
$\downarrow$
$\sqrt{4}$ output 2

Thus,

$$f(0) = \sqrt{0 + 4} = \sqrt{4} = 2,$$
$$f(5) = \sqrt{5 + 4} = \sqrt{9} = 3,$$
$$f(8) = \sqrt{8 + 4} = \sqrt{12} = \sqrt{4 \cdot 3} = \sqrt{4}\sqrt{3} = 2\sqrt{3},$$
$$f(12) = \sqrt{12 + 4} = \sqrt{16} = 4.$$

■

Dependent and Independent Variables

Since the value of the variable y in $y = f(x)$ always depends on the choice of x, we say that y is the **dependent variable**. By contrast, the choice of x is independent of y, and hence x is called the **independent variable**. We note that this use of the letters f, x, and y is standard practice. However, other letters are sometimes employed. For example, in $z = h(t)$, h denotes the function, t is the independent variable and z is the dependent variable.

Example 3   If $f(x) = x^2 - x + 1$, find $f(-1)$, $f(x + h)$, and $f(x^2 + 1)$.

Solution   The independent variable x is replaced by -1, $x + h$, and $x^2 + 1$, respectively. To further emphasize this replacement, the original function can be written as

$$f(\ \ ) = (\ \ )^2 - (\ \ ) + 1.$$

Thus, for the given inputs, we have

$$f(-1) = (-1)^2 - (-1) + 1$$
$$= 1 + 1 + 1 = 3,$$
$$f(x + h) = (x + h)^2 - (x + h) + 1$$
$$= x^2 + 2xh + h^2 - x - h + 1,$$
$$f(x^2 + 1) = (x^2 + 1)^2 - (x^2 + 1) + 1$$
$$= x^4 + 2x^2 + 1 - x^2 - 1 + 1$$
$$= x^4 + x^2 + 1.$$

Example 4   If $h(t) = 2\sqrt{t-1}$, find $h(1)$, $h(s)$, and $h(v+1)$.

Solution   The independent variable t is replaced by 1, s, and $v + 1$, respectively, to obtain

$$h(1) = 2\sqrt{1-1} = 2\sqrt{0} = 0$$
$$h(s) = 2\sqrt{s-1}$$
$$h(v + 1) = 2\sqrt{(v+1) - 1} = 2\sqrt{v}.$$

Domain and Range

In our machine analogy the **domain** of a function is the set of all real inputs that result in real outputs. The set of outputs is called the **range** of the function. Formally we define the range of a function f with domain X to be the set $\{f(x) \mid x \in X\}$. For example, the range of the function $f(x) = x^2$ is the set of nonnegative real numbers.

When a function is defined by a formula, the domain is understood to be the set of all real numbers for which the corresponding function values are real. This set is sometimes referred to as the "natural domain" of the function. For example, the domain of the function $f(x) = \sqrt{x}$ is the set of all nonnegative real numbers.

Example 5   Find the domain of the function $f(x) = \sqrt{x+4}$.

Solution   Since the radicand $x + 4$ must always be nonnegative, the domain is determined from the inequality

$$x + 4 \geq 0.$$

That is, the domain is the set $\{x \mid x \geq -4\}$.

In the previous example, note that if x is chosen less than -4, say $x = -6$, then

$$f(-6) = \sqrt{-6 + 4} = \sqrt{-2},$$

which is not a real output. Thus, $x = -6$ is not an allowable input and hence is not in the domain of the function $f(x) = \sqrt{x+4}$.

Example 6   Find the domain of the function $f(s) = \dfrac{s}{s^2 - 4}$.

Solution   A quotient of real numbers is a real number unless the denominator is zero. Therefore, the domain of the function f is every real number s except those numbers satisfying

$$s^2 - 4 = 0.$$

The solutions to this equation are $s = 2$ and $s = -2$. Thus, the domain of the function is all real numbers except $s = \pm 2$.   ■

Example 7   The domain of the function $g(x) = x/(x^2 + 4)$ is the entire set of real numbers, since there is no real number x for which $x^2 + 4 = 0$.   ■

The reader should note that many algebraic concepts from Chapter 0 are being utilized in this discussion of functions.

Example 8   Find the domain and range of

$$g(x) = 5 + \sqrt{x - 3}.$$

Solution   The domain is determined by the requirement that $x - 3 \geq 0$. Hence, the domain is $\{x \mid x \geq 3\}$. Now, since $\sqrt{x - 3} \geq 0$ for $x \geq 3$, we have $5 + \sqrt{x - 3} \geq 5$ for these same values of x. Thus, the range is $\{y \mid y \geq 5\}$.   ■

In the preceding examples we have determined the "natural" domain of the given function. However, the domain of a function can be explicitly specified provided that the resulting function values are real. For example, we can restrict the domain of the function $f(x) = x^2$ to be the set of nonnegative real numbers; that is,

$$f(x) = x^2, \qquad x \geq 0.$$

This is not the same function as

$$f(x) = x^2,$$

where the domain is understood to be the set of all real numbers.

Example 9   Find the range of $f(x) = \sqrt{9 - x^2}$.

Solution   The range consists of all the function values

$$f(x) = \sqrt{9 - x^2},$$

for which $9 - x^2 \geq 0$. Therefore, we consider the equation

$$y = \sqrt{9 - x^2}. \tag{1}$$

This is equivalent to

$$y^2 = 9 - x^2 \quad \text{or} \quad x^2 + y^2 = 9,$$

provided that $y \geq 0$. Hence, we see that (1) represents the set of all points (x, y) on the upper half of the circle of radius 3 centered at the origin. See Figure 2.2. Thus the y values satisfying (1) are $0 \leq y \leq 3$, and so the range is $\{y \mid 0 \leq y \leq 3\}$.

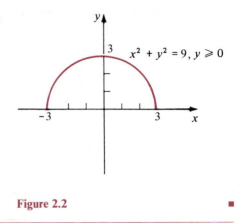

$x^2 + y^2 = 9, y \geq 0$

Figure 2.2

As the next example illustrates, the problem of determining whether a *single* number r is in the range of the function $y = f(x)$ is equivalent to solving the equation $f(x) = r$.

Example 10   For what value of x is $f(x) = \sqrt{x - 1}$ equal to 5?

Solution   We solve $f(x) = 5$ or

$$\sqrt{x - 1} = 5$$
$$x - 1 = 25$$
$$x = 26.$$

To verify that 26 is the desired value, we compute

$$f(26) = \sqrt{26 - 1} = 5.$$

Graph of a Function

The **graph of a function** $y = f(x)$ is the graph of the relation

$$\{(x, y) \mid y = f(x), x \text{ in the domain of } f\}.$$

Example 11   The graph of the function defined by

x	y
1 →	3
3 →	5
6 →	7
8	

is shown in Figure 2.3.

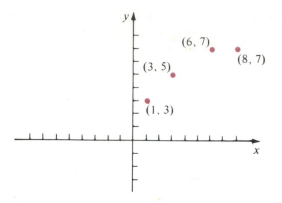

Figure 2.3

Example 12 Find the y coordinate of the point on the graph of the function $f(x) = x^3$ whose x coordinate is 4.

Solution Since any point on the graph of $y = f(x)$ has coordinates $(x, f(x))$, the point with x coordinate equal to 4 is $(4, f(4))$. Thus, the y coordinate of this point is

$$y = f(4) = 4^3 = 64.$$

The graph of a function provides a geometric illustration of the domain and range of the function. The domain is the set of all x coordinates of points on the graph, and the range is the set of all y coordinates of points on the graph.

Example 13 The graph of the function $f(x) = x^2$, shown in Figure 2.4, is the graph of the equation $y = x^2$ that was sketched in Figure 1.11 in Section 1.1. The graph indicates that the domain of f is the set of real numbers and the range is $\{y \mid y \geq 0\}$.

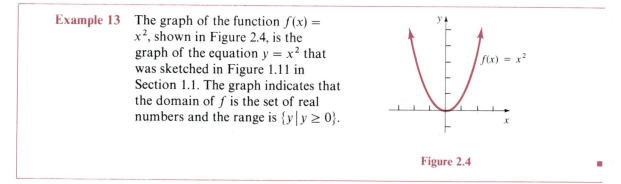

$f(x) = x^2$

Figure 2.4

Piecewise-Defined Functions It is important to note that a function can be defined in several pieces. For example,

$$g(x) = \begin{cases} x^2, & x < 0 \\ x + 1, & x \geq 0 \end{cases}$$

is called a **piecewise-defined function**. It is *not* two functions, but rather one function in which the rule is given in two parts: one for the negative real numbers and one for the nonnegative real numbers. Thus, since -4 is negative,

$$g(-4) = (-4)^2 = 16;$$

and since 6 is nonnegative,

$$g(6) = (6) + 1 = 7.$$

Example 14 The graph of the piecewise-defined function

$$f(x) = \begin{cases} -1, & x < 0 \\ 0, & x = 0 \\ 1, & x > 0 \end{cases}$$

is shown in Figure 2.5. The graph indicates that the domain of f is the set of real numbers and the range is $\{-1, 0, 1\}$.

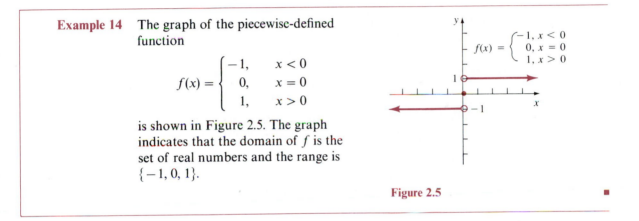

Figure 2.5

Vertical Line Test From the definition of a function we know that for each x in the domain of f there corresponds a *unique* value $f(x)$ in the range. Since the graph of f is the set $\{(x, f(x)) \mid x \text{ in the domain of } f\}$, we conclude that any vertical line can intersect the graph in at most one point. Conversely, if each vertical line intersects the graph of a relation in at most one point, then the relation is a function. This last statement is called the **vertical line test** for a function.

Example 15 From Figure 2.6 we see that any vertical line intersects the graph of the relation defined by $y = x^2$ in at most one point. Thus, by the vertical line test, the relation defines a function $y = f(x) = x^2$.

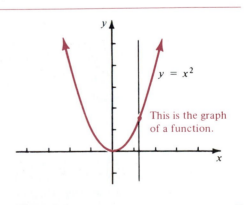

$y = x^2$

This is the graph of a function.

Figure 2.6

Example 16　As shown in Figure 2.7 a vertical line can intersect the graph of the relation defined by $x^2 + y^2 = 4$ in more than one point. Thus the relation does not define a function $y = f(x)$.

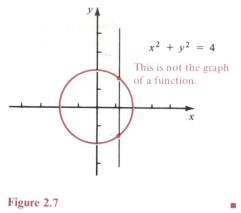

$x^2 + y^2 = 4$

This is not the graph of a function.

Figure 2.7

Very often in calculus in order to solve a problem, it is necessary to first set up a function and determine its domain. The following example is typical of the kind of reasoning required.

Example 17　A rectangular plot of land that is to contain 2500 square meters will be fenced and divided into two equal parts by an additional fence that is parallel to two sides. Express the total amount F of fencing required as a function of the length x as shown in Figure 2.8. Find the domain of F.

Solution　If we denote the length of the field by x and the width by y, then, as shown in Figure 2.8, the total amount of fencing required is

$$F = 2x + 3y. \qquad (2)$$

We also have the condition that the area enclosed is 2500 m^2, so that

$$xy = 2500. \qquad (3)$$

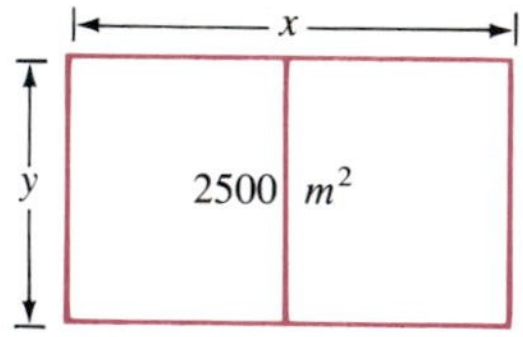

2500 m^2

Figure 2.8

Solving (3) for $y = 2500/x$ and substituting into the expression (2) for F enables us to write F as a function of the single variable x:

$$F(x) = 2x + 3\left(\frac{2500}{x}\right) \qquad (4)$$

Two considerations enter into determining the domain of F. From (4) we see that $x \neq 0$. In addition, from the context of the word problem where x represents a length, we see that x cannot be negative. Therefore the domain of F is $\{x \mid x > 0\}$. ∎

Exercise 2.1

In Problems 1–6, determine whether the correspondence $x \to y$ given by the set of ordered pairs (x, y) is a function.

1. $\{(1, 2), (2, -3), (3, 4), (-4, -1), (1, 5)\}$
2. $\{(-1, 5), (7, 2), (3, -4)\}$
3. $\{(0, 0), (1, 1), (2, 2)\}$
4. $\{(4, 2), (-4, 3), (8, 6), (5, 4)\}$
5. $\{(0, 1), (1, 1), (2, 1)\}$
6. $\{(3, 2), (-6, 2), (-3, 9), (-6, 9)\}$
7. If $f(x) = x^2 - 1$, find $f(0)$; $f(1)$; $f(\sqrt{2})$; $f(-2)$.
8. If $f(x) = x^2 + x$, find $f(0)$; $f(1)$; $f(\sqrt{2})$; $f(-2)$.
9. If $h(x) = \sqrt{x + 1}$, find $h(0)$; $h(3)$; $h(-1)$; $h(5)$.
10. If $r(x) = \sqrt{2x + 4}$, find $r(0)$; $r(4)$; $r(\frac{1}{2})$; $r(-\frac{1}{2})$.
11. If $f(x) = \dfrac{3x}{x^2 + 1}$, find $f(0)$; $f(1)$; $f(\sqrt{2})$; $f(-1)$.
12. If $f(x) = \dfrac{x^2}{x^3 - 1}$, find $f(0)$; $f(-1)$; $f(\sqrt{2})$; $f(\frac{1}{2})$.
13. If $f(t) = 3t^3 - t$, find $f(a)$; $f(a + 1)$; $f(a^2)$; $f\left(\dfrac{1}{a}\right)$; $f(-a)$.
14. If $p(z) = 2z^4 - z^2$, find $p(b)$; $p(b + 1)$; $p(b^2)$; $p\left(\dfrac{1}{b}\right)$; $p(-b)$.
15. If $f(x) = \begin{cases} x^2 + 2x, & x \geq 0 \\ -x^3, & x < 0 \end{cases}$, find $f(1)$; $f(0)$; $f(\sqrt{2})$; $f(-2)$.
16. If $f(x) = \begin{cases} -10, & x < 0 \\ 0, & x = 0 \\ 10, & x > 0 \end{cases}$, find $f(1)$; $f(0)$; $f(\sqrt{2})$; $f(-2)$.
17. If $f(x) = 3x - 4$, find
$$\frac{f(1 + h) - f(1)}{h}, \qquad \text{where } h \neq 0 \text{ is a constant.}$$
18. If $f(x) = x^2 + 2x - 1$, find
$$\frac{f(3 + h) - f(3)}{h}, \qquad \text{where } h \neq 0 \text{ is a constant.}$$

In Problems 19–22, find $\dfrac{f(x + h) - f(x)}{h}$, where $h \neq 0$ is a constant.

19. $f(x) = x^2$
20. $f(x) = x^3$
21. $f(x) = \dfrac{1}{x}$
22. $f(x) = \sqrt{x + 1}$

In Problems 23–26, find $\dfrac{f(x) - f(a)}{x - a}$, where $x \neq a$.

23. $f(x) = 2x + 1$
24. $f(x) = x^2 + x$
25. $f(x) = \sqrt{x}$
26. $f(x) = \dfrac{1}{x}$

In Problems 27–32, find the domain of the given function.

27. $f(x) = \dfrac{x^2}{x^3 - 1}$

28. $g(t) = \dfrac{3}{t^2 + 6t + 5}$

29. $f(x) = \dfrac{x}{x^2 + 25}$

30. $f(x) = \dfrac{1}{\sqrt{x - 2}}$

31. $f(x) = \dfrac{\sqrt{x + 1}}{x^2}$

32. $f(x) = \dfrac{x}{x^2 - 1}$

In Problems 33–40, find the domain and the range of the given function.

33. $f(x) = 6$

34. $f(x) = 3x - 15$

35. $f(x) = x^2 + 2x + 1$

36. $f(x) = x^3 + x$

37. $f(x) = \sqrt{x - 5}$

38. $f(x) = \sqrt{4x - 3}$

39. $f(x) = \sqrt{3x} + 2$

40. $f(x) = \sqrt{1 - x^2}$

41. For what values of x is $f(x) = \sqrt{x - 4}$ equal to 4? to 0?

42. For what values of x is $f(x) = \sqrt{x^2 - 1}$ equal to 0? to -1?

43. For what values of x is $f(x) = \begin{cases} x^2, & x \geq 0 \\ x^3, & x < 0 \end{cases}$ equal to -8? to 4?

44. For what values of x is $f(x) = \begin{cases} x - 1, & x < 0 \\ 0, & x = 0 \\ x^2, & x > 0 \end{cases}$ equal to 1? to -4?

In Problems 45–53, graph the given function. State the domain and the range.

45. $f(x) = 3$

46. $f(x) = x$

47. $f(x) = x^3$

48. $f(x) = \begin{cases} 0, & x < 0 \\ x, & x \geq 0 \end{cases}$

49. $f(x) = \begin{cases} -x, & x \neq 1 \\ 1, & x = 1 \end{cases}$

50. $g(s) = \begin{cases} s - 1, & x \leq 0 \\ s + 1, & x > 0 \end{cases}$

51. $f(x) = \begin{cases} 3, & x > 3 \\ x, & -3 \leq x \leq 3 \\ -3, & x < -3 \end{cases}$

52. $h(t) = \begin{cases} 0, & t < 0 \\ t, & 0 \leq t < 1 \\ t + 1, & t \geq 1 \end{cases}$

53. $z(x) = \begin{cases} x^2, & x < 0 \\ 0, & x = 0 \\ -x^2, & x > 0 \end{cases}$

In Problems 54 and 55, find the y coordinate of the point on the graph of the given function whose x coordinate is (a) 3; (b) 0; (c) a.

54. $f(x) = x^2 - x$

55. $f(x) = \sqrt{x + 5}$

In Problems 56–61, graph the given relation and use the vertical line test to determine whether the relation defines y as a function of x.

56. $\{(x, y) \mid x^2 + y^2 = 9\}$

57. $\{(x, y) \mid xy > 0\}$

58. $\{(x, y) \mid y = 3x - 2\}$

59. $\{(x, y) \mid y = \sqrt{x}\}$

60. $\{(x, y) \mid y - x^2 = 0\}$

61. $\{(x, y) \mid x - y^2 = 0\}$

In Problems 62–65 use the vertical line test to determine whether the relation defines y as a function of x. If it does, state the domain and range.

62.

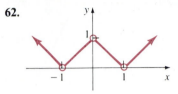

Figure 2.9

63.

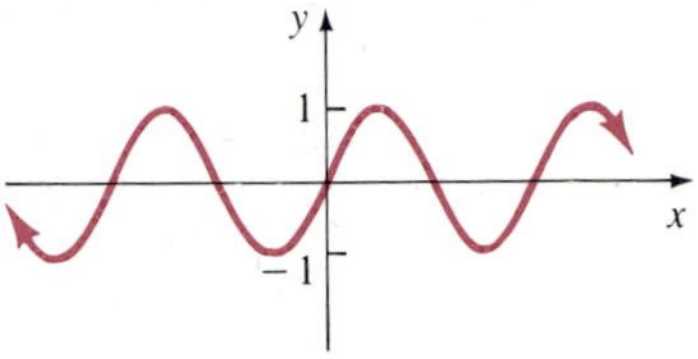

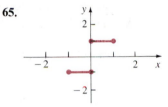

Figure 2.10

64.

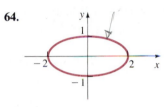

Figure 2.11

65.

Figure 2.12

66. Express the volume V of a cube as a function of the area A of its base.

67. Express the area A of a circle as a function of its diameter d; as a function of its circumference C.

68. Express the surface area S of a solid cylinder of volume 1 cubic meter as a function of its radius r.

69. The height of a rectangular box is five times the width and the length is two times the width. Express the volume of the box as a function of the width; as a function of the length.

70. A rectangular box with no top is to be constructed with a square base x cm on a side and a volume of 64,000 cm^3. Express the surface area S as a function of x. Find the domain of S.

71. Solve Problem 70 if the box has a top.

72. A rancher wishes to build a rectangular corral of 144,000 ft^2 with one side along a vertical cliff. The fencing along the cliff costs \$1.00 per foot and the fencing along the other 3 sides costs \$2.50 per foot. Express the total cost of the fencing T as a function of the length x of the fence along the vertical cliff. Find the domain of T.

73. First-class postage requires 22¢ for the first ounce or fraction thereof and 17¢ for each additional ounce or fraction thereof. Graph the function. State the domain and range.

74. A direct-dialed long distance call from Los Angeles to San Francisco costs 62¢ for the first minute and 45¢ for each additional minute or fraction thereof. Graph the function. State the domain and range.

In Problems 75–78 express the area of the shaded region as a function of the x coordinate of the point P(x, y).

75.

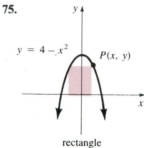

y
y = 4 − x²
P(x, y)
x
rectangle

Figure 2.13

76.

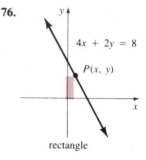

y
4x + 2y = 8
P(x, y)
x
rectangle

Figure 2.14

77.

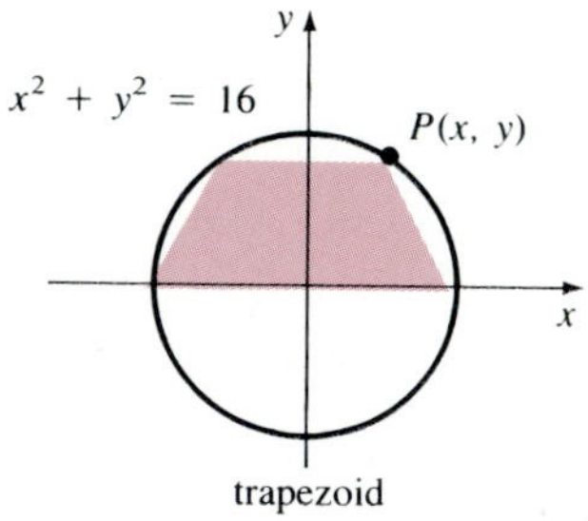

y
x² + y² = 16
P(x, y)
x
trapezoid

Figure 2.15

78.

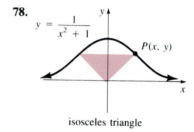

y
y = 1/(x² + 1)
P(x, y)
x
isosceles triangle

Figure 2.16

79. Find an expression for the slope of the line through the points A and B in Figure 2.17.

y
B
A
y = f(x)
1
1 + h
x

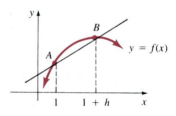

Figure 2.17

80. Find an expression for the slope of the line through the points A and B in Figure 2.18.

y
y = g(x)
B
A
a
a + h
x

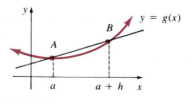

Figure 2.18

81. Find the y coordinates of the points A, B, C, and D in Figure 2.19.

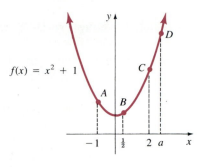

$f(x) = x^2 + 1$

Figure 2.19

82. If the graph of $f(x) = x^2 + a$ is as shown in Figure 2.20, find a.

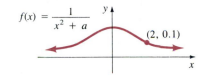

$f(x) = \dfrac{1}{x^2 + a}$

$(2, 0.1)$

Figure 2.20

Calculator Problems

83. If $f(x) = x^{100}$, find $f(2)$, $f(0.9)$, $f(1.01)$.

84. If $f(x) = x^{2.5}$, find $f(2.5)$, $f(10)$, $f(1.01)$.

85. If $g(r) = \sqrt[5]{r + 1}$, find $g(1)$, $g(0.1)$, $g(-11)$.

86. If $h(t) = \sqrt{0.95t}$, find $h(0.95)$, $h(1.01)$, $h(2)$.

2.2 Linear and Quadratic Functions

Linear Functions

One of the simplest and yet most important types of functions is the **linear function**

$$f(x) = ax + b, \qquad a \neq 0, \tag{1}$$

where a and b are constants. Since $f(x)$ is a real number for any choice of x, we conclude that the domain of (1) is the set of real numbers. If we write (1) in the form

$$y = ax + b,$$

we have the equation of a straight line with slope a and y intercept b.

Example 1 Graph $f(x) = \frac{1}{2}x - \frac{3}{2}$.

Solution The y intercept of the graph is $b = -\frac{3}{2}$. We note that this corresponds to the value of the function at $x = 0$; that is, $f(0) = -\frac{3}{2}$. Recall that to graph a straight line, only two points are required. While we could substitute any value of x into $f(x)$ to obtain another point, it is good practice to try to determine the x intercept of the graph. Setting $f(x) = 0$, we have

$$\tfrac{1}{2}x - \tfrac{3}{2} = 0,$$

which gives $x = 3$. Thus the graph in Figure 2.21 is drawn through the points $(0, -\frac{3}{2})$ and $(3, 0)$. ∎

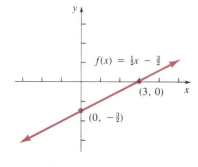

Figure 2.21

Example 2 If t represents temperature measured in degrees Fahrenheit and T represents temperature measured in degrees Celsius, the equation

$$T = \tfrac{5}{9}(t - 32)$$

represents the conversion from the Fahrenheit to the Celsius scale. Using function notation, we can write

$$T(t) = \tfrac{5}{9}(t - 32).$$

This emphasizes the fact that Celsius temperature T is a function of Fahrenheit temperature t. Thus for $t = 32°$F we have

$$T(32) = 0°\text{C},$$

and for $t = 212°$F we have

$$T(212) = \tfrac{5}{9}(212 - 32) = 100°\text{C}.$$ ∎

Quadratic Functions

A **second-degree** or **quadratic function** is any function of the form

$$f(x) = ax^2 + bx + c, \qquad a \neq 0, \tag{2}$$

where a, b, and c are constants. The graph of a quadratic function is called a **parabola***. It can be shown that every quadratic function has a graph that is similar in shape to the graph of $f(x) = x^2$. If $a > 0$, the parabola will open upward as in Figure 2.22.

If $a < 0$ in (2), the parabola will open downward, as in the next example.

---

* For a geometric discussion of the parabola see Section 6.2.

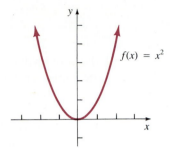

Figure 2.22

Example 3   To obtain the graph of $f(x) = -x^2$ in Figure 2.23 we observe that the y coordinate of any point on the graph is the negative of the corresponding y coordinate on the graph of $f(x) = x^2$.

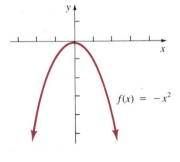

Figure 2.23   ∎

Intercepts

The **y intercept** of the graph of the quadratic function (2) is the value
$$f(0) = c.$$
To determine whether the graph has any **x intercepts**, we solve the equation $f(x) = 0$; that is,
$$ax^2 + bx + c = 0. \tag{3}$$
Applying the quadratic formula to (3), we can see that the graph of a quadratic function may or may not have x intercepts. The following table summarizes the three possibilities.

$ax^2 + bx + c = 0,$ $\qquad x = \dfrac{-b \pm \sqrt{b^2 - 4ac}}{2a}$	
$b^2 - 4ac > 0$	Two real roots; two x intercepts. The graph crosses the x axis twice.
$b^2 - 4ac = 0$	One real root at $x = -b/2a$; one x intercept. The graph is tangent to the x axis.
$b^2 - 4ac < 0$	No real roots; no x intercepts. The graph is either entirely above or entirely below the x axis.

Example 4   Graph $f(x) = x^2 - 2x - 3$.

Solution   Since f is a quadratic function with $a = 1 > 0$, the graph will be a parabola opening upward. The y intercept is $f(0) = -3$. To find the x intercepts, we solve

$$x^2 - 2x - 3 = 0$$

and find that $x = -1$ or $x = 3$. By plotting points from the accompanying table and joining them with a smooth curve, we obtain the graph in Figure 2.24.

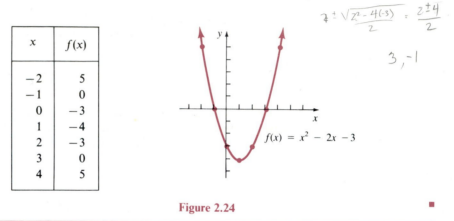

x	$f(x)$
-2	5
-1	0
0	-3
1	-4
2	-3
3	0
4	5

$$\frac{2 \pm \sqrt{2^2 - 4(-3)}}{2} = \frac{2 \pm 4}{2}$$

$$3, -1$$

$f(x) = x^2 - 2x - 3$

Figure 2.24

Vertex

If the graph of a quadratic function opens upward, then the minimum or lowest point on the parabola is called the **vertex**; if the graph opens downward, then the maximum or highest point on the graph is called the **vertex**. See Figure 2.25.

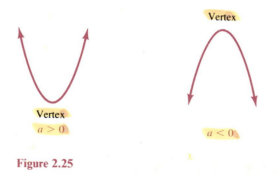

Vertex
$a > 0$

Vertex
$a < 0$

Figure 2.25

In Example 4 it *appears* that the vertex of the parabola is located at $(1, -4)$. Being able to accurately determine the vertex of a parabola would be a considerable aid in graphing a quadratic function. In order to do this, we consider the following.

By completing the square on the quadratic expression $ax^2 + bx + c$, we may write

$$f(x) = ax^2 + bx + c$$

as

$$f(x) = a\left(x + \frac{b}{2a}\right)^2 + \frac{4ac - b^2}{4a}. \tag{4}$$

We note that, regardless of what value is substituted for x in (4), the term $(4ac - b^2)/4a$ remains unchanged. Thus the first term $a[x + (b/2a)]^2$ determines the magnitudes of the function values $f(x)$. If $a > 0$, then $a[x + (b/2a)]^2 \geq 0$. It follows that $f(x)$ has its minimum value when

$$\left(x + \frac{b}{2a}\right)^2 = 0;$$

that is, for $x = -b/2a$. Similarly, if $a < 0$, then $a[x + (b/2a)]^2 \leq 0$, and f has its maximum value when $x = -b/2a$. Therefore the vertex of the graph of a quadratic function $f(x) = ax^2 + bx + c$ is located at

$$\left(\frac{-b}{2a}, f\left(\frac{-b}{2a}\right)\right). \tag{5}$$

Example 5 For $f(x) = x^2 - 2x - 3$ we identify $a = 1$ and $b = -2$. Then the vertex of the graph is located at

$$\left(\frac{-(-2)}{2 \cdot 1}, f\left(\frac{-(-2)}{2 \cdot 1}\right)\right) = (1, f(1))$$

$$= (1, -4),$$

as previously indicated in Figure 2.24. ∎

Axis

The graph of a quadratic function is symmetric with respect to the vertical line through the vertex of the parabola. This line is called the **axis** of the parabola. From (5) we conclude that the axis of the graph of $f(x) = ax^2 + bx + c$ has the equation

$$x = \frac{-b}{2a}. \tag{6}$$

Example 6   Graph $f(x) = -4x^2 + 12x - 9$.

Solution   The graph of this quadratic function is a parabola that opens downward since $a = -4 < 0$. Identifying $a = -4$ and $b = 12$, we have from (5) that the vertex is located at

$$\left(\tfrac{3}{2}, f\left(\tfrac{3}{2}\right)\right) = \left(\tfrac{3}{2}, 0\right).$$

The y intercept is $f(0) = -9$. Solving $-4x^2 + 12x - 9 = 0$, we find that there is only one x intercept, namely $x = \tfrac{3}{2}$. Of course, this was to be expected, since the vertex $\left(\tfrac{3}{2}, 0\right)$ is on the x axis. As shown in Figure 2.26, a rough sketch can be obtained from these two points alone, since the parabola must be symmetric with respect to the axis $x = \tfrac{3}{2}$.

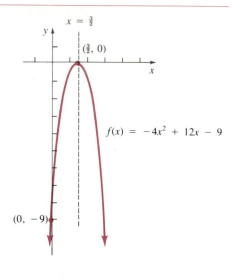

Figure 2.26

Example 7   Graph $f(x) = x^2 + 2x + 4$.

Solution   The graph is a parabola that opens upward. Identifying $a = 1$ and $b = 2$, we see that the vertex is located at

$$(-1, f(-1)) = (-1, 3).$$

The y intercept is $f(0) = 4$. Solving $x^2 + 2x + 4 = 0$, we find no real solutions. Therefore the graph has no x intercepts. Since the vertex is above the x axis, the graph must lie entirely above the x axis. Using symmetry with respect to the axis $x = -1$, we obtain the graph as shown in Figure 2.27.

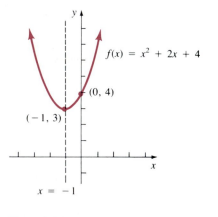

Figure 2.27

As indicated in the following examples, many applied problems involve finding the maximum (or minimum) value of a quadratic function.

Example 8   A rancher wishes to build a rectangular corral with 1000 feet of fencing. What should the dimensions of the corral be if the area enclosed is to be maximum?

Solution   As shown in Figure 2.28, we denote the width and length of the corral by w and l, respectively. The area is given by

$$A = wl, \qquad (7)$$

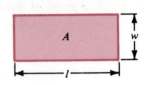

Figure 2.28

and the perimeter equals 1000 feet. Thus

$$2w + 2l = 1000.$$

Solving this equation for l,

$$l = \tfrac{1}{2}(1000 - 2w) = 500 - w;$$

substituting the result into (7), we have

$$A = w(500 - w).$$

Thus we have expressed the area A as a function of the single variable w:

$$A(w) = w(500 - w) = 500w - w^2.$$

This function is graphed in Figure 2.29. Since the parabola opens downward, the maximum value of A occurs at the vertex, where

$$w = \frac{-500}{2(-1)} = 250.$$

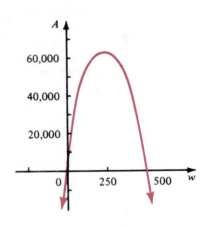

Figure 2.29

Since the corresponding length is

$$l = 500 - w = 250,$$

the desired dimensions are 250 feet by 250 feet.   ∎

Example 9   The distance s in feet above the ground of a certain toy rocket t seconds after being fired from the top of a building is given by $s(t) = -16t^2 + 96t + 256$. Find (a) the maximum height attained by the rocket; (b) the time when it strikes the ground; and (c) the height of the building from which the rocket was fired.

Solution   We see that $s(t) = -16t^2 + 96t + 256$ is a quadratic function in the form (2) with $a = -16$, $b = 96$, $c = 256$, and t in the place of x. Therefore the graph of this function will 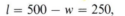 be a parabola that opens downward.

a. The maximum value of s occurs when

$$t = \frac{-b}{2a} = \frac{-96}{-32} = 3.$$

Thus the maximum height attained is

$$s(3) = -16(3)^2 + 96(3) + 256 = 400 \text{ ft.}$$

b. The rocket strikes the ground when $s(t) = 0$:

$$-16t^2 + 96t + 256 = 0$$
$$-16(t^2 - 6t - 16) = 0$$
$$-16(t - 8)(t + 2) = 0$$
$$t = 8 \quad \text{or} \quad t = -2$$

Since t represents the number of seconds after the rocket is fired, we discard the solution $t = -2$. Therefore the rocket strikes the ground $t = 8$ seconds after being fired.

c. The height of the building from which the rocket was fired is given by its initial position

$$s(0) = 256 \text{ ft.}$$

The graph of $s(t) = -16t^2 + 96t + 256$ is sketched in Figure 2.30. The domain of the function is restricted to $0 \le t \le 8$ since s represents feet above the ground. We observe that the maximum height attained is the ordinate of the vertex; the time at which the rocket strikes the ground corresponds to the t intercept; and the initial height corresponds to the s intercept.

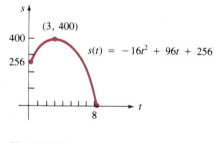

Figure 2.30

Exercise 2.2

In Problems 1–6, find the x and y intercepts of the given linear function. Graph.

1. $f(x) = 2x - 8$

2. $f(x) = -\frac{2}{3}x - 4$

3. $f(x) = -3x + 6$

4. $f(x) = 2x + 5$

5. $f(x) = 3(2 - \frac{1}{2}x)$

6. $f(x) = -\frac{1}{10}(2x - 5) + \frac{1}{2}$

7. If $f(x) = ax + b$, evaluate $\dfrac{f(x + h) - f(x)}{h}$ where $h \ne 0$ is a constant.

8. One meter is approximately 3.28 feet. Determine formulas for converting meters into feet, and feet into meters.

9. If $T = \frac{5}{9}(t - 32)$ is the formula converting t degrees Fahrenheit to T degrees Celsius, determine the formula for converting degrees Celsius to degrees Fahrenheit. Convert 77°F to degrees Celsius. Convert 20°C to degrees Fahrenheit.

10. The value of a computer, in dollars, is given by

$$V(x) = 500,000\left(1 - \frac{x}{40}\right), \qquad 0 \le x \le 40,$$

   where x is the time in years. What is the initial value of the computer? At what point in time is the computer worth $\frac{1}{2}$ its initial value? At what point in time will the computer have lost $\frac{3}{4}$ of its initial value? When is it worth nothing?

11. Straight-line, or linear, depreciation consists of an item losing *all* its initial worth of A dollars over a period of n years by an amount A/n each year. If an item costing $20,000 when new is depreciated linearly over 25 years, determine a formula giving its value V after x years $(0 \le x \le 25)$. What is its value after 10 years?

12. In simple interest, the amount accrued S is a linear function of time measured in years: $S = P + Prt$. Compute S after 15 years if the principal is $P = 100$ and the annual rate of interest is $r = 4\%$. At what time is $S = 220$?

13. A copy machine makes y copies in t minutes. The relationship between y and t is linear. If the machine makes 10 copies after running for 1 minute and 1190 copies after running for 1 hour, express y as a function of t. How much time must the machine run to make the first copy?

In Problems 14–19, determine whether the graph of the given quadratic function opens upward or downward. Find the x and y intercepts.

14. $f(x) = -x^2 - x + 6$
15. $f(x) = 2x^2 - 4x + 1$
16. $f(x) = 4 - x^2$
17. $f(x) = -(3 - x)^2$
18. $f(x) = 5x^2 - 2x + 4$
19. $f(x) = (3 - x)(x + 1)$

In Problems 20–25, graph the given quadratic function. Determine the vertex, the x and y intercepts, and the axis of symmetry.

20. $f(x) = x^2 - 3x + 2$
21. $f(x) = \frac{1}{2}x^2 + x + 1$
22. $f(x) = -x^2 + 6x - 5$
23. $f(x) = x^2 - 2x - 7$
24. $f(x) = -2x^2 - 16x$
25. $f(x) = -(4 - 3x^2)$

26. If $f(x) = ax^2 + bx + c$, evaluate $\dfrac{f(x + h) - f(x)}{h}$, where $h \ne 0$ is a constant.

27. Is it possible to find two real numbers whose sum is 30 and whose product is a maximum? A minimum?

28. The sum of two real numbers is 6. Find the two numbers such that their product is a maximum. Can you generalize this result?

29. Find the point on the line $y = 2x$ closest to $(5, 0)$. What is the minimum distance? [*Hint:* Consider the square of the distance.]

30. A rancher wants to enclose a corral alongside a straight stream as shown in Figure 2.31. If the total length of fencing on hand is 3000 feet, find the maximum value of the area function $A(x)$. What are the dimensions of the corral of maximum area? (No fencing is required along the stream.)

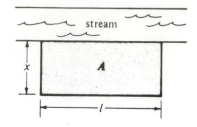

Figure 2.31

31. A long rectangular sheet of metal 10 inches wide is to be made into a rain gutter by bending up two sides perpendicular to the sheet as shown in Figure 2.32. How many inches should be bent up on each side in order to make the capacity of the gutter a maximum?

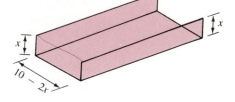

Figure 2.32

32. A Norman window consists of a rectangle surmounted by a semicircle as shown in Figure 2.33. Find the dimensions of the window with the largest area if its perimeter is 12 meters.

Figure 2.33

33. A piece of wire 40 inches long is to be bent into a rectangle. Show that the rectangle will have the greatest area if it is a square.

34. If an object is thrown straight upward from the ground with an initial velocity of 48 ft/sec, the height s in feet above the ground after t seconds is given by $s(t) = 48t - 16t^2$. Find (a) the maximum height the object attains; (b) when it hits the ground.

35. If an egg is dropped from a height of 144 feet, its distance s in feet above the ground after t seconds is given by $s(t) = -16t^2 + 144$. When does it hit the ground?

36. A school determines that 100 copies of a yearbook will be sold if the price is $5.00 and that the number of copies sold will decrease by 10 for each $1.00 added to the price. What price will yield the maximum gross income from sales and what is the maximum gross income?

In Problems 37–44, identify each of the given functions as linear or quadratic. Graph.

37. $f(x) = 4 - 5x$

38. $s(t) = -16t^2 + 48t$

39. $s(x) = 2x(x - 4)$ **40.** $g(x) = 2 - (x - 2)$

41. $h(t) = (4 - 3t)(2 + t)$ **42.** $f(v) = 4v - 2(v - 1)$

43. $g(x) = x^2 - x(x - 1)$ **44.** $f(x) = 4x^2 - 8x + 2$

45. Graph $f(x) = 6x - 2$, then answer true or false for each of the following statements:
  a. The x intercept is -2.
  b. The graph passes through $(2, 10)$.
  c. $f(1) = 4$
  d. $\dfrac{f(x) - f(8)}{x - 8} = 6$

46. Graph $f(x) = 2x^2 + 16x - 2$, then answer true or false for each of the following statements:
  a. The graph opens downward.
  b. The minimum function value is $f(x) = -34$.
  c. There are no x intercepts.
  d. The graph passes through the point $(-2, -26)$.
  e. The graph is symmetric with respect to the line $x = 4$.
  f. $\dfrac{f(x) - f(1)}{x - 1} = 2x + 18$.

In Problems 47–50, find the linear function $f(x) = ax + b$ which satisfies the given conditions.

47. The graph of f has y intercept 2 and x intercept -0.5.

48. The graph of f is parallel to the line $7x - 4y + 1 = 0$ and passes through $(7, 2)$.

49. The graph of f is perpendicular to the graph of $g(x) = 2.4x + 3$ and $f(4) = 2$.

50. The y intercept of the graph of f is 1.7 and $f(10) = 0.7$.

In Problems 51–54, find the quadratic function $f(x) = ax^2 + bx + c$ which satisfies the given conditions.

51. The function f has the values: $f(0) = 5$, $f(1) = 10$, and $f(-1) = 4$.

52. The vertex of the graph of f is located at $(1, 2)$, and the y intercept is 4.

53. The maximum value of f is 10; the graph of f is symmetric with respect to the line $x = -1$ and has y intercept 8.

54. The graph of f has y intercept -1 and x intercepts 1 and 3.

Calculator Problems

55. When money is deposited in an account earning simple interest, the amount accrued S is a linear function of time t measured in years: $S = P + Prt$. Compute S after 15 years if the principal is $P = 500$ and the annual interest rate r is 16.5%. At what time does $S = 550$?

56. The electric resistance R in ohms of a certain resistor as a function of the temperature x in degrees Centigrade is given by $R(x) = 0.006x + 5.005$. Find the resistance at $27°C$. Find the temperature if the resistance is 4.911 ohms.

57. Under the condition of constant acceleration the velocity v of an object is a linear function of time t. Given $v(t) = 17.5t - 22.3$, find when the velocity is 0.

58. The voltage E of a certain thermocouple as a function of temperature t is given by $E(t) = 2.8t + 0.008t^2$. Find the voltage when the temperature is $t = 48$.

59. The number of units demanded of a certain commodity that costs x dollars per unit is given by the function $D(x) = 1101x^2 - 6022x + 9025$. Calculate $D(2)$, $D(3)$, and $D(4)$. When is the demand a minimum? Why does such a demand function not make usual economic sense for $x > 2.73$. [*Hint:* Normally demand decreases when price increases.]

Review Problems

60. For $f(x) = 3/(1 + x^2)$, find $f(2)$, $f(-4)$, and $f(x + 1)$. What is the domain of this function?

61. Graph

$$f(x) = \begin{cases} x + 1, & x > 1 \\ 1, & x = 1 \\ x - 1, & x < 1 \end{cases}$$

62. If the point $(1, 5)$ lies on the graph of a function $y = f(x)$, can $f(1) = 3$? Justify your answer.

2.3 Techniques for Graphing Functions

The graph of a function provides us with a geometric interpretation of the function's behavior.

Constant Functions

A function is said to be **constant** if its range consists of a single real number. For example,

$$g(x) = -1 \quad \text{and} \quad h(x) = 3$$

are constant functions. Since the "output" of a constant function is the same for all "inputs" x, the graph of a constant function is a horizontal line. See Figure 2.34.

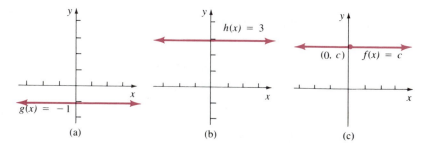

(a)　　　(b)　　　(c)

Figure 2.34

**Increasing and
Decreasing
Functions**

As shown in Figure 2.35(a), for $a > 0$, the values of $f(x) = ax + b$ *increase* as x increases (from left to right the graph rises); whereas for $a < 0$, we see in Figure 2.35(b) that the function values *decrease* as x increases (the graph falls). In the first case we say that f is increasing and in the second that f is decreasing. In general a function f is said to be **increasing on an interval** if $x_1 < x_2$ implies $f(x_1) < f(x_2)$ for any x_1 and x_2 in the interval. Similarly, a function f is said to be **decreasing on an interval** if $x_1 < x_2$ implies $f(x_1) > f(x_2)$ for any x_1 and x_2 in the interval.

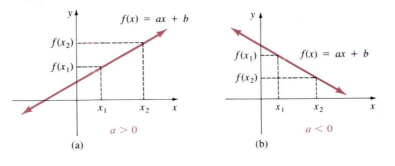

Figure 2.35

Example 1 For the functions graphed in Figure 2.36, determine the intervals on which each is increasing, decreasing, or constant.

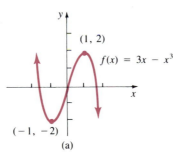 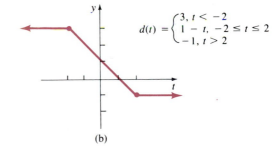

Figure 2.36

Solution   **a.** The function $y = f(x)$ is

decreasing on $(-\infty, -1)$

increasing on $(-1, 1)$

decreasing on $(1, +\infty)$.

b. The function $y = d(t)$ is

constant on $(-\infty, -2)$

decreasing on $(-2, 2)$

constant on $(2, +\infty)$. ∎

Example 2 The linear function

$$f(x) = 2x + 3$$

is increasing on the interval $(-\infty, \infty)$, since $a = 2 > 0$. ∎

Example 3 Find the interval on which the function

$$f(x) = -x^2 - 4x + 5$$

is increasing, and the interval on which it is decreasing.

Solution The graph of the function is a parabola that opens downward with
vertex at $(-2, 9)$. As shown in
Figure 2.37, the function is increasing on the interval $(-\infty, -2)$ and
decreasing on the interval $(-2, \infty)$.

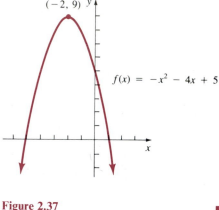

$$f(x) = -x^2 - 4x + 5$$

Figure 2.37 ∎

Locating the points where a function changes its behavior, for example from
increasing to decreasing, is essential to obtaining an accurate graph. In addition these
points often correspond to maximum or minimum values of the function. These
values are important to find in many applications.

Intercepts

Recall from Section 1.1 that the graph of a function $y = f(x)$ is the graph of the
relation

$$\{(x, y) \mid y = f(x), x \text{ in the domain of } f\}.$$

We have seen that by plotting a number of ordered pairs (x, y) where
$y = f(x)$ we may be able to determine the configuration of the graph. We now consider several techniques that enable us to obtain more accurate graphs with greater
ease than by simply plotting points. In particular, it is good practice to find the
intercepts. The y intercept of the graph is $f(0)$, and the x intercepts are the solutions of
the equation $f(x) = 0$.

Example 4 The function $f(x) = \sqrt{x} - 4$ has y intercept $f(0) = -4$. Solving $f(x) = 0$, we have

$$\sqrt{x} - 4 = 0$$

$$\sqrt{x} = 4$$
$$x = 16.$$

Therefore the x intercept is $x = 16$. Plotting the points from the table, we sketch the graph in Figure 2.38.

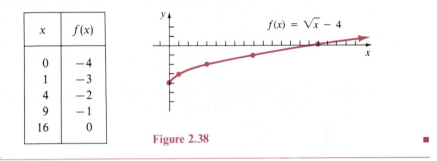

x	$f(x)$
0	-4
1	-3
4	-2
9	-1
16	0

Figure 2.38

Even and Odd Functions

In Section 1.1 we discussed different types of symmetry for the graphs of relations. These concepts can be applied to graphs of functions. A function is said to be **even** if its graph is symmetric with respect to the y axis; and a function is **odd** if its graph is symmetric with respect to the origin. From the tests for symmetry in Section 1.1 we obtain the following tests for even and odd functions.

> The function $y = f(x)$ is **even** if $f(-x) = f(x)$.
>
> The function $y = f(x)$ is **odd** if $f(-x) = -f(x)$.

Example 5 Determine whether the following functions are even, odd, or neither.

    a. $f(x) = x^3$ **b.** $g(x) = x^2 + 4$ **c.** $h(x) = x^3 + 1$.

Solution   **a.** Since $f(-x) = (-x)^3 = -x^3 = -f(x)$, this function is odd.

    b. This function is even, since

$$g(-x) = (-x)^2 + 4 = x^2 + 4 = g(x).$$

    c. Substituting $-x$ for x, we obtain

$$h(-x) = (-x)^3 + 1 = -x^3 + 1.$$

Since $h(x) = x^3 + 1$ and $-h(x) = -x^3 - 1$, we see that the function h is neither even nor odd. ∎

As shown in the next example, determining whether a function is even or odd can be an aid in graphing.

Example 6   Graph $f(x) = \dfrac{1}{x}$.

Solution   *y intercept*: Since $f(0)$ is not defined, the function has no y intercept.

x intercept: If we set $f(x) = 0$, we have $1/x = 0$. Since $1/x$ cannot equal zero (a quotient is zero only when the numerator is zero), the graph has no x intercepts.

Symmetry: Since $f(-x) = -1/x = -f(x)$, we see that the function is odd and hence the graph is symmetric with respect to the origin.

The graph is sketched in Figure 2.39 by plotting points for $x > 0$ from the table below, and then symmetry is used to complete the graph for $x < 0$.

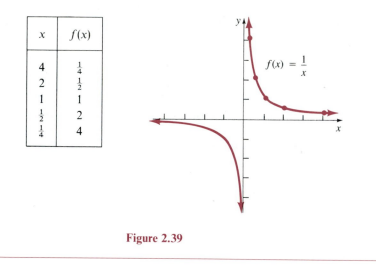

x	$f(x)$
4	$\frac{1}{4}$
2	$\frac{1}{2}$
1	1
$\frac{1}{2}$	2
$\frac{1}{4}$	4

Figure 2.39

Shifted Graphs

Certain graphs have the same shape but different positions relative to the coordinate axes. Compare the graph of $f(x) = x^2$ with the companion graphs shown in Figure 2.40. These graphs are **vertical** and **horizontal shifts** of the graphs of $f(x) = x^2$ and are special cases of the following general rules.

Shifted Graphs $c > 0$

Original graph: $y = f(x)$

Vertical shift c units upward: $y = f(x) + c$

Vertical shift c units downward: $y = f(x) - c$

Horizontal shift c units to the left: $y = f(x + c)$

Horizontal shift c units to right: $y = f(x - c)$

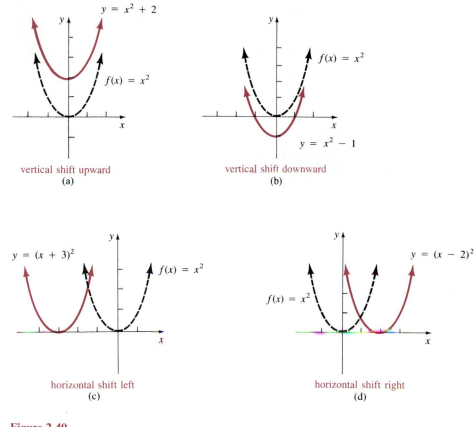

Figure 2.40

Example 7   Sketch the graph of $f(x) = |x|$, then use the technique of shifting to graph

   a. $g(x) = |x + 1|$ **b.** $h(x) = |x| - 3$ **c.** $k(x) = |x - 2| + 1$.

Solution   First we obtain the graph of f as follows.

y intercept: $f(0) = |0| = 0$.

x intercepts: Setting $f(x) = 0$, we obtain $|x| = 0$ or $x = 0$.

Symmetry: $f(-x) = |-x| = |x| = f(x)$. Therefore f is an even function and its graph is symmetric with respect to the y axis.

 With the above information and the values in the table below we obtain the graph in Figure 2.41.

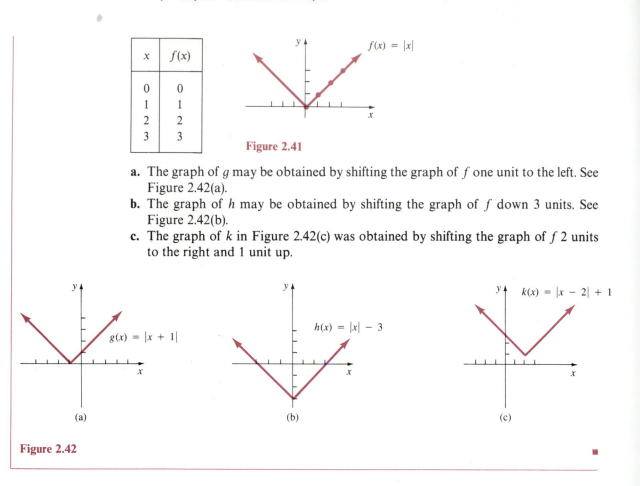

x	$f(x)$
0	0
1	1
2	2
3	3

$f(x) = |x|$

Figure 2.41

a. The graph of g may be obtained by shifting the graph of f one unit to the left. See Figure 2.42(a).
b. The graph of h may be obtained by shifting the graph of f down 3 units. See Figure 2.42(b).
c. The graph of k in Figure 2.42(c) was obtained by shifting the graph of f 2 units to the right and 1 unit up.

$g(x) = |x + 1|$

$h(x) = |x| - 3$

$k(x) = |x - 2| + 1$

(a)

(b)

(c)

Figure 2.42

Reflected Graphs

The relationship between the graphs of $y = f(x)$ and $y = -f(x)$ is seen from the following observation. For each point (x, y) on the graph of f, the point $(x, -y)$ will lie on the graph of $y = -f(x)$. It follows that the graph of $y = -f(x)$ may be obtained by **reflecting** the graph of f through the x axis.

Example 8 The graph of $g(x) = -x^2$ may be obtained, as shown in Figure 2.43, by reflecting the graph of $f(x) = x^2$ through the x axis.

$f(x) = x^2$

$g(x) = -x^2$

Figure 2.43

Stretching Graphs

To obtain the graph of $y = cf(x)$ for $c > 0, c \neq 1$, we can multiply the y coordinates of points on the graph of $y = f(x)$ by c. The result is a stretching of the original graph.

Example 9 Graph **a.** $g(x) = 3x^2$ **b.** $h(x) = \frac{1}{3}x^2$.

Solution **a.** To sketch the graph of $f(x) = 3x^2$ we consider the graph of $f(x) = x^2$, shown as the dashed curve in Figure 2.44, and multiply the y coordinate of each point by 3. The result is the narrower parabola drawn in color.

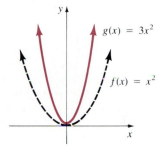

Figure 2.44

b. The graph of $h(x) = \frac{1}{3}x^2$ may be obtained by multiplying the y coordinates of points on the graph of $f(x) = x^2$ by $\frac{1}{3}$. The result is a wider parabola. Both graphs are shown in Figure 2.45.

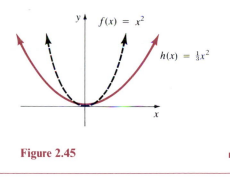

Figure 2.45

■

Exercise 2.3

In Problems 1–8, graph the given function and determine the intervals on which the given function is increasing, decreasing, or constant.

1. $h(x) = 2$

2. $g(x) = x - (x + 4)$

3. $f(x) = 12x - 11$

4. $h(x) = 3x^2 - 8x + 1$

5. $g(x) = x(x - 10)$

6. $f(x) = -3x + 5$

7. $f(x) = -2x^2 - 6x + 3$

8. $g(x) = |x|/x$

In Problems 9–16, determine whether the given function is even, odd, or neither.

9. $f(x) = 4x - x^3$

10. $g(x) = x^3 + 6x^2 + 1$

11. $h(x) = (x - 1)^2$

12. $r(x) = 6x + 2$

13. $f(x) = 3$

14. $h(x) = x^2 + 4$

15. $g(x) = 6x^5 - x^3$

16. $f(x) = x(x - 1)$

In Problems 17–24, use shifting, reflecting, or stretching to sketch on the same coordinate axes the graphs of f for the given values of a.

17. $f(x) = x + a; \quad a = 0, a = 1, a = -2$

18. $f(x) = 4x + a; \quad a = 0, a = -1, a = 3$

19. $f(x) = x^2 + a; \quad a = 0, a = -1, a = 4$

20. $f(x) = |x| + a; \quad a = 0, a = 1, a = -1$

21. $f(x) = (x + a)^2; \quad a = 0, a = 2, a = -1$

22. $f(x) = |x + a|; \quad a = 0, a = -1, a = 5$

23. $f(x) = ax^2; \quad a = 1, a = -1, a = 4$

24. $f(x) = a|x|; \quad a = 1, a = -1, a = \frac{1}{2}$

In Problems 25–32, use the graph of $f(x) = \sqrt[3]{x}$ *shown in Figure 2.46 to graph the given function.*

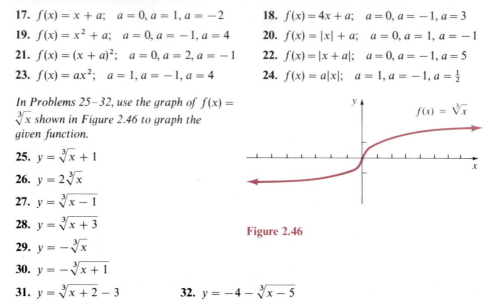

25. $y = \sqrt[3]{x} + 1$

26. $y = 2\sqrt[3]{x}$

27. $y = \sqrt[3]{x} - 1$

28. $y = \sqrt[3]{x + 3}$

Figure 2.46

29. $y = -\sqrt[3]{x}$

30. $y = -\sqrt[3]{x} + 1$

31. $y = \sqrt[3]{x + 2} - 3$ **32.** $y = -4 - \sqrt[3]{x - 5}$

In Problems 33–36, each graph was obtained by shifting or reflecting the graph of $f(x) = |x|$, *as seen in Figure 2.41. Determine a formula for each function graphed.*

33.

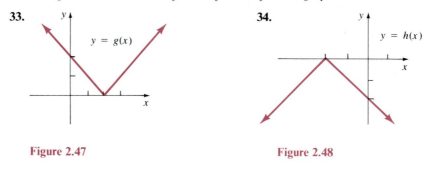

$y = g(x)$

Figure 2.47

34.

$y = h(x)$

Figure 2.48

35.

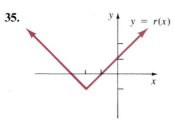

$y = r(x)$

Figure 2.49

36.

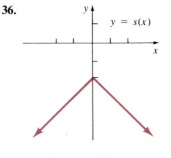

$y = s(x)$

Figure 2.50

In Problems 37–42, find intercepts, determine whether the function is even or odd, sketch the graph (use shifting, reflecting, or stretching, if appropriate), and determine the intervals on which f is increasing, decreasing, or constant.

37. $f(x) = 3 - x^2$ **38.** $f(x) = 4 - |x|$ **39.** $f(x) = |x^2 - 4|$

40. $f(x) = (1/10)x^2$ **41.** $f(x) = 2(x^2 + 2)$ **42.** $f(x) = -4(x + 1)^2$

In Problems 43–46, find the x and y intercepts of the graph of the given function.

43. $f(x) = x^4 - 6x^2 + 5$

44. $f(x) = 8x^3 + 12x^2 + 6x + 1$

45. $f(x) = 6x^4 - 25x^3 + 14x^2 + 27x - 18$

46. $f(x) = 4x^5 + 4x^4 + 3x^3 + 3x^2 - x - 1$

47. The graph of $f(x) = ax^2 + bx + c$ may be obtained from the graph of $y = x^2$ by what shifting and stretching? [*Hint:* Determine d, h, and k so that $f(x) = d(x + h)^2 + k$.]

48. The **greatest integer function** f is defined as follows: If $n \le x < n + 1$, then $f(x) = n$. Graph this function. On what intervals is f constant?

49. The symbol $[\![x]\!]$ is used to denote the largest integer n for which $n \le x$. Thus the greatest integer function defined in Problem 48 may be denoted $f(x) = [\![x]\!]$. Sketch the graph of
(a) $f(x) = [\![x - 2]\!]$; (b) $f(x) = -[\![x]\!]$; (c) $f(x) = 3[\![x]\!]$.

Review Problems

50. Find the y coordinate of the point on the graph of $f(x) = \pi x^3 - 2x + \sqrt{2}$ if the x coordinate of the point is (a) 1; (b) $\sqrt{2}$; (c) a.

51. Find k so that the graph of $f(x) = 2x + k + 3$ passes through $(1, \sqrt{2})$.

52. Graph $g(x) = \begin{cases} 1, & x \le 0 \\ x^2, & 0 < x < 1 \\ -x + 2, & x \ge 1. \end{cases}$

53. Find the domain of $f(x) = \dfrac{1}{\sqrt{x + 5}}$.

54. If $f(x) = 1/x$, find $f(1/x)$.

55. Find the maximum value of the quadratic function $f(x) = 3 + 4x - x^2$.

2.4   Special Functions and Their Graphs

We have already considered three special functions in some detail—namely, constant, linear, and quadratic functions. In this section we shall discuss several additional classes of functions and their graphs.

**2.4.1
Polynomial
Functions**

A **polynomial function** of degree n is any function of the form

$$P(x) = a_n x^n + a_{n-1} x^{n-1} + \cdots + a_1 x + a_0, \qquad a_n \neq 0, \qquad (1)$$

where the coefficients $a_0, a_1, \ldots, a_n$ are constants and n is a nonnegative integer. In the preceding section we considered linear and quadratic functions, which correspond to the cases $n = 1$ and $n = 2$, respectively. When $n = 3$, we obtain a third-degree, or *cubic*, polynomial function.

Example 1   Graph the cubic polynomial function $f(x) = x^3$.

Solution   *y intercept*: Since $f(0) = 0$, the y intercept is at the origin.

x intercept: Solving

$$f(x) = x^3 = 0,$$

we have $x = 0$, and we see that the origin is also the only x intercept.

Symmetry: Since $f(-x) = (-x)^3 = -x^3 = -f(x)$, f is an odd function. Thus the graph of f will be symmetric with respect to the origin.

The graph is obtained by plotting the points from the accompanying table and using symmetry. See Figure 2.51.

x	$f(x)$
2	8
1	1
$\frac{1}{2}$	$\frac{1}{8}$
0	0

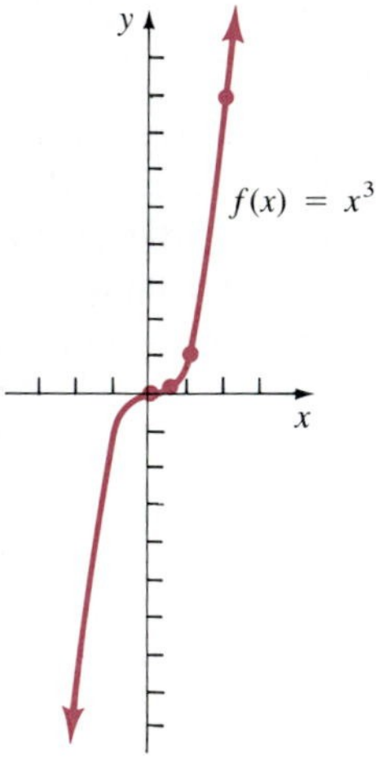

$f(x) = x^3$

Figure 2.51   ∎

The graphs of polynomial functions of degrees greater than or equal to 3 that are not shifts or reflections of simpler functions, are somewhat tedious to obtain by simply plotting points. Fortunately, calculus provides a very powerful technique which is a considerable aid in graphing polynomial functions.

2.4.2
Rational Functions

The quotient of two polynomial functions

$$f(x) = \frac{P(x)}{Q(x)} = \frac{a_n x^n + a_{n-1} x^{n-1} + \cdots + a_1 x + a_0}{b_m x^m + b_{m-1} x^{m-1} + \cdots + b_1 x + b_0} \tag{2}$$

is called a **rational function**. The domain of a rational function is the set of all real numbers except those for which the denominator is zero. For example, $f(x) = (2x^3 - 1)/(x^2 - 9)$ is a rational function with domain $\{x \mid x \neq \pm 3\}$.

Example 2 Graph the function $f(x) = \dfrac{2}{x - 1}$.

Solution The graph of this function could be obtained by a horizontal shift and a stretching of the graph of $y = 1/x$. However, we shall learn more about the graphs of rational functions if we examine the behavior of f as follows.

y intercept: $f(0) = -2$.

x intercepts: Since the numerator of the function is never 0, the graph has no x intercepts.

Symmetry: It is left to the reader to verify that f is neither even nor odd.

Domain: Setting the denominator equal to 0, we see that $x = 1$ is not in the domain of the function.

Plotting points: As the table indicates, for $|x|$ large, the corresponding function values are near 0. That is, the graph of the function approaches the x axis as $|x|$ becomes large. Similarly, for values of x near 1, the corresponding function values are large in absolute value. Thus the graph of the function approaches the vertical line $x = 1$ as x approaches 1. The graph is shown in Figure 2.52.

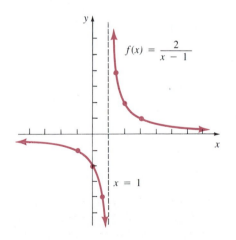

Figure 2.52

x	$f(x)$
-999	-0.002
-99	-0.02
-9	-0.2
-1	-1
0	-2
0.5	-4
0.9	-20
0.99	-200
0.999	-2000

x	$f(x)$
1.001	2000
1.01	200
1.1	20
1.5	4
2	2
3	1
11	0.2
101	0.02
1001	0.002

■

Asymptotes Suppose that

$$f(x) = \frac{P(x)}{Q(x)}$$

is a rational function and a is a real number such that $Q(a) = 0$ and $P(a) \neq 0$. Then the values of $|f(x)|$ increase without bound as x approaches a, and we say that the line $x = a$ is a **vertical asymptote**. Now suppose that $|x|$ increases without bound. If the corresponding function values approach a constant c, then the horizontal line $y = c$ is called a **horizontal asymptote**. In the preceding example the line $x = 1$ is a vertical asymptote and the horizontal line $y = 0$ (the x axis) is a horizontal asymptote.

Example 3 Graph the function $f(x) = \dfrac{1}{x^2}$.

Solution *Intercepts*: The function has no x or y intercepts.

Symmetry: $f(-x) = \dfrac{1}{(-x)^2} = \dfrac{1}{x^2} = f(x)$. Therefore f is even, and the graph of f will be symmetric with respect to the y axis.

x	$f(x)$
$\frac{1}{10}$	100
$\frac{1}{2}$	4
1	1
2	$\frac{1}{4}$
5	$\frac{1}{25}$
10	$\frac{1}{100}$

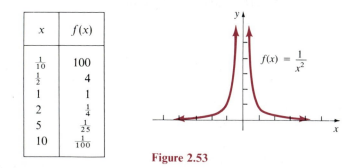

Figure 2.53

Asymptotes: Since the denominator is 0 when $x = 0$ and the numerator is never 0, the line $x = 0$ (the y axis) is a vertical asymptote. As $|x|$ becomes large, the function values approach 0, and the line $y = 0$ (the x axis) is a horizontal asymptote. See Figure 2.53.

■

A method for determining whether the graph of a rational function has a horizontal asymptote is to divide the numerator and denominator by the highest power of x *occurring in the denominator*. We then examine the behavior of the resulting quotient for $|x|$ large.

Example 4   Graph the function $f(x) = \dfrac{3 - x}{2 + x}$.

Solution   The y intercept is $f(0) = \frac{3}{2}$ and the x intercept is $x = 3$. The function is neither even nor odd. Since the denominator is 0 when $x = -2$, but the numerator is not 0, the line $x = -2$ is a vertical asymptote. To find the horizontal asymptote, we divide the numerator and denominator by x:

$$f(x) = \frac{\dfrac{3}{x} - \dfrac{x}{x}}{\dfrac{2}{x} + \dfrac{x}{x}} = \frac{\dfrac{3}{x} - 1}{\dfrac{2}{x} + 1}.$$

For $|x|$ large, the terms $3/x$ and $2/x$ are near 0 and the function value is near $\frac{-1}{1} = -1$. Therefore the line $y = -1$ is a horizontal asymptote. The graph is shown in Figure 2.54 with the dashed lines indicating the asymptotes.

x	$f(x)$
-5	$-\frac{8}{3}$
-3	-6
-1	4
0	$\frac{3}{2}$
1	$\frac{2}{3}$
3	0

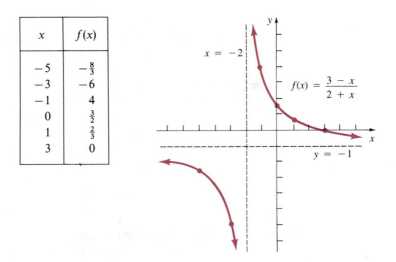

Figure 2.54

■

Example 5 Graph the function $f(x) = \dfrac{1}{1 - x^2}$.

Solution The y intercept is $f(0) = 1$, and there are no x intercepts. Since $f(-x) = f(x)$, the function is even, and its graph will be symmetric with respect to the y axis. As $|x|$ becomes large, the function values approach 0. Thus the line $y = 0$ is a horizontal asymptote. Solving $1 - x^2 = 0$, we find that $x = -1$ and $x = 1$ are vertical asymptotes. With this information it is possible to give a rough sketch of the graph. See Figure 2.55. As in the preceding example, it is a good idea to plot a few points on either side of the vertical asymptotes.

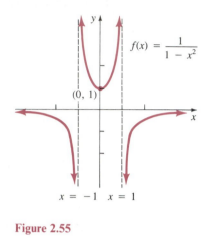

Figure 2.55

2.4.3 Algebraic Functions

Both polynomial and rational functions belong to a larger class known as **algebraic functions**. An algebraic function of x is obtained by performing a finite number of additions, subtractions, multiplications, divisions, and extractions of roots on the variable x and constants. For example, $f(x) = x^{1/3}$, $f(x) = 1/\sqrt{x} + 2$, $f(x) = x^2 - 5x$, and $f(x) = 7x/(5x + 8)$ are all algebraic functions.

In general, the following procedure is useful when graphing a function:

---

1. Try to identify or categorize the function. Is it linear? Quadratic? Rational? Can the graph be obtained by shifting, reflecting, or stretching the graph of a simpler function?

2. Find any intercepts.

3. Check for symmetry.

4. Determine the domain.

5. If the function is rational, find any vertical or horizontal asymptotes.

6. Plot points.

Example 6 Graph the function $g(x) = \sqrt{x + 1}$.

Solution The algebraic function $g(x) = \sqrt{x + 1}$ is not linear, quadratic, or rational. However, its graph could be obtained by shifting the graph of the simpler function $f(x) = \sqrt{x}$.

Thus, we will first graph $f(x) = \sqrt{x}$. The y intercept of f is $f(0) = 0$, and $x = 0$ is the only x intercept. Since the domain of f is $x \geq 0$, there is no need to check symmetry. From the table of values below for $f(x) = \sqrt{x}$ we obtain the dashed curve in Figure 2.56. Shifting this graph 1 unit to the left yields the solid curve in Figure 2.56 as the graph of $g(x) = \sqrt{x + 1}$.

x	$f(x)$
0	0
1	1
4	2

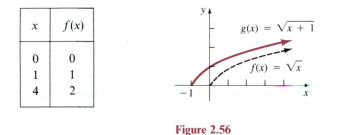

Figure 2.56

Power functions   For any real numbers k and n it can be shown that

$$f(x) = kx^n \tag{3}$$

is a function, called the **power function**. The meaning of the expression kx^n for irrational numbers n will be discussed in Chapter 3. For n a nonnegative integer the power function (3) reduces to a polynomial function. If n is rational, the power function belongs to the class of algebraic functions. For example, if $k = 1$ and $n = 2$ in (3) we have the simple quadratic function $f(x) = x^2$ (graphed in Figure 2.4); and if $k = 1$ and $n = -1$ we have $f(x) = x^{-1} = 1/x$, (graphed in Figure 2.39).

Example 7   Graph the power function $f(x) = x^{2/3}$.

Solution   The x and y intercepts are the origin. We note that the function can also be written in the form

$$f(x) = (x^{1/3})^2 = (\sqrt[3]{x})^2.$$

Since it is possible to find the cube root of any real number, the domain of the function is the set of real numbers. Since $f(-x) = f(x)$, the function is even and the graph will be symmetric with respect to the y axis. Plotting the points from the table and using symmetry, we obtain the graph in Figure 2.57.

x	$f(x)$
0	0
$\frac{1}{8}$	$\frac{1}{4}$
1	1
4	2.52
8	4

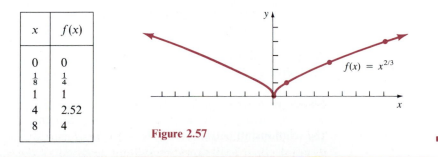

Figure 2.57

It is interesting to note that for *even integer* values of n greater than two, the graphs of the power function $f(x) = x^n$ all have shapes similar to the parabola $y = x^2$. See Figure 2.58(a). However, these graphs are not true parabolas. In addition, for *odd integer* values of n greater than three, the graphs of $f(x) = x^n$ resemble the graph of $y = x^3$. See Figure 2.58(b).

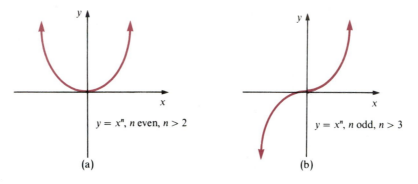

(a) $y = x^n$, n even, $n > 2$ (b) $y = x^n$, n odd, $n > 3$

Figure 2.58

Variation   The power function occurs frequently in application in physics, biology, business, psychology, economics, and many other fields. It arises whenever two variables are directly or inversely proportional. If the variables s and t are directly proportional we say s **varies directly** as t (or simply s varies as t), and we write $s = kt$ where k is a constant. If s and t are inversely proportional, we say s **varies inversely** as t and write $s = k/t$ where k is a constant. For example, if y varies directly as x^2 we have the power function $y = kx^2$; or if in another situation y is found to vary inversely as x^3 we have the power function $y = k/x^3 = kx^{-3}$. The constant k is referred to as the **constant of proportionality**.

Example 8   Assume y varies directly as x^3 and $y = -16$ when $x = 2$. Find y when $x = 3$.

Solution   Since y varies directly as x^3, x and y are related by $y = kx^3$ where the proportionality constant k is yet to be determined. If $y = -16$ when $x = 2$, we must have $-16 = k(2)^3$ or $k = -2$. Thus the exact relationship between y and x is $y = -2x^3$. Hence, when $x = 3$ we have $y = -2(3)^3 = -54$.   ∎

Example 9   Assume y varies inversely as x and $y = 4/5$ when $x = 10$. Find the relationship between x and y.

Solution   Since y varies inversely as x, we have $y = k/x$ for some constant k. From the fact that $y = 4/5$ when $x = 10$ we find $4/5 = k/10$. Solving for k we obtain $k = 8$. Thus the relationship between x and y is $y = 8/x$.   ∎

Example 10   In certain biological studies it is assumed that the weight W in pounds of a person is directly proportional to the cube of the person's height h in inches. Find the constant of proportionality for a person 60 inches tall who weighs 100 pounds. Find the weight corresponding to a height of 66 inches.

Solution   The weight W varies directly as h^3, the cube of the height, so $W = kh^3$. Since $W = 100$ when $h = 60$, we find $100 = k(60)^3$ or $k = 1/2160$ as the constant of proportionality. Thus the relationship between h and W is $W = (1/2160)h^3$. Substituting $h = 66$ we find the corresponding weight $W = (1/2160)(66)^3 = 133.1$ pounds.   ■

Exercise 2.4

In Problems 1–6, graph the given polynomial function.

1. $f(x) = -x^3$
2. $f(x) = 1 + x^3$
3. $f(x) = 1 - x^3$
4. $f(x) = x^3 - x$
5. $f(x) = x^4$
6. $f(x) = x^5$

In Problems 7–18, graph the given rational function. Determine intercepts and asymptotes.

7. $f(x) = \dfrac{2}{x}$
8. $f(x) = \dfrac{1}{x-2}$
9. $f(x) = \dfrac{1}{x^3}$

10. $f(x) = \dfrac{x^2}{x^2-4}$
11. $f(x) = \dfrac{4x-9}{3+2x}$
12. $f(x) = \dfrac{x}{x+1}$

13. $f(x) = \dfrac{1-x^2}{x^2}$
14. $f(x) = \dfrac{x+1}{x-1}$
15. $f(x) = \dfrac{x}{x^2-1}$

16. $f(x) = \dfrac{2-3x}{x}$
17. $f(x) = \dfrac{x^2-9}{x}$
18. $f(x) = \dfrac{1}{x^2-16}$

19. Does the function $f(x) = (x^2 - 1)/(x - 1)$ have a vertical asymptote? Graph.

20. Does the function $f(x) = (x^2 - x)/x$ have a vertical asymptote? Graph.

In Problems 21–29, graph the given function.

21. $f(x) = x^{1/3}$
22. $f(x) = x^{2/5}$
23. $f(x) = x^{2/3} - 1$

24. $f(x) = \dfrac{1}{\sqrt{x}}$
25. $f(x) = \sqrt{x^2 - 1}$
26. $f(x) = (x - 1)^{2/3}$

27. $f(x) = \sqrt{x - 3}$
28. $f(x) = \dfrac{1}{|x|}$
29. $f(x) = 2 + \sqrt{x + 1}$.

30. If y varies directly as x^2, and $y = 16$ when $x = 2$, find y when $x = 6$.

31. If y varies directly as $\sqrt{x + 1}$, and $y = 4$ when $x = 0$, find y when $x = 8$.

32. If y varies inversely as x^3, and $y = 4$ when $x = 2$, find y when $x = 10$.

33. If y varies inversely as $2x + 1$, and $y = 15$ when $x = 7$, find y when $x = 1$.

34. According to Boyle's Law, if the temperature is constant the pressure that a gas exerts varies inversely as its volume. If a gas has a volume of 72 cubic centimeters when its pressure is 12 kilograms per square centimeter, find the volume when the pressure is 60 kilograms per square centimeter. Find the pressure when the volume is 48 cubic centimeters.

35. The weight of a certain copper pipe varies directly as its length. If 2 feet of pipe weighs 1.4 pounds, how much will 10 feet weigh?

36. The amount of paint needed to paint some toy blocks varies directly as the square of the length of an edge. If one pint of paint is needed to paint 100 blocks with 2-inch edges, how many pints of paint are required to paint 1000 blocks with 3-inch edges?

Calculator Problems

37. The weight of an object varies inversely as the square of its distance from the center of the earth. If an object weighs 200 pounds on the surface of the earth, how much would it weigh at a distance of 6 miles above the earth's surface? (Assume that the radius of the earth is 4000 miles.)

38. The distance an object falls under the influence of gravity varies directly as the square of time. If an object falls 9.80 meters in 1 second, how far does it fall in 1.25 seconds?

39. The kinetic energy K of an object with velocity v and mass m is given by $K = \frac{1}{2}mv^2$. Find the kinetic energy of 0.11 kilogram of blood expelled from the heart into an aorta with a velocity of 0.14 meters per second.

40. The cost of producing x units of a certain product is found to be directly proportional to $x^{1/3}$ with a proportionality constant of 320. Find the cost of producing 248 units.

41. Hooke's Law states that the force required to stretch a spring x units beyond its natural length is proportional to x. If it requires a force of 12.1 pounds to stretch a spring 7 inches, what force is required to stretch it 1 foot?

Review Problems

42. For $f(x) = 2x^3 + x$, find $f(-1)$, $f(3)$, $f(2z)$.

43. Find the interval on which the function $f(x) = (x - 2)(x + 3)$ is decreasing and the interval on which it is increasing.

44. What shifting must be performed on the graph of $y = f(x)$ to produce the graph of $g(x) = f(x - 4) + 2$?

45. Graph $h(t) = \begin{cases} t^2, & t < 0 \\ -t, & 0 \leq t \leq 1 \\ 3, & t \geq 3 \end{cases}$

46. What are the domain and range of the function h in Problem 45?

2.5 The Algebra of Functions

Given two functions f and g, we can construct four additional functions by forming their **sum**, **difference**, **product**, and **quotient**. We define

$$(f + g)(x) = f(x) + g(x) \tag{1}$$

$$(f - g)(x) = f(x) - g(x) \tag{2}$$

$$(fg)(x) = f(x)g(x) \tag{3}$$

$$(f/g)(x) = \frac{f(x)}{g(x)}, \qquad (g(x) \neq 0). \tag{4}$$

Example 1 Given

$$f(x) = x^2 + 4x \quad \text{and} \quad g(x) = x^2 - 9,$$

we have

$$(f + g)(x) = (x^2 + 4x) + (x^2 - 9) = 2x^2 + 4x - 9,$$

$$(f - g)(x) = (x^2 + 4x) - (x^2 - 9) = 4x + 9,$$

$$(fg)(x) = (x^2 + 4x)(x^2 - 9) = x^4 + 4x^3 - 9x^2 - 36x,$$

$$(f/g)(x) = \frac{x^2 + 4x}{x^2 - 9}.$$ ∎

In the preceding example we see that $x = 3$ and $x = -3$ are not in the domain of $(f/g)(x)$.

Domains

The domain of each function given in (1)–(4) is the *intersection* of the domain of f with the domain of g. In the case of (4) we must also exclude values of x for which the denominator $g(x)$ is 0. For example, if we add

$$f(x) = \sqrt{1 - x}, \qquad x \leq 1, \tag{5}$$

and

$$g(x) = \sqrt{x + 2}, \qquad x \geq -2, \tag{6}$$

we obtain

$$(f + g)(x) = \sqrt{1 - x} + \sqrt{x + 2}.$$

The domain of this new function is the interval $-2 \leq x \leq 1$ *common* to both domains.

Example 2 For the functions given in (5) and (6), the domain of

$$(f/g)(x) = \frac{f(x)}{g(x)} = \frac{\sqrt{1-x}}{\sqrt{x+2}} = \sqrt{\frac{1-x}{x+2}}$$

is $\{x \mid -2 < x \leq 1\}$. Here the value $x = -2$ must be excluded to avoid having 0 in the denominator. ∎

Graph of a Sum   The sum of two functions can be easily interpreted graphically. Specifically the graph of $(f + g)(x) = f(x) + g(x)$ can be obtained by adding the y coordinates of the graphs of the original two functions.

Example 3 Graph $(f + g)(x)$ if

$$f(x) = x \quad \text{and} \quad g(x) = \sqrt{x}.$$

Solution First, we note that the domain of

$$(f + g)(x) = x + \sqrt{x}$$

is $\{x \mid x \geq 0\}$. Figure 2.59 shows the graphs of f and g for $x \geq 0$. As indicated, at a given value of x we can obtain the y coordinate of a point on the graph of $f + g$ by adding y_1 and y_2.

The graph of the sum is given in Figure 2.60.

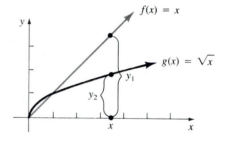

Figure 2.59

x	$f(x)$	$g(x)$	$f(x) + g(x)$
$\frac{1}{4}$	$\frac{1}{4}$	$\frac{1}{2}$	0.75
$\frac{1}{3}$	$\frac{1}{3}$	0.58	0.91
$\frac{1}{2}$	$\frac{1}{2}$	0.71	1.21
$\frac{3}{4}$	$\frac{3}{4}$	0.87	1.62
1	1	1	2
2	2	1.41	3.41
3	3	1.73	4.73
4	4	2	6

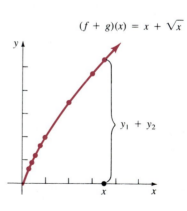

Figure 2.60 ∎

Example 4   The domain of

$$f(x) = \sqrt{x}$$

is $\{x \mid x \geq 0\}$. Note that the domain of the product

$$(ff)(x) = [f(x)]^2 = x$$

is still $\{x \mid x \geq 0\}$, even though the new function appears to be defined for all real numbers. The reason for this continued restriction is that each factor in the product

$$f(x)f(x) = \sqrt{x}\sqrt{x} = x$$

must be meaningful in the set of real numbers.   ∎

Example 5   Suppose f and g are the piecewise-defined functions

$$f(x) = \begin{cases} x, & x < 0, \\ x + 1, & x \geq 0, \end{cases}$$

$$g(x) = \begin{cases} x^2, & x \leq -1, \\ x - 2, & x > -1. \end{cases}$$

Find $f + g$, $f - g$, and fg.

Solution   We can combine the parts of each function only on common subintervals of the x axis. To accomplish this, we first rewrite f and g as

$$f(x) = \begin{cases} x, & x \leq -1, \\ x, & -1 < x < 0, \\ x + 1, & x \geq 0, \end{cases}$$

$$g(x) = \begin{cases} x^2, & x \leq -1, \\ x - 2, & -1 < x < 0, \\ x - 2, & x \geq 0. \end{cases}$$

It follows that

$$(f + g)(x) = \begin{cases} x + x^2, & x \leq -1, \\ 2x - 2, & -1 < x < 0, \\ 2x - 1, & x \geq 0, \end{cases}$$

$$(f - g)(x) = \begin{cases} x - x^2, & x \leq -1, \\ 2, & -1 < x < 0, \\ 3, & x \geq 0, \end{cases}$$

$$(fg)(x) = \begin{cases} x^3, & x \leq -1, \\ x^2 - 2x, & -1 < x < 0, \\ x^2 - x - 2, & x \geq 0. \end{cases}$$
                                   ∎

Function Composition

We now consider another important method for combining two functions, f and g, called **function composition**. We define the **composition of f and g**, denoted by $f \circ g$, to be the function

$$(f \circ g)(x) = f(g(x)). \tag{7}$$

In (7) it is understood that the values $g(x)$ are in the domain of f. Symbolically, using our machine analogy, we can represent the composition of f and g as shown in Figure 2.61.

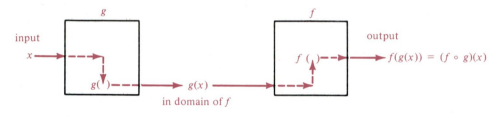

Figure 2.61

Similarly, the **composition of g and f**, denoted $g \circ f$, is

$$(g \circ f)(x) = g(f(x)). \tag{8}$$

Of course, in (8) $f(x)$ must be a number in the domain of g. The functions $f \circ g$ and $g \circ f$ are called **composite functions**.

Example 6 Evaluate $f \circ g$ if

$$f(x) = x^2 + 3x - 1 \quad \text{and} \quad g(x) = 2x^2 + 1.$$

Solution For emphasis we write f in the form

$$f(\ ) = (\ )^2 + 3(\ ) - 1.$$

Thus to evaluate $(f \circ g)(x) = f(g(x))$, we can substitute $g(x)$ into each set of parentheses. We find

$$(f \circ g)(x) = f(g(x)) = f(2x^2 + 1)$$
$$= (2x^2 + 1)^2 + 3(2x^2 + 1) - 1$$
$$= 4x^4 + 4x^2 + 1 + 6x^2 + 3 - 1$$
$$= 4x^4 + 10x^2 + 3.$$

∎

Example 7   For the functions given in the previous example evaluate $g \circ f$.

Solution   Using equation (8), we find

$$(g \circ f)(x) = g(x^2 + 3x - 1)$$
$$= 2(x^2 + 3x - 1)^2 + 1$$
$$= 2(x^4 + 6x^3 + 7x^2 - 6x + 1) + 1$$
$$= 2x^4 + 12x^3 + 14x^2 - 12x + 3.$$

■

Note that in general $f \circ g \neq g \circ f$.

Example 8   Let $f(x) = x^2$ and $g(x) = 2x + 1$. Find

  a. $(f \circ g)(3)$ **b.** $(g \circ f)(3)$ **c.** $(f \circ f)(x)$

Solution   **a.** $(f \circ g)(3) = f(g(3)) = f(7) = 49.$
b. $(g \circ f)(3) = g(f(3)) = g(9) = 19.$
c. $(f \circ f)(x) = f(f(x)) = f(x^2) = x^4.$

■

Domain of the Composition

The domain of the composition $f(g(x))$ consists of those values of x in the domain of g such that $g(x)$ is in the domain of f. Of course, this does not preclude the possibility that the domain of $f \circ g$ may be the entire domain of g.

Example 9   The domains of

$$f(x) = x^2 - \frac{3}{x} \quad \text{and} \quad g(x) = \sqrt{x}$$

are $\{x \mid x \neq 0\}$ and $\{x \mid x \geq 0\}$, respectively. The domain of

$$(f \circ g)(x) = f(g(x)) = f(\sqrt{x})$$
$$= (\sqrt{x})^2 - \frac{3}{\sqrt{x}} = x - \frac{3}{\sqrt{x}}$$

is $\{x \mid x > 0\}$.

■

Example 10   In order to determine the domain of $f \circ g$ if

$$f(x) = \sqrt{x - 3},$$

and

$$g(x) = x^2 + 2,$$

we note that $g(x) \geq 2$ for all x. To guarantee that the numbers represented by $g(x)$ are in the domain of f (namely, those numbers greater than or equal to 3), we must restrict the domain of g to be

$$\{x \,|\, x \leq -1 \quad \text{or} \quad x \geq 1\},$$

so that $g(x) \geq 3$. See Figure 2.62. Thus

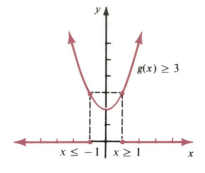

$$(f \circ g)(x) = f(g(x))$$
$$= f(x^2 + 2)$$
$$= \sqrt{(x^2 + 2) - 3}$$
$$= \sqrt{x^2 - 1}$$

is defined only for $x \geq 1$ or $x \leq -1$.

Figure 2.62

■

Example 11   Observe that $g \circ f$ is undefined for the functions

$$g(x) = 1 + \sqrt{x} \quad \text{and} \quad f(x) = -x^2 - 1.$$

Since $f(x) < 0$ for all x, these function values are not in the domain of g. On the other hand,

$$(f \circ g)(x) = f(g(x)) = f(1 + \sqrt{x})$$
$$= -(1 + \sqrt{x})^2 - 1 = -2 - 2\sqrt{x} - x$$

is defined for $x \geq 0$.

■

In calculus it is sometimes necessary to express a given function as a composition of two other functions.

Example 12   Express $h(x) = \sqrt{x^2 + 25}$ as the composition $h = f \circ g$ of two functions f and g.

Solution   If $h(x) = (f \circ g)(x)$, then $h(x) = f(g(x))$. Since g is the first function applied to the variable x, we let

$$g(x) = x^2 + 25.$$

Then, if

$$f(x) = \sqrt{x},$$

we have

$$(f \circ g)(x) = f(g(x)) = f(x^2 + 25) = \sqrt{x^2 + 25} = h(x),$$

as desired.

■

In certain applications a quantity y is given as a function of a variable x which in turn is a function of another variable t. By means of function composition we can express y as a function of t. An example follows.

Example 13 A weather balloon is being inflated. If the radius is increasing at the rate of 5 cm/sec, express the volume of the balloon as a function of time t in seconds.

Solution If r denotes the radius of the balloon, then

$$r(t) = 5t.$$

Since the volume of a sphere of radius r is

$$V = \frac{4}{3}\pi r^3,$$

the composition is

$$(V \circ r)(t) = V(r(t))$$

$$= V(5t) = \frac{4}{3}\pi(5t)^3 = \frac{500}{3}\pi t^3.$$

Thus the volume of the balloon as a function of time is given by

$$V(t) = \frac{500}{3}\pi t^3. \qquad \blacksquare$$

Exercise 2.5

In Problems 1–10, find the indicated functions and give their domains.

1. $f(x) = 2x^2 - x + 3, g(x) = x^2 + 1; f + g, fg$

2. $f(x) = x, g(x) = \sqrt{x - 1}; fg, f/g$

3. $f(x) = 2x - \dfrac{1}{\sqrt{x}}, g(x) = 2x + \dfrac{1}{\sqrt{x}}; f + g, fg$

4. $f(x) = 3x^3 - 4x^2 + 5x - 6, g(x) = (1 - x)^3; f + g, f - g$

5. $f(x) = x^2 - 4, g(x) = x + 2; fg, f/g$

6. $f(x) = (2x + 3)^{1/2}, g(x) = 2x + 3 + (2x + 3)^{1/2}; f - g, fg$

7. $f(x) = \sqrt{1 - x}, g(x) = \sqrt{x + 2}; f + g, fg$

8. $f(x) = 2 + \sqrt{x + 2}, g(x) = \sqrt{5x + 5}; f - g, f/g$

9. $f(x) = \begin{cases} 2x - 1, & x < 0 \\ \sqrt{x}, & x \geq 0 \end{cases}$, $g(x) = \begin{cases} x + 3, & x \leq 2 \\ \sqrt{x} - 4, & x > 2 \end{cases}; f + g, f - g, fg$

10. $f(x) = \begin{cases} x^2 + 1, & x < 1 \\ x, & 1 \leq x < 2 \\ 2x^2, & x \geq 2 \end{cases}$, $g(x) = \begin{cases} x^2 - 1, & x < 0 \\ 4x + 5, & 0 \leq x < 1 \\ x, & x \geq 1 \end{cases}; f + g, f - g, fg$

In Problems 11–20, for the given functions f and g, graph the function f + g.

11. $f(x) = -2x + 6, g(x) = 3x + 1$

12. $f(x) = x^2, g(x) = x^2 - 8$

13. $f(x) = -1, g(x) = \sqrt{x}$

14. $f(x) = 2, g(x) = \sqrt{x - 1}$

15. $f(x) = x, g(x) = |x|$

16. $f(x) = x, g(x) = \dfrac{1}{x}$

17. $f(x) = \dfrac{1}{x^2}, g(x) = -1$

18. $f(x) = \dfrac{1}{x^2}, g(x) = x$

19. $f(x) = (x + 1)^2, g(x) = -x^2$

20. $f(x) = \begin{cases} x^2 - x, & x < 0 \\ x, & x \ge 0 \end{cases}, g(x) = x$

In Problems 21–24, graph the given function using one or more of the following techniques: addition of y coordinates, stretching, reflecting, or shifting.

21. $f(x) = x - |x|$

22. $f(x) = x^2 - 2|x| + 1$

23. $f(x) = x^2 - 2|x - 2| - 3$

24. $f(x) = x - [\![x]\!]$, where $[\![x]\!]$ is the greatest integer function. (See Problem 49 in Section 2.3.)

25. Let $f(x) = x^2 - 6x + 5$. Graph $y = \frac{1}{2}\{f(x) + |f(x)|\}$.

26. Describe how the graph of $y = \frac{1}{2}\{f(x) + |f(x)|\}$ is related to the graph of $y = f(x)$.

In Problems 27–42, find $f \circ g$ and $g \circ f$.

27. $f(x) = 1 + x^2, g(x) = \sqrt{x - 1}$

28. $f(x) = x^2 - x + 5, g(x) = -x + 4$

29. $f(x) = \dfrac{1}{2x - 1}, g(x) = x^2 + 1$

30. $f(x) = \dfrac{x + 1}{x}, g(x) = \dfrac{1}{x}$

31. $f(x) = 2x - 3, g(x) = \dfrac{x + 3}{2}$

32. $f(x) = x - 1, g(x) = x^3$

33. $f(x) = x + \dfrac{1}{x^2}, g(x) = \dfrac{1}{x}$

34. $f(x) = \sqrt{x - 4}, g(x) = x^2$

35. $f(x) = x + 1, g(x) = x + \sqrt{x - 1}$

36. $f(x) = x^3 - 4, g(x) = \sqrt[3]{x + 4}$

37. $f(x) = 2, g(x) = x + 1$

38. $f(x) = x^3 + x, g(x) = 4$

39. $f(x) = \begin{cases} x - 1, & x \le 0 \\ x + 1, & x > 0 \end{cases}, g(x) = x^2$

40. $f(x) = 2x + 1, g(x) = \begin{cases} x^2, & x < 0 \\ -x^2, & x \ge 0 \end{cases}$

41. $f(x) = \begin{cases} x^2 - x, & x < 0 \\ x, & 0 \le x \le 1, \\ 2x - x^2, & x > 1 \end{cases} g(x) = x^2 - 1$

42. $f(x) = \begin{cases} -10, & x < 0 \\ 0, & x = 0 \end{cases}, g(x) = x^3$

43. If $f(x) = \sqrt{x}$, find $f \circ f$ and $f \circ (1/f)$.

44. If $f(x) = 2x + 6$, find $f \circ f$ and $f \circ (1/f)$.

45. If $f(x) = x^2 + 1$, find $f \circ f$ and $f \circ (1/f)$.

In Problems 46–49, find (a) $(f \circ g)(4)$; (b) $(g \circ f)(4)$; (c) $(f \circ f)(-1)$ for the given function.

46. $f(x) = x + 1, g(x) = x^3$

47. $f(x) = x^2 + 1, g(x) = 2x^2$

48. $f(x) = 4x, g(x) = \sqrt{x}$

49. $f(x) = -x^2 + x, g(x) = |x|$

In Problems 50–52, find $(f \circ g \circ h)(x) = f(g(h(x)))$ for the given functions.

50. $f(x) = \frac{1}{4}(x^{-1} - 1), g(x) = \dfrac{1}{x^2}, h(x) = 2x + 1$

51. $f(x) = \sqrt{2x - 3}, g(x) = x^2 + 3, h(x) = \sqrt{5x - 7}$

52. $f(x) = \sqrt{x}, g(x) = x^2, h(x) = x - 1$

In Problems 53–56, express the given function h as the composition $h = f \circ g$ of two functions f and g.

53. $h(x) = (3x - 5)^4$

54. $h(x) = (2x + 1)^{-2}$

55. $h(x) = \sqrt{x + x^{-1}}$

56. $h(x) = \left(\dfrac{x + 1}{x - 1}\right)^2$

57. Find functions f and g satisfying $h(x) = (g \circ f)(x) = |\sqrt{x} - 4|$.

58. Find functions f and g satisfying $h(x) = (g \circ f)(x) = (x + 2)^2 - 3\sqrt{x + 2}$.

59. For $f(x) = x^2 - 1, g(x) = x^2 + 1$, verify $f \circ g \neq g \circ f$.

60. Compare $f \circ g$ and $g \circ f$ for $f(x) = x^2 - 1, x > 0$ and $g(x) = \sqrt{x + 1}$.

61. Answer true or false: $f \circ (g + h) = (f \circ g) + (f \circ h)$.

62. Express the area A of a circle as a function of the circumference C of the circle. [*Hint:* First express the radius r as a function of the circumference C.]

63. The diameter d of a cube is the distance between opposite vertices. Express the diameter as a function of the length s of a side, by first expressing the diagonal y (Figure 2.63) as a function of s.

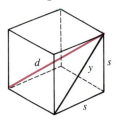

Figure 2.63

64. A birdwatcher sights a bird 100 feet due east of his position. If the bird is flying due south (Figure 2.64) at 500 ft/min, express the distance d from the birdwatcher to the bird as a function of time t. Find the distance 5 minutes after the sighting.

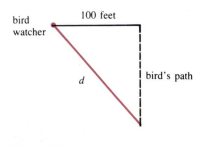

Figure 2.64

65. A certain bacteria when cultured grows in a circular shape. If the radius in centimeters is given by

$$r(t) = 4 - \frac{4}{t^2 + 1}$$

at time t hours, express the area covered by the bacteria as a function of time.

66. Express the circumference of the bacteria culture in Problem 65 as a function of time.

Review Problems

67. Assume v varies directly as t and $v = 4$ when $t = 1$. Find v when $t = 4$.

68. Graph $f(x) = \dfrac{x - 1}{x}$.

69. For what value(s) of a does $(a, 2a)$ lie on the graph of $f(x) = x^3 + x$?

70. For $f(x) = 1/x^2$, find $f(3)$, $(1/f)(3)$, and $(f \circ f)(3)$.

71. Find an expression for the slope of the line passing through the points A and B in Figure 2.65.

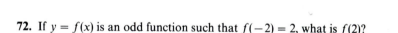

Figure 2.65

72. If $y = f(x)$ is an odd function such that $f(-2) = 2$, what is $f(2)$?

2.6 Inverse Functions

In this section we shall discuss the inverse of a function. This will be a rule of correspondence that "reverses" the original function. For example, consider the function f defined by the table

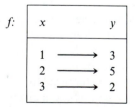

By reversing the columns in this table we obtain a new rule of correspondence, which is also a function, denoted by f^{-1}.

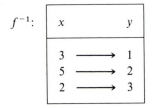

The symbol f^{-1} is read "f inverse." It is important to note that "-1" in f^{-1} is *not* an exponent; that is,

$$f^{-1} \neq \frac{1}{f};$$

rather, f^{-1} denotes the inverse of f.

Now consider another function g defined by the table

g:

x	y
1 →	4
2 →	6
3 →	6

If we reverse the roles of x and y in this table, we obtain the correspondence

x	y
4 →	1
6 →	2
	3

which is not a function since there are two values of y, the numbers 2 and 3, associated with $x = 6$.

One-to-One Functions

We wish to determine what property a function must have in order that the "reversed rule" is also a function. Note that in the first example above, each element in the range of f was associated with *only one* element in the domain, while in the second example an element in the range of g (namely, 6) corresponded to *more than one* element in the domain.

If we require that each element in the range of f be associated with a different element of the range, then the function is said to be **one-to-one**. It is precisely this property that is required for the "reversed rule" to be a function. A formal definition of a one-to-one function is given by the following.

Definition 2.2   *A function is **one-to-one** if and only if $f(x_1) = f(x_2)$ implies $x_1 = x_2$ for all x_1 and x_2 in the domain of f.*

From Definition 2.2 it follows that a function f is *not* one-to-one if distinct elements $x_1 \neq x_2$ can be found in the domain of f such that $f(x_1) = f(x_2)$.

Example 1   The function $f(x) = x^2$ is not one-to-one, since $f(-3) = f(3) = 9$.   ■

Example 2   Determine whether $f(x) = \dfrac{1}{x+1}$ is one-to-one.

Solution   We assume that $f(x_1) = f(x_2)$. Then

$$\frac{1}{x_1 + 1} = \frac{1}{x_2 + 1}$$

$$x_2 + 1 = x_1 + 1$$

$$x_2 = x_1.$$

Since the assumption $f(x_1) = f(x_2)$ leads to the conclusion that $x_1 = x_2$, we have shown that f is one-to-one.   ■

Inverse Function

Let f be a one-to-one function. The **inverse function** f^{-1} is obtained by associating each element v in the range of f with the unique element u in the domain of f such that $f(u) = v$. That is,

$$f^{-1}(v) = u \quad \text{if and only if} \quad f(u) = v.$$

Example 3   Find the inverse of the function f defined by

f:	x		y
	1	$\longrightarrow$	2
	2	$\longrightarrow$	4
	3	$\longrightarrow$	6

Solution   It is clear from an examination of the table that f is one-to-one, and hence has an inverse. The domain of f^{-1} will be the set $\{2, 4, 6\}$ and the range of f^{-1} will be $\{1, 2, 3\}$. Now

$$f^{-1}(2) = \text{that number } u \text{ such that } f(u) = 2.$$

Since $f(1) = 2$, we have $f^{-1}(2) = 1$. Similarly, $f^{-1}(4)=2$ [since $f(2)=4$], and $f^{-1}(6) = 3$. In Figure 2.66 we illustrate the fact that the correspondence f^{-1} is the reversal of f.

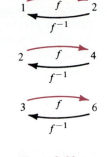

Figure 2.66

Before attempting to find the inverse of a function, one should first determine whether the given function is one-to-one. In addition to the algebraic method discussed above, we now present a graphical technique for determining whether a function is one-to-one. (See Problems 21 and 22 for a third method.)

Horizontal Line Test

Let $y = f(x)$ be a one-to-one function and consider its graph

$$\{(x, y)\,|\,y = f(x), x \text{ in the domain of } f\}.$$

Since each y value corresponds to at most one x value, any horizontal line intersects the graph of $y = f(x)$ in at most one point. Conversely, if each horizontal line intersects the graph of a function in at most one point, then the function is one-to-one. In Figure 2.67 we illustrate this **horizontal line test** for one-to-one functions.

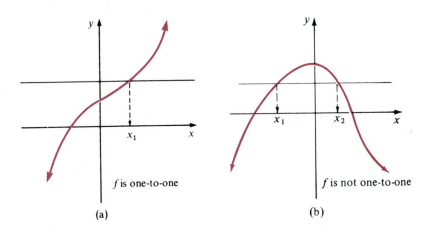

f is one-to-one

(a)

f is not one-to-one

(b)

Figure 2.67

Example 4   Applying the horizontal line test to the function $f(x) = x^2 - 2x$, we see from Figure 2.68 that f is not one-to-one.

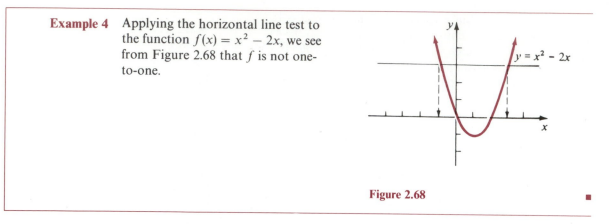

Figure 2.68

■

Finding the Inverse

Now that we can determine whether a particular function is one-to-one (and hence has an inverse), we consider a method for finding its inverse. Recall from the definition of f^{-1} that

$$f^{-1}(v) = u \quad \text{if} \quad f(u) = v.$$

It then follows that

$$(f^{-1} \circ f)(u) = f^{-1}(f(u))$$
$$= f^{-1}(v) = u.$$

In fact, it can be shown that the existence of the inverse function f^{-1} is equivalent to

$$\boxed{(f^{-1} \circ f)(x) = x \quad \text{and} \quad (f \circ f^{-1})(x) = x.} \tag{1}$$

The second of these equations may be used to compute the inverse of a function.

Example 5   Verify $f(x) = 5x - 7$ is a one-to-one function and find f^{-1}.

$$f(x) = 5x - 7.$$

Solution   Since the graph of $y = 5x - 7$ is a nonhorizontal straight line, we have from the horizontal line test that f is one-to-one. To find f^{-1} we use the second equation in (1),

$$(f \circ f^{-1})(x) = x \quad \text{or} \quad f(f^{-1}(x)) = x. \tag{2}$$

Now the function f is the rule given by

$$f(\ ) = 5(\ ) - 7,$$

so that (2) may be written as

$$5(f^{-1}(x)) - 7 = x,$$

which we can solve for $f^{-1}(x)$:

$$5f^{-1}(x) = x + 7$$

$$f^{-1}(x) = \frac{1}{5}x + \frac{7}{5}.$$

$\blacksquare$

Example 6 Verify that $f(x) = \dfrac{2}{x^3 + 1}$ is one-to-one and find f^{-1}.

Solution From the assumption that $f(x_1) = f(x_2)$ we obtain:

$$\frac{2}{x_1^3 + 1} = \frac{2}{x_2^3 + 1}$$

$$2(x_2^3 + 1) = 2(x_1^3 + 1)$$

$$x_2^3 + 1 = x_1^3 + 1$$

$$x_2^3 = x_1^3$$

$$x_2 = x_1.$$

Thus, by Definition 2.2, f is one-to-one. To compute the inverse we use the requirement that $f(f^{-1}(x)) = x$ and the fact that for this function

$$f(\ \ ) = \frac{2}{(\ \ )^3 + 1}.$$

It follows that $\dfrac{2}{(f^{-1}(x))^3 + 1} = x$ or

$$2 = x[(f^{-1}(x))^3 + 1]$$

$$2 = x(f^{-1}(x))^3 + x$$

$$2 - x = x(f^{-1}(x))^3$$

$$(f^{-1}(x))^3 = \frac{2 - x}{x}.$$

Thus, we obtain $f^{-1}(x) = \sqrt[3]{\dfrac{2 - x}{x}}.$

$\blacksquare$

In Figure 2.69 we illustrate the relationship between the graphs of the function

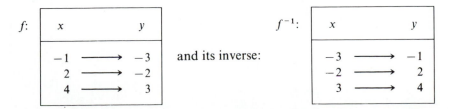

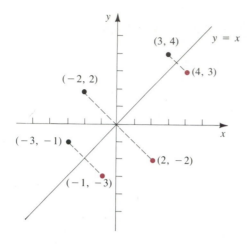

Figure 2.69

We note that the graph of f^{-1} is the reflection of the graph of f in the line $y = x$. In fact, it follows from the definition of f^{-1} that whenever the point (u, v) is on the graph of the function f, then (v, u) is on the graph of f^{-1}. (For further discussion of this reflection property, see Problem 36.)

Inverse Relation

Finally we note that interchanging the variables x and y in *any* function defined by a formula results in a relation called the **inverse relation**. For example, the inverse relation of

$$f = \{(x, y) \mid y = x^2\}$$

is

$$\{(x, y) \mid x = y^2\}. \tag{3}$$

The preceding discussion of the relationship between the graphs of a function and its inverse applies in this case as well. That is, the graph of an inverse relation is the reflection of the graph of the function in the line $y = x$. As we see from Figure 2.70(b), the inverse relation (3) is *not* a function since it fails the vertical line test for functions.

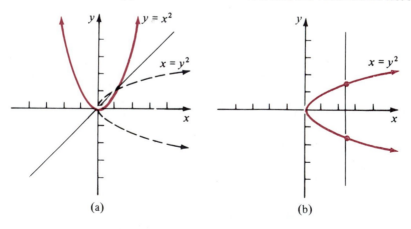

(a)   (b)

Figure 2.70

Example 7   Graph the inverse relation of $y = x^3$ and determine whether it is a function.

Solution   The graph of $y = x^3$ and its inverse relation $x = y^3$ are shown in Figure 2.71. We see from the vertical line test that the graph of $x = y^3$ defines a function:

$$f^{-1} = \{(x, y) \mid x = y^3\}$$
$$= \{(x, y) \mid y = \sqrt[3]{x}\}.$$

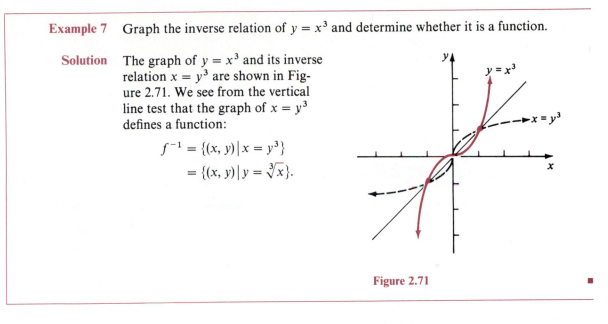

Figure 2.71

∎

Alternate Method to Find f^{-1}

The discussion of inverse relation has provided us with an alternate method for finding the inverse of a one-to-one function. Observe that in the preceding example the given function $f(x) = x^3$, described by the equation $y = x^3$, is one-to-one. By interchanging the variables x and y in the equation, we obtain an equation $x = y^3$ which describes the inverse function. By solving this latter equation for y, we obtain the rule for f^{-1}, that is, $x = y^3$ gives $y = \sqrt[3]{x}$ or $f^{-1}(x) = \sqrt[3]{x}$.

Example 8   We can find the inverse of the one-to-one function

$$f(x) = \frac{2}{x^3 + 1}$$

by first writing

$$y = \frac{2}{x^3 + 1} \tag{4}$$

Then if we interchange the variables x and y in (4) we obtain (5), the equation describing the inverse function. The final step is to solve (5) for y.

$$x = \frac{2}{y^3 + 1} \tag{5}$$

$$y^3 + 1 = \frac{2}{x}$$

$$y^3 = \frac{2}{x} - 1$$

$$y = \sqrt[3]{\frac{2}{x} - 1} = \sqrt[3]{\frac{2 - x}{x}}.$$

Thus, $$f^{-1}(x) = \sqrt[3]{\frac{2-x}{x}}.$$

This, of course, is the same as the inverse found by the composition property in Example 6.

■

Exercise 2.6

In Problems 1–6, determine whether the given function is one-to-one by any method.

1. $f(x) = 3x$ **2.** $f(x) = -2x + 1$ **3.** $f(x) = x^4$

4. $f(x) = x^2 - 2$ **5.** $f(x) = x^3$ **6.** $f(x) = (x - 2)(x + 1)$

In Problems 7–12, use Definition 2.2 to verify that the given function is one-to-one.

7. $f(x) = 3x - 5$ **8.** $f(x) = 2x + 6$ **9.** $f(x) = \dfrac{1}{x}$

10. $f(x) = \dfrac{3}{x + 2}$ **11.** $f(x) = \sqrt{x}$ **12.** $f(x) = \dfrac{x^2 - 1}{x}, x > 0$

In Problems 13–18, determine whether the given function is one-to-one by the horizontal-line test.

13. $f(x) = x^2 - 6x$ **14.** $f(x) = \dfrac{2}{x}$ **15.** $f(x) = \dfrac{1}{x - 3}$

16. $f(x) = \dfrac{x}{x^2 + 1}$ **17.** $f(x) = \dfrac{2}{x^2 - 1}$ **18.** $f(x) = x^3 + 3x^2 + 3x + 1$

19. Find conditions on the constants a and b such that $f(x) = ax + b$ is a one-to-one function.

20. Find conditions on the constants a, b, and c such that $f(x) = ax^2 + bx + c$ is a one-to-one function.

An equivalent definition of a one-to-one function is given by the following: The function f is one-to-one if and only if $x_1 \neq x_2$ implies that $f(x_1) \neq f(x_2)$ for x_1 and x_2 in the domain of f.

21. Use the definition above to prove that an increasing function is one-to-one. [Recall that a function f is *increasing* if $x_1 < x_2$ implies that $f(x_1) < f(x_2)$.]

22. Use the definition given above to prove that a decreasing function is one-to-one. [Recall that a function f is *decreasing* if $x_1 < x_2$ implies that $f(x_1) > f(x_2)$.]

In Problems 23–34, verify that the given function is one-to-one and find its inverse using the composition property $(f \circ f^{-1})(x) = x$.

23. $f(x) = 3x - 9$ **24.** $f(x) = \frac{1}{2}x - 2$ **25.** $f(x) = \dfrac{1}{x - 3}$ **26.** $f(x) = \dfrac{2}{3x - 5}$

27. $f(x) = \dfrac{1}{x}$ **28.** $f(x) = \dfrac{x}{x + 2}$ **29.** $f(x) = \dfrac{3x}{2x - 1}$ **30.** $f(x) = x^3$

31. $f(x) = x^3 + 2$ **32.** $f(x) = x^{1/2}$

33. $f(x) = \sqrt[3]{x + 1}$ **34.** $f(x) = \sqrt{x - 1}$

35. Use the definition of f^{-1} to show that $(f \circ f^{-1})(v) = v$.

36. In this problem we will prove that the line $y = x$ is the perpendicular bisector of the line segment joining the points (u, v) and (v, u). Refer to Figure 2.72.

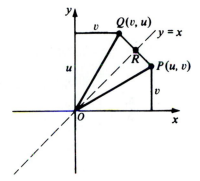

    a. Prove that triangles ORP and ORQ are right triangles by showing that the line through (u, v) and (v, u) is perpendicular to the line $y = x$. [*Hint:* Recall from Section 1.4 the relationship between the slopes of perpendicular lines.]

    b. Prove that the line segment joining O and P has the same length as the line segment joining O and Q.

    c. Use congruent triangles to prove that the line $y = x$ bisects the segment joining P and Q.

Figure 2.72

In Problems 37–42, graph the given function and its inverse relation. Determine whether the inverse relation is a function.

37. $f(x) = x^2 - 2x + 1$ **38.** $f(x) = (x + 1)^3$ **39.** $f(x) = \dfrac{2}{x^2 + 1}$

40. $f(x) = \dfrac{1}{x^3 - 1}$ **41.** $f(x) = \dfrac{x}{x + 1}$ **42.** $f(x) = |x + 1|$

In Problems 43–51, verify that the given function is one-to-one and find its inverse by interchanging the variables.

43. $f(x) = 2 - 4x$ **44.** $f(x) = \dfrac{1}{3 - x}$ **45.** $f(x) = \dfrac{5}{2x + 1}$

46. $f(x) = (3x + 1)^3$ **47.** $f(x) = \dfrac{x}{x - 1}$ **48.** $f(x) = \dfrac{2x}{1 - x}$

49. $f(x) = \sqrt[3]{x - 1}$ **50.** $f(x) = \sqrt[5]{2x + 1}$ **51.** $f(x) = 3x^5 - 2$

52. Find the inverse of the function $f(x) = x^2 - 5$, $x \geq 0$.

53. Verify that the function $f(x) = 2(x - 4)^2$ does not have an inverse. Restrict the domain of f so that the restricted function has an inverse and find that inverse.

Review Problems

54. Find $f \circ g$ and $g \circ f$ if $f(x) = x^3 + 6x$ and $g(x) = \sqrt{x - 1}$.

55. Graph $f(x) = \begin{cases} -x + 2, & x < 1 \\ 3, & x = 1 \\ x^2, & x > 1 \end{cases}$

56. Find the vertical and horizontal asymptotes for $f(x) = \dfrac{x}{x^2 - 25}$.

57. Find $\dfrac{f(1 + h) - f(1)}{h}$ if $f(x) = \dfrac{1}{x + 1}$.

58. Find the vertex, x and y intercepts, and axis symmetry of $f(x) = x^2 + x - 30$.

59. Find the value(s) of x for which the product $(x + 4)(x - 1)$ is the smallest possible.

Chapter Review

A **function** f is a rule of correspondence that associates with a real number x one and only one real number y. The number y associated with x by f is usually written

$$y = f(x).$$

We call y the **dependent variable** and x the **independent variable**. The set of all values of the independent variable x that result in real function values is called the **domain** of the function. The set of corresponding values for the dependent variable y is called the **range** of the function.

The **linear function**

$$f(x) = ax + b$$

is an **increasing** function if $a > 0$ and a **decreasing** function if $a < 0$.

The maximum or minimum value, of the **quadratic function**

$$f(x) = ax^2 + bx + c,$$

is the y coordinate

$$f\left(-\frac{b}{2a}\right)$$

of the **vertex** of the parabola.

The **graph of a function** $y = f(x)$ is the set of ordered pairs

$$\{(x, y) \mid y = f(x), x \text{ in the domain of } f\}.$$

The **y intercept** is the value $f(0)$, provided that 0 is in the domain of f. The **x intercepts** are the real solutions of the equation

$$f(x) = 0.$$

The graph of $y = f(x + c) + k$ may be obtained by **shifting** the graph of $y = f(x)$ c units to the left and k units upward if c and k are both positive. The graph of $y = -f(x)$ may be obtained by **reflecting** the graph of $y = f(x)$ through the x axis.

If c is a real number such that $f(x)$ approaches c as x (or $-x$) becomes large, then the horizontal line

$$y = c$$

is called a **horizontal asymptote** for the graph of f. If the values of $f(x)$ become unbounded as x is taken close to a real number a, we say that the vertical line

$$x = a$$

is a **vertical asymptote**. A knowledge of the asymptotes and intercepts of a rational function is frequently sufficient to sketch its graph.

Given two functions f and g, we define four additional functions, $f + g$, $f - g$, fg, and f/g, by forming their **sum**, **difference**, **product**, and **quotient**. The domain of $f + g$, $f - g$, or fg is the intersection of the domain of f with the domain of g. The domain of f/g is the intersection of the domains of f and g, excluding values of x for which $g(x) = 0$.

Another important method for combining two functions, f and g, is **function composition**. The **composition of f and g**, denoted by $f \circ g$, is the function defined by

$$(f \circ g)(x) = f(g(x)).$$

The domain of $f \circ g$ consists of all values of x in the domain of g such that $g(x)$ is in the domain of f.

A function f is said to be **one-to-one** if and only if each element in the range of f is associated with exactly one element in the domain. If each horizontal line intersects the graph of $y = f(x)$ in at most one point, then f is a one-to-one function.

For a one-to-one function f, the **inverse function** is defined to be

$$f^{-1}(v) = u \quad \text{if and only if} \quad f(u) = v.$$

To find this inverse function, we can use the property

$$(f \circ f^{-1})(x) = f(f^{-1}(x)) = x.$$

The graph of f^{-1} is obtained by reflecting the graph of f in the line $y = x$.

Chapter 2 Review Exercises

In Problems 1–10, answer true or false.

1. The correspondence given by the set of ordered pairs $\{(1, 2),\ (2, 3),\ (3, 3)\}$ is a function. _____

2. If $f(x) = 2x + 1$, then $f(a + 1) = 2a + 2$. _____

3. The line $y = 5$ is a horizontal asymptote for the graph of $f(x) = \dfrac{5}{x + 1}$. _____

4. The domain of $f(x) = \dfrac{1}{x - 1}$ is $\{x \mid x \geq 1\}$. _____

5. The graph of $f(x) = x^2 + 2x + 2$ has no x intercepts. _____

6. The y intercept of $f(x) = 4(x + 5)$ is 20. _____

7. If $y = 10x^{-4}$, then y varies inversely as x^4. _____

8. The graph of $f(x) = \dfrac{1}{x^2 + 1}$ has vertical asymptotes $x = \pm 1$. _____

9. If $f(x) = 3x$ and $g(x) = x + 1$, then $(f \circ g)(x) = 3x + 1.$ _____

10. The horizontal line test is used to determine if a function is one-to-one. _____

11. If $f(x) = x^3 + x^2 - 2$, find $f(0)$; $f(1)$; $f(-2)$; $f(\tfrac{1}{2})$.

12. If $f(x) = \begin{cases} -1, & x < 0 \\ 0, & x = 0, \text{ find } f(0);\ f(1);\ f(\sqrt{2});\ f(-2) \\ 1, & x > 0 \end{cases}$

In Problems 13 and 14, find the domain and range of the given function.

13. $f(x) = \sqrt{2 - x}$ 14. $f(x) = x^2 + 2x + 1$

15. For what value of x is $f(x) = \sqrt{3 - x}$ equal to 1?

In Problems 16 and 17, (a) graph the given function; (b) determine the x coordinate(s) of the point(s) on the graph whose y coordinate is 3.

16. $f(x) = x^2 + 1$ 17. $f(x) = \begin{cases} -x, & x < -1 \\ x^2, & x \geq -1 \end{cases}$

In Problems 18 and 19, graph the given relation and determine whether the relation defines y as a function of x.

18. $\{(x, y) \mid y^2 = x\}$ 19. $\{(x, y) \mid x = 3y + 4\}$

In Problems 20 and 21, find the x and y intercepts of the given linear function. Graph.

20. $f(x) = 4 - 7x$ 21. $f(x) = \tfrac{3}{2}x - \tfrac{5}{2}$

In Problems 22–24, find the x and y intercepts and vertex of the given quadratic function. Graph.

22. $f(x) = 3x^2 + 2x$ 23. $f(x) = -x^2 + 4x - 4$ 24. $f(x) = 2x^2 - 2x + 1$

25. A ball is thrown upward. Suppose that its position in feet above the ground after t seconds is given by $s(t) = -16t^2 + 64t$.

a. When is the maximum height attained?
b. What is the maximum height reached by the ball?
c. When does the ball hit the ground?

26. The velocity (in feet per second) after t seconds of a rock thrown vertically downward from the top of a building is given by

$$v(t) = 32t + 16, \qquad \text{for } 0 \le t \le 4.$$

a. Find the velocity of the rock after 1 second.
b. Find the initial velocity of the rock.

In Problems 27 and 28, determine whether the given function is even or odd.

27. $f(x) = 2x - x^3$ **28.** $f(x) = x^2 + |x|$

In Problems 29–32, use one or more of the techniques of shifting, reflecting, or stretching to graph the given function.

29. $f(x) = 4 - (x + 1)^2$ **30.** $f(x) = -\sqrt{x - 5}$

31. $f(x) = \dfrac{4}{x} + 2$ **32.** $f(x) = (x + 5)^3 - 3$

In Problems 33–36, each graph was obtained by shifting or reflecting the graph of $f(x) = x^{2/3}$ as seen in Figure 2.57. Determine a formula for each function graphed.

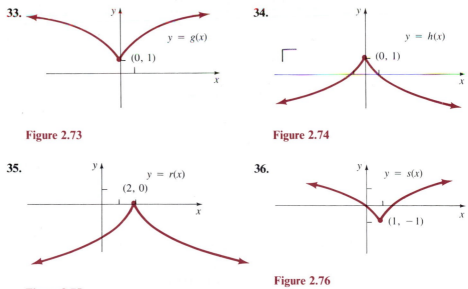

33.

Figure 2.73

$y = g(x)$

$(0, 1)$

34.

Figure 2.74

$y = h(x)$

$(0, 1)$

35.

Figure 2.75

$y = r(x)$

$(2, 0)$

36.

Figure 2.76

$y = s(x)$

$(1, -1)$

In Problems 37–39, graph the given function. Determine intercepts and asymptotes.

37. $f(x) = \dfrac{2x - 5}{1 + 3x}$ **38.** $f(x) = \sqrt{2 - x}$ **39.** $f(x) = \dfrac{x}{x^2 - 4}$

In Problems 40 and 41, find the indicated functions and give their domains.

40. $f(x) = \begin{cases} x, & x < 0 \\ \sqrt{x}, & x \ge 0 \end{cases}$, $g(x) = \begin{cases} x^2, & x < 0 \\ 2x, & x \ge 0 \end{cases}$; $f + g, f - g, fg, \dfrac{f}{g}, f \circ g, g \circ f$

41. $f(x) = x^3 + 1$, $g(x) = \dfrac{1}{x-1}$; $f + g$, $f - g$, fg, $\dfrac{f}{g}$, $f \circ g$, $g \circ f$

In Problems 42 and 43, for the given functions f and g graph (a) f + g and (b) f − g.

42. $f(x) = x^3$, $g(x) = x$

43. $f(x) = 1 - \sqrt{x}$, $g(x) = 1 + \sqrt{x}$

In Problems 44 and 45, verify that the given function is one-to-one and find the inverse function.

44. $f(x) = 3 - \frac{1}{2}x$

45. $f(x) = \dfrac{5}{x-2}$

In Problems 46–48, graph the given function and its inverse relation. Determine whether the inverse relation is a function.

46. $f(x) = \dfrac{1}{x-1}$

47. $f(x) = (x-2)^3$

48. $f(x) = x^2 + 4x + 4$

49. For what value(s) of a does $(a, 16a)$ lie on the graph of

$$f(x) = \begin{cases} 4x - 3, & x < 0 \\ x^3, & 0 \le x \le 1 \\ x^2 + 64, & x > 1 \end{cases}$$

50. If $(2, 3)$ is on the graph of $f(x) = \sqrt[3]{c(x+1)}$, find c.

51. Find the quadratic function $f(x) = ax^2 + bx + c$ if the graph of f has 7 as its y intercept and -2 as its only x intercept.

52. Find the linear function $f(x) = ax + b$ if the graph of f has x intercept 2 and y intercept 1.2.

53. Determine a quadratic function that describes the parabolic archway shown in Figure 2.77.

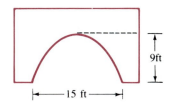

Figure 2.77

54. An 8-ohm resistor and a variable resistor are placed in parallel as shown in Figure 2.78. The total resistance R (ohms) is related to the resistance r (ohms) of the variable resistor by the equation

$$R = \dfrac{8r}{8 + r}$$

Sketch the graph of R as a function of r for $r \ge 0$.

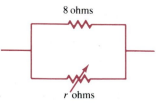

Figure 2.78

3 Exponential and Logarithmic Functions

3.1 Exponential Functions

Irrational Exponents

In Section 0.2 we defined b^r for any positive base b and any rational exponent r. For example,

$$4^{1/5} = \sqrt[5]{4} \quad \text{and} \quad 4^{1.4} = 4^{14/10} = 4^{7/5} = (\sqrt[5]{4})^7.$$

For any irrational number r, b^r can be defined, but a precise definition is beyond the scope of this text. We can, however, suggest one possible procedure for defining a number such as $4^{\sqrt{2}}$. Since

$$\sqrt{2} = 1.414213562\ldots, \tag{1}$$

the rational numbers

$$1, 1.4, 1.41, 1.414, 1.4142, 1.41421, \ldots$$

give successively better approximations to $\sqrt{2}$. This suggests that the numbers

$$4^1, 4^{1.4}, 4^{1.41}, 4^{1.414}, 4^{1.4142}, 4^{1.41421}, \ldots$$

give successively better approximations to the value of $4^{\sqrt{2}}$. In fact, this can be shown to be true with a precise definition of b^r for irrational r. Using the $\boxed{y^x}$ key on a scientific calculator one can find that a nine decimal place *approximation* to $4^{\sqrt{2}}$ is 7.102993294.

We shall accept the following statement as fact:

For $b > 0$ and any real number r, the expression b^r represents a unique real number, and in addition, the laws of exponents hold for all real numbers.

It follows from the preceding statement that the equation $y = 4^x$ describes a function with the set of real numbers as its domain. Plotting points obtained from the accompanying tables and then connecting the points by a smooth curve yields the graph in Figure 3.1. Notice that $y = 4^x$ is an increasing function.

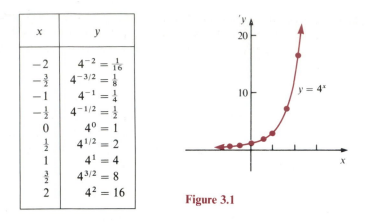

x	y
-2	$4^{-2} = \frac{1}{16}$
$-\frac{3}{2}$	$4^{-3/2} = \frac{1}{8}$
-1	$4^{-1} = \frac{1}{4}$
$-\frac{1}{2}$	$4^{-1/2} = \frac{1}{2}$
0	$4^0 = 1$
$\frac{1}{2}$	$4^{1/2} = 2$
1	$4^1 = 4$
$\frac{3}{2}$	$4^{3/2} = 8$
2	$4^2 = 16$

Figure 3.1

Example 1   Graph the function $y = 4^{-x}$.

Solution

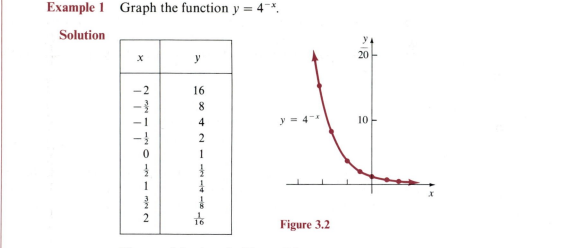

x	y
-2	16
$-\frac{3}{2}$	8
-1	4
$-\frac{1}{2}$	2
0	1
$\frac{1}{2}$	$\frac{1}{2}$
1	$\frac{1}{4}$
$\frac{3}{2}$	$\frac{1}{8}$
2	$\frac{1}{16}$

Figure 3.2

The graph is given in Figure 3.2.   ■

Notice in the preceding example $y = 4^{-x}$ is a decreasing function. Also, using the laws of exponents, we can write

$$y = 4^{-x} = (4^{-1})^x = \left(\frac{1}{4}\right)^x.$$

Both $y = 4^x$ and $y = 4^{-x}$ are examples of **exponential functions**.

> **Definition 3.1**   *Any function of the form*
> $$f(x) = b^x,$$
> $b > 0, b \neq 1, x$ *real, is said to be an* **exponential function**. *The number b is said to be the* **base** *of the function.*

In Definition 3.1 we restrict the base b to be a positive number to avoid a situation such as $(-4)^{1/2}$. Also, for $b = 1$ we obtain the constant function $f(x) = 1^x = 1$. As indicated in Figure 3.3, an exponential function increases for increasing x when $b > 1$, and decreases for increasing x when $0 < b < 1$.

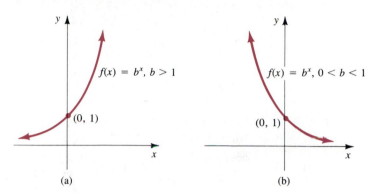

Figure 3.3

Properties of the Exponential Function

From the sketches in Figure 3.3 we can observe the following properties of the exponential function with base b.

- The domain is all real numbers.
- The range is all positive real numbers.
- The graph has y intercept 1, but there is no x intercept.
- The x axis is an asymptote.
- The function is increasing if $b > 1$ and is decreasing if $0 < b < 1$.
- The function is one-to-one.

Example 2   The population P in a community after t years is given by the formula

$$P(t) = 1000(\tfrac{3}{2})^t.$$

Does the population increase or decrease for increasing time? What is the initial population? What is the population after 1, 2, and 5 years?

Solution   We can think of $P(t)$ as a constant multiple of $(\tfrac{3}{2})^t$. Since $1000 > 0$ and $b = \tfrac{3}{2} > 1$, the population increases for increasing time. The initial population occurs when $t = 0$:

$$P(0) = 1000(\tfrac{3}{2})^0 = 1000.$$

Also,
$$P(1) = 1000(\tfrac{3}{2})^1 = 1500,$$
$$P(2) = 1000(\tfrac{3}{2})^2 = 1000(\tfrac{9}{4}) = 2250,$$
$$P(5) = 1000(\tfrac{3}{2})^5 = 1000(\tfrac{243}{32}) \approx 7594.$$

■

Example 3   A function such as $f(x) = 4^{x-2}$ is a multiple of an exponential function since

$$f(x) = 4^{x-2} = (4^{-2})4^x = (\tfrac{1}{16})4^x.$$

■

Other Exponents

As the following examples show, when the exponent of the base is an algebraic expression involving x, the graph of the function may not resemble either of those given in Figure 3.3.

Example 4   Graph $f(x) = 4^{x^2}$.

Solution

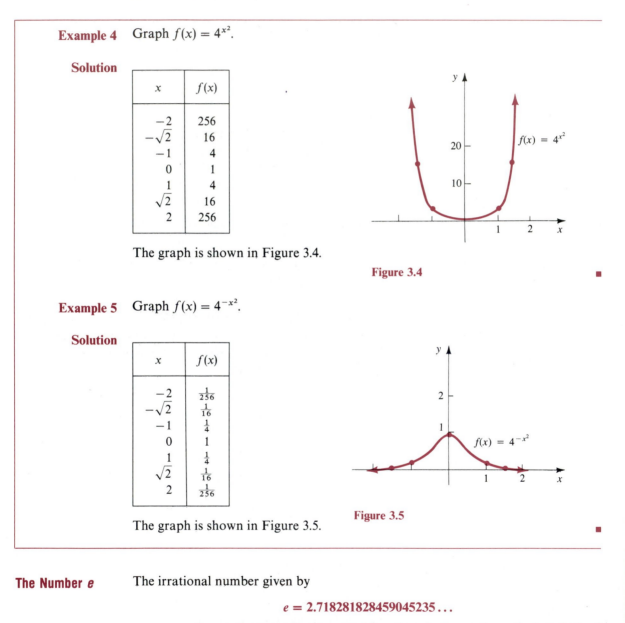

x	$f(x)$
-2	256
$-\sqrt{2}$	16
-1	4
0	1
1	4
$\sqrt{2}$	16
2	256

The graph is shown in Figure 3.4.

Figure 3.4

Example 5   Graph $f(x) = 4^{-x^2}$.

Solution

x	$f(x)$
-2	$\frac{1}{256}$
$-\sqrt{2}$	$\frac{1}{16}$
-1	$\frac{1}{4}$
0	1
1	$\frac{1}{4}$
$\sqrt{2}$	$\frac{1}{16}$
2	$\frac{1}{256}$

Figure 3.5

The graph is shown in Figure 3.5.

The Number e

The irrational number given by

$$e = 2.718281828459045235\ldots$$

occurs so often as the base of exponential functions in the mathematical analysis of certain physical, sociological, and economic phenomena that it merits special consideration. Although it is not our purpose here to give a detailed discussion of the number e, we will examine one approach to its definition. We consider the function f

defined by

$$f(n) = \left(1 + \frac{1}{n}\right)^n,$$

where the domain is restricted to the set of positive integers. As the following table shows, the values of $f(n)$ (rounded to five decimals) are getting closer together;

n	$f(n)$
1	2
2	2.25
3	2.37037
4	2.44141
5	2.48832
10	2.59374
100	2.70481
1000	2.71692
10,000	2.71815
100,000	2.71827
1,000,000	2.71828
10,000,000	2.71828

in fact, the terms in the sequence of function values $f(1), f(2), \ldots, f(10,000), \ldots$ do not increase without bound. In calculus it is shown that these values are approaching e. The number e is said to be the **limit** of the sequence and this is written

$$\lim_{n \to \infty} \left(1 + \frac{1}{n}\right)^n = e. \tag{2}$$

The reader is encouraged to duplicate the calculation of $f(n)$ using the $\boxed{y^x}$ key on a calculator.

Since $e > 1$ and $1/e < 1$, the graphs of the exponential functions $f(x) = e^x$ and $f(x) = e^{-x} = \left(\dfrac{1}{e}\right)^x$ have the form of those in Figure 3.3. See Figure 3.6.

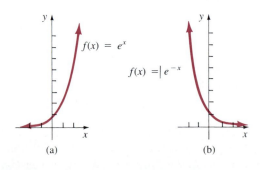

(a)   (b)

Figure 3.6

Example 6 Compute **a.** e^4 **b.** $\dfrac{1}{e^3}$.

Solution Because of its importance, many scientific calculators have an $\boxed{e^x}$ key. If not, then we enter x and press, in succession, $\boxed{\text{INV}}$ $\boxed{\text{ln}}$ (or $\boxed{f^{-1}}$ $\boxed{\text{ln}}$ on some calculators.)
a. With the aid of a calculator we find

$$e^4 \approx 54.5982$$

where the symbol $\approx$ indicates that the answer is an approximation to the true value.

b. To compute $\dfrac{1}{e^3}$ we rewrite the expression as e^{-3}. Entering -3 and proceeding as above gives

$$e^{-3} \approx 0.0498.$$ ∎

Note: Although most calculators are capable of displaying eight or nine decimal places, from here on we shall generally write a number with at most four rounded decimal places. Also, a short table of values of the exponential functions e^x and e^{-x} can be found at the back of this text (Table III).

Exercise 3.1

In Problems 1–20, graph the given function.

1. $f(x) = 2^x$ **2.** $f(x) = 2^{-x}$ **3.** $f(x) = 2^{x^2}$ **4.** $f(x) = 2^{x+1}$

5. $f(x) = 2^{|x|}$ **6.** $f(x) = 2^{-x^2}$ **7.** $f(x) = x2^x$ **8.** $f(x) = 3(2^x)$

9. $f(x) = -2^x$ **10.** $f(x) = -2^{-x}$ **11.** $f(x) = (\tfrac{3}{4})^x$ **12.** $f(x) = (\tfrac{4}{3})^x$

13. $f(x) = 3^x - 3$ **14.** $f(x) = 3^{-x} + 1$ **15.** $f(x) = 3^{-x+1}$ **16.** $f(x) = 3^x + 3^{-x}$

17. $f(x) = e^{-2x}$ **18.** $f(x) = 2 + e^{-x}$ **19.** $f(x) = 1 - e^x$ **20.** $f(x) = e^{-(x-2)^2}$

21. The function $f(x) = \dfrac{e^x + e^{-x}}{2}$ is called the **hyperbolic cosine**. Sketch the graph of f.

22. The function $f(x) = \dfrac{e^x - e^{-x}}{2}$ is called the **hyperbolic sine**. Sketch the graph of f.

In Problems 23–40, state whether the given equation is true or false.

23. $2^{2x} = 4^x$ **24.** $2^{-x} = (\tfrac{1}{2})^x$ **25.** $2^{x-1} = \dfrac{1}{2}(2^x)$

26. $2^{-x} = (2^x)^{-1}$ **27.** $2^x \cdot 2^y = 4^{x+y}$ **28.** $3^x \cdot 4^x = 12^x$

29. $2^{3x} \cdot 2^{2x} = 2^{5x}$ **30.** $2^x + 2^{-x} = (2 + 2^{-1})^x$ **31.** $2^{x^2} = (2^x)^2$

32. $2^{-x^2} = \left(\dfrac{1}{2^x}\right)^2$ **33.** $\dfrac{2^{x^2}}{2^x} = 2^x$ **34.** $(\tfrac{3}{2})^x = 3^x \cdot 2^{-x}$

35. $4^{x/2} = 2^x$ **36.** $2^{-x} = 2^{1/x}$ **37.** $2^{x-1} = (2^x)^{-1}$

38. $2^{|x|} = |2^x|$ **39.** $2^{3+3x} = 8^{1+x}$ **40.** $2^x + 2^x = 2^{x+1}$

41. If $f(x) = b^x$, show that $f(x_1 + x_2) = f(x_1)f(x_2)$.

42. Give two negative and two positive values of x not in the domain of the function $f(x) = (-10)^x$. Are these values in the domain of $f(x) = (-10)^{x^2}$?

43. What is the range of the function $f(x) = 5 + 10^{-x}$? $f(x) = 5 - 10^{-x}$? [*Hint:* Sketch the graphs.]

44. Determine, if any, the point(s) of intersection for $y = 2^x + 4$ and $y = 3^{1-x} + 2$. [*Hint:* Inspect the graph of each function.]

45. Determine by inspection whether the graphs of $y = 3^x - 1$ and $y = 2^x$ intersect. If so, find the point(s) of intersection.

Calculator Problems

46. The number of bacteria present in a culture after t minutes is given by $N(t) = (200)4^{t/2}$. Find the initial number of bacteria present. Also find the number present after 2 minutes, 4 minutes, and 10 minutes. Estimate the number after 1 hour.

47. Suppose a radioactive substance is known to decay at a rate such that at the end of any day there is only one-half as much present as there was at the beginning of the day.
   a. If there were 100 grams of the substance initially, how much will be left after t days? [*Hint:* If $f(t)$ is the number of grams left at the end of the tth day, then $f(0) = 100$, $f(1) = \frac{1}{2}(100)$, $f(2) = \frac{1}{2}(\frac{1}{2})(100)$, and so on.]
   b. How much of the substance is left at the end of one week? Two weeks?

48. The function $h(t) = 79.04 + 6.39t - e^{3.26 - 0.99t}$ is sometimes used to predict the height (in cm) of preschool-aged children. If t represents time (in yrs), find the height predicted for a two-year-old.

In Problems 49 and 50 use a calculator to fill in the table for the given function. Round your answers to five decimal places.

49. $f(h) = (1 + h)^{1/h}$ **50.** $f(n) = (1 + 2/n)^n$

h	$f(h)$
1	
0.1	
0.01	
0.001	
0.0001	
0.00001	
0.000001	
0.0000001	

n	$f(n)$
1	
10	
100	
1000	
10,000	
100,000	
1,000,000	
10,000,000	

51. From inspection of Problem 49, conjecture the value of $\lim_{h \to 0}(1 + h)^{1/h}$.

52. From inspection of Problem 50, conjecture the value of $\lim_{n \to \infty}(1 + 2/n)^n$.

53. The following approximation is established in calculus:

$$e^x \approx 1 + x + \frac{x^2}{2} + \frac{x^3}{6} + \frac{x^4}{24}.$$

Use this result to approximate

a. e^{-1} **b.** e^3 **c.** $\sqrt{e}$.

3.2 Logarithmic Functions

Logarithms

Since the exponential function $y = b^x$ ($b > 0$, $b \neq 1$) is one-to-one, it follows from Section 2.6 that the inverse relation $\{(x, y) \mid x = b^y\}$ defines y as a function of x. One way to express the role of the variable y in the equation $x = b^y$ is to say

y is that exponent of the base b which yields the number x.

It is common practice to replace the word "exponent" with the word "logarithm" and to say that y is "the logarithm to the base b of x." This is written as $y = \log_b x$. Thus we have the following definition.

Definition 3.2 $y = \log_b x$ if and only if $b^y = x$.

Example 1   From Definition 3.2,
a. $\log_3 9 = 2$ is equivalent to $3^2 = 9$,
b. $\log_{10}(0.0001) = -4$ is equivalent to $10^{-4} = 0.0001$,
c. $\log_{16} \frac{1}{4} = -\frac{1}{2}$ is equivalent to $16^{-1/2} = \frac{1}{4}$.   ■

Example 2 $\log_{1/3} 27 = -3$ since $\left(\frac{1}{3}\right)^{-3} = (3^{-1})^{-3} = 3^3 = 27$.   ■

Example 3 $8^{-5/3} = \frac{1}{32}$ can be written equivalently as $\log_8 \frac{1}{32} = -\frac{5}{3}$.   ■

From Definition 3.2 we see

$$\log_b b = 1 \tag{1}$$

and

$$\log_b 1 = 0 \tag{2}$$

because $b^1 = b$ and $b^0 = 1$, respectively. Also, it should be noted that $\log_b x$ is meaningless for $x < 0$, since there is no exponent for which $b^y < 0$.

Example 4 Solve for the unknown.
a. $\log_2 8 = y$. **b.** $\log_4 x = -\frac{1}{2}$. **c.** $\log_b 25 = 2$.

Solution In each case we use Definition 3.2.
a. $\log_2 8 = y$ is equivalent to

$$2^y = 8 \quad \text{or} \quad 2^y = 2^3.$$

By equating exponents of the same base, we conclude that $y = 3$.
b. $\log_4 x = -\frac{1}{2}$ is equivalent to

$$4^{-1/2} = x,$$

so $x = \left(\frac{1}{4}\right)^{1/2} = \frac{1}{2}$.
c. $\log_b 25 = 2$ is equivalent to

$$b^2 = 25 \quad \text{or} \quad b^2 = 5^2,$$

so by equating bases to the same power we find that $b = 5$. ■

We have the following definition.

Definition 3.3 *The* **logarithmic function with base** b,

$$f(x) = \log_b x,$$

is the inverse of the exponential function with base b.

Recall from Section 2.6 that the graph of an inverse function can be obtained by reflecting the graph of the original function in the line $y = x$. This technique is used in Figure 3.7 to obtain the graph of $f(x) = \log_b x$ for $b > 1$. The graph of $f(x) = \log_b x$ for $0 < b < 1$ is left as an exercise. See Problem 78.

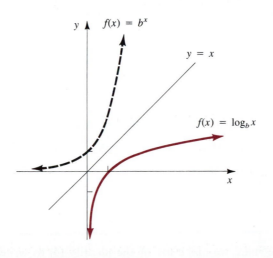

Figure 3.7

Properties of the Logarithmic Function

As Figure 3.7 illustrates, the logarithmic function with base $b(b > 1)$ has the following properties:

- The domain is the set of positive real numbers.
- The range is the set of real numbers.
- The graph has x intercept 1, but there is no y intercept.
- The y axis is a vertical asymptote.
- The function is increasing on the interval $(0, \infty)$.
- The function is one-to-one.

By substituting $y = \log_b x$ into the equivalent equation $x = b^y$, we obtain the important identity

$$x = b^{\log_b x}. \tag{3}$$

Example 5   **a.** $3^{\log_3 7} = 7$ **b.** $10^{\log_{10} 5^2} = 5^2$ ■

By substituting $x = b^y$ into $y = \log_b x$ we obtain another identity

$$y = \log_b b^y. \tag{4}$$

Example 6   **a.** $\log_{10} 10^2 = 2$ **b.** $\log_8 8^{\sqrt{2}} = \sqrt{2}$ ■

Rules of Logarithms

The following three properties of logarithms are simply a reformulation of the laws of exponents.

For any positive real numbers M and N:

I. $\log_b MN = \log_b M + \log_b N$,

II. $\log_b \dfrac{M}{N} = \log_b M - \log_b N$,

III. $\log_b N^c = c \log_b N$,   for c any real number.

To verify these rules, we make use of the identity (4) to write any two positive numbers M and N as, respectively,

$$M = b^{\log_b M} \quad \text{and} \quad N = b^{\log_b N}.$$

Therefore, $$MN = b^{\log_b M} \cdot b^{\log_b N}$$

$$= b^{\log_b M + \log_b N}.$$

From Definition 3.2 the latter statement is the same as rule I:

$$\log_b MN = \log_b M + \log_b N.$$

Similarly, $$\frac{M}{N} = \frac{b^{\log_b M}}{b^{\log_b N}}$$

$$= b^{\log_b M - \log_b N}$$

and $$N^c = (b^{\log_b N})^c$$

$$= b^{c \log_b N}$$

are equivalent to rules II and III, respectively.

Example 7   There are now two ways to simplify an expression such as $\log_5 5^3$:

$$\log_5 5^3 = 3 \log_5 5 \qquad \text{(rule III)}$$

$$= 3, \qquad\qquad \text{[equation (1)]}$$

or alternatively,

$$\log_5 5^3 = 3. \qquad \text{[equation (4)]}$$

Example 8   Solve for c: $\log_{10} 100^c = 3$.

Solution   By rule III and equation (1), we can write

$$\log_{10} 100^c = \log_{10} (10^2)^c$$

$$= \log_{10} 10^{2c}$$

$$= 2c \log_{10} 10$$

$$= 2c.$$

Thus, we have $\log_{10} 100^c = 3 = 2c$, from which it follows that $c = \frac{3}{2}$.

Example 9   Simplify $\frac{1}{2} \log_9 36 + 2 \log_9 4 - \log_9 4$.

Solution   There are several possible ways to approach this problem. Note, for example, that the second and third terms can be combined as

$$2 \log_9 4 - \log_9 4 = \log_9 4.$$

Alternatively, we can use rule III followed by rule I to combine these terms:

$$2 \log_9 4 - \log_9 4 = \log_9 4^2 - \log_9 4$$
$$= \log_9 16 - \log_9 4$$
$$= \log_9 \tfrac{16}{4}$$
$$= \log_9 4.$$

Hence
$$\tfrac{1}{2} \log_9 36 + 2 \log_9 4 - \log_9 4 = \log_9 (36)^{1/2} + \log_9 4$$
$$= \log_9 6 + \log_9 4$$
$$= \log_9 24. \qquad \blacksquare$$

Example 10   If $\log_b 3 = 0.4771$ and $\log_b 5 = 0.6990$, find

a. $\log_b 15$ **b.** $\log_b 0.6$ **c.** $\log_b 27$ **d.** $\dfrac{\log_b 3}{\log_b 5}$.

Solution   In the first three parts of this example, we shall use rules I–III to express the given logarithm in terms of $\log_b 3$ and $\log_b 5$.

a. From rule I,

$$\log_b 15 = \log_b 3 \cdot 5$$
$$= \log_b 3 + \log_b 5$$
$$= 0.4771 + 0.6990 = 1.1761.$$

b. From rule II,

$$\log_b 0.6 = \log_b \tfrac{3}{5}$$
$$= \log_b 3 - \log_b 5$$
$$= 0.4771 - 0.6990$$
$$= -0.2219.$$

c. From rule III,

$$\log_b 27 = \log_b 3^3$$
$$= 3 \log_b 3$$
$$= 3(0.4771) = 1.4313.$$

d. Notice that none of the rules apply in this case. In particular, *a quotient of logarithms is not the difference of logarithms.* Thus, with the aid of a calculator we simply *divide* the two numbers:

$$\frac{\log_b 3}{\log_b 5} = \frac{0.4771}{0.6990} \approx 0.6825. \qquad \blacksquare$$

Rules I and II are often misinterpreted. The basic idea in part **d** of Example 10 is that

$$\frac{\log_b M}{\log_b N} \neq \log_b M - \log_b N.$$

Also, the *logarithm of a sum is not the sum of logarithms.* That is,

$$\log_b(M + N) \neq \log_b M + \log_b N.$$

Graphs

The next three examples illustrate some graphs of logarithmic functions with base 10.

Example 11 Graph $f(x) = \log_{10} x$.

Solution The following table shows the corresponding values of y for selected values of x.

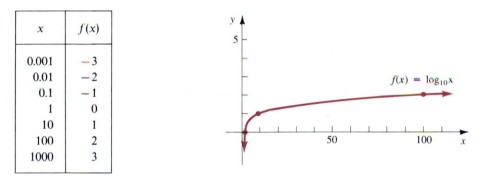

x	$f(x)$
0.001	-3
0.01	-2
0.1	-1
1	0
10	1
100	2
1000	3

The graph is shown in Figure 3.8.   **Figure 3.8**

Example 12 Graph $f(x) = \log_{10}(x + 10)$.

Solution From our discussion of shifted graphs in Section 2.7 we know that the graph of $y = \log_{10}(x + 10)$ is the graph of $y = \log_{10} x$ shifted 10 units to the left. See Figure 3.9.

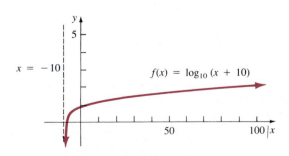

Figure 3.9

Example 13   Graph $f(x) = \log_{10}|x|$.

Solution   The absolute value extends the domain of the logarithmic function to all real numbers except $x = 0$. For example,

$$\log_{10}|-10| = \log_{10}10 = 1.$$

Furthermore, the function is even since

$$f(-x) = \log_{10}|-x| = \log_{10}|x| = f(x).$$

Hence the graph will be symmetric with respect to the y axis. The graph is sketched in Figure 3.10 by plotting points for $x > 0$ as in Example 11 and then using symmetry to complete the graph for $x < 0$.

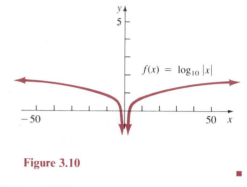

Figure 3.10

Exercise 3.2

In Problems 1–12, rewrite the given exponential statement in the equivalent logarithmic form.

1. $4^{-1/2} = \frac{1}{2}$ **2.** $7^2 = 49$ **3.** $9^0 = 1$

4. $(3^2)^{-2} = \frac{1}{81}$ **5.** $e^y = x$ **6.** $8^{2/3} = 4$

7. $\left(\frac{1}{64}\right)^{-1/2} = 8$ **8.** $(a+b)^2 = a^2 + 2ab + b^2$ **9.** $36^{-3/2} = \frac{1}{216}$

10. $5^{-4} = 0.0016$ **11.** $6^{-0.6131} = \frac{1}{3}$ **12.** $6^{2.2662} = 58$

In Problems 13–24, rewrite the given logarithmic statement in the equivalent exponential form.

13. $\log_3 81 = 4$ **14.** $\log_2 32 = 5$ **15.** $\log_{10} 10 = 1$ **16.** $\log_{17} 17^5 = 5$

17. $\log_5 \frac{1}{25} = -2$ **18.** $\log_{\sqrt{2}} 2 = 2$ **19.** $\log_{16} 2 = \frac{1}{4}$ **20.** $\log_9 \frac{1}{3} = -\frac{1}{2}$

21. $\log_b b^2 = 2$ **22.** $\log_b u = v$ **23.** $\log_b 9 = 2$ **24.** $3\log_b 4 = 2$

In Problems 25–36, find the value of the given logarithm.

25. $\log_8(0.125)$ **26.** $\log_9(729)^2$ **27.** $\log_2(2^2 + 2^2)$ **28.** $\log_7 \sqrt[3]{49}$

29. $\log_{64} \frac{1}{32}$ **30.** $\log_{\sqrt{3}} 9$ **31.** $\log_{1/2} 16$ **32.** $\log_8 \frac{1}{4}$

33. $\log_{5/2} \frac{8}{125}$ **34.** $\log_6 216$ **35.** $\log_4 \frac{1}{64}$ **36.** $\log_2(2 \cdot 4 \cdot 8)$

In Problems 37–48, solve for the unknown.

37. $\log_b 125 = 3$ **38.** $\log_4 N = -2$ **39.** $\log_7 343 = y$ **40.** $\log_5 25^c = 4$

41. $\log_2\left(\dfrac{1}{N}\right) = 5$ **42.** $\log_b 6 = -1$ **43.** $2\log_9 N = 1$ **44.** $\log_2 4^{-3} = x$

45. $\log_3 \frac{1}{27} = k$ **46.** $\log_{10}\left(\frac{1}{1000}\right)^c = 1$ **47.** $\dfrac{\log_2 16}{\log_2 8} = r$ **48.** $-3\log_b 2 = 1$

In Problems 49–60, use $\log_b 4 = 0.6021$ *and* $\log_b 5 = 0.6990$ *to evaluate the given logarithm.*

49. $\log_b 2$ **50.** $\log_b 20$ **51.** $\log_b 64$ **52.** $\log_b 625$

53. $\log_b \sqrt{40}$ **54.** $\log_b \frac{5}{4}$ **55.** $\log_b \sqrt[3]{5}$ **56.** $\log_b 80$

57. $\log_b 0.8$ **58.** $\log_b 2.5$ **59.** $\log_b \frac{1}{16}$ **60.** $\log_b(4^2 + 4^3)$

In Problems 61–66, simplify and write as one logarithm.

61. $\log_{10} 2 + \log_{10} 5$

62. $\frac{1}{2}\log_5 49 - \frac{1}{3}\log_5 8 + 13\log_5 1$

63. $\log_{10}(x^4 - 4) - \log_{10}(x^2 + 2)$

64. $\log_{10}\left(\dfrac{x}{y}\right) - 2\log_{10}x^3 + \log_{10}y^{-4}$

65. $\log_2 5 + \log_2 5^2 + \log_2 5^3 - \log_2 5^6$

66. $3\log_{10}5 + 2\log_{10}8 - 2\log_{10}4 - 4\log_{10}2 - \frac{1}{2}\log_{10}25$

In Problems 67–76, graph the given function.

67. $f(x) = \log_2 x$

68. $f(x) = \log_2 |x + 3|$

69. $f(x) = \log_2(x - 3)$

70. $f(x) = \log_2 \dfrac{1}{x}$

71. $f(x) = \log_2 \dfrac{1}{|x|}$

72. $f(x) = \log_2 2x$ [*Hint:* Apply rule I before graphing.]

73. $f(x) = \log_2 \sqrt{x}$ [*Hint:* Apply rule III before graphing.]

74. $f(x) = |\log_2 x|$

75. $f(x) = -1 + \log_2 x$

76. $f(x) = 1 - \log_2 x$

77. On the same axes, compare the graphs of

$$f(x) = \log_3 x, \quad f(x) = \log_4 x, \quad f(x) = \log_5 x.$$

78. Sketch the graph of $f(x) = \log_b x$ for $0 < b < 1$.

In Problems 79 and 80, the figures illustrate the shifted graph of the given function. Find an equation of the graph.

79. $f(x) = \log_{10} x$

80. $f(x) = \log_{10}|x|$

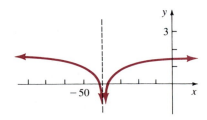

Figure 3.11   **Figure 3.12**

Review Problems

81. Sketch the graph of $f(x) = 2^{-3-x}$.

82. True or false: $e^x + e^{-x} = e^0$.

83. Consider the function $f(h) = \dfrac{e^{2+h} - e^2}{h}$. Compute $f(0.001)$ and $f(-0.0001)$.

84. Simplify the expression $\left(\dfrac{e^x + e^{-x}}{2}\right)^2 - \left(\dfrac{e^x - e^{-x}}{2}\right)^2$.

3.3 Common and Natural Logarithms

Logarithms utilizing the bases $b = 10$ and the irrational number $b = e$ have been in use since the sixteenth century. Since the number 10 is the base of our number system, we saw in Chapter 0 that any real number N could be expressed as a multiple of a power of 10: $N = n \times 10^c$. Because of this ability to express numbers in scientific notation, rules I–III of Section 3.2 make it possible to use the base 10 logarithm as an aid in performing arithmetic calculations involving complicated products, quotients, and powers. These time-consuming calculations involve tables and a correcting procedure known as linear interpolation. The innovation of the scientific calculator has single-handedly destroyed the need for these tedious hand calculations. For example, to calculate the number

$$\sqrt{1776}$$

with a typical scientific calculator, we simply enter 1776 and press the $\boxed{\sqrt{}}$ key to obtain the answer correct to eight decimal places:

$$42.1426\,1501$$

The reader should not jump to the conclusion from this introductory remark that logarithms are no longer important. They are extremely important! But in this chapter we can show you only a few of the many applications of logarithms; the true importance of logarithms will be seen as you advance through courses in calculus, science, and engineering.

Common and Natural Logarithms

Logarithms with the base 10 are called **common logarithms**, and logarithms with the base e are called **natural logarithms**. It is usual practice to write the natural logarithm

$$\log_e x \quad \text{as} \quad \ln x.$$

The symbol $\ln x$ is read either "the natural logarithm of x" or phonetically as "ell-en of x." Bear in mind that all the properties and rules of logarithms discussed in Section

3.2 pertain to the natural logarithm as well. In particular, we note that Definition 3.2 and equations (1), (2), and (3) of Section 3.2 with base $b = e$ give, in turn,

- $y = \ln x$ if and only if $x = e^y$
- $\ln e = 1$
- $\ln 1 = 0$
- $x = e^{\ln x}$.

Also, since $e > 1$, the graph of the natural logarithmic function $f(x) = \ln x$ is similar to the graph of $f(x) = \log_b x$ given in Figure 3.7. In Figure 3.13 we have compared the graph of $y = \ln x$ with its inverse function $y = e^x$.

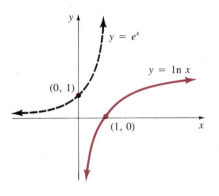

Figure 3.13

To compute the common logarithm of a positive number x we use the $\boxed{\log}$ key on a scientific calculator; the natural logarithm of x is obtained by using the $\boxed{\ln}$ key. Values of common and natural logarithms are also tabulated in Tables I and II respectively.

Example 1   Use a calculator to find the values of

$$\textbf{a. } \log_{10} 257.48 \qquad \textbf{b. } \log_{10} 0.0000624 \qquad \textbf{c. } \ln 5932$$

Solution   **a.** After entering the number 257.48 in the calculator, we press the $\boxed{\log}$ key:

$$\log_{10} 257.48 \approx 2.4107.$$

b. Proceeding as in part a we use the $\boxed{\log}$ key:

$$\log_{10} 0.0000624 \approx -4.2048.$$

c. In this last case we enter 5932 in the calculator and press the $\boxed{\ln}$ key:

$$\ln 5932 \approx 8.6881.$$

■

Example 2   Solve for x: **a.** $\log_{10}x = 1.65$ **b.** $\ln x = 3.2$.

Solution   **a.** We know that $\log_{10}x = 1.65$ is equivalent to $x = 10^{1.65}$. Some calculators may have a $\boxed{10^x}$ key, in which case simply enter 1.65 and press that key. For many calculators, however, enter 1.65 and press $\boxed{\text{INV}}$ $\boxed{\log}$ in succession. We find

$$x \approx 44.6684.$$

b. The logarithmic statement $\ln x = 3.2$ is equivalent to $x = e^{3.2}$. Using either $\boxed{e^x}$ or $\boxed{\text{INV}}$ $\boxed{\ln}$ it follows that

$$x \approx 24.5325. \qquad\blacksquare$$

Example 3   Find the value of $\log_4 70$.

Solution   There is no direct way of computing this number, since calculators only utilize common or natural logarithms. However, if we let x denote the value of $\log_4 70$, then we know that the statement

$$x = \log_4 70 \quad \text{is equivalent to} \quad 4^x = 70.$$

If we now take the common logarithm of both sides of the last equation, then

$$\log_{10}4^x = \log_{10}70$$
$$x \log_{10}4 = \log_{10}70$$
$$x = \frac{\log_{10}70}{\log_{10}4}.$$

Dividing logarithms gives

$$\log_4 70 = \frac{\log_{10}70}{\log_{10}4} \approx 3.0646. \qquad\blacksquare$$

In Example 3, it is important to note that in the division $(\log_{10}70)/\log_{10}4$ we computed $\log_{10}70$ and $\log_{10}4$, stored these values in the calculator, and then divided. Intermediate steps in a calculation should not, as a rule, be written down. For example, if we write

$$\frac{\log_{10}70}{\log_{10}4} \approx \frac{1.8451}{0.6021}$$

and then re-enter the numbers 1.8451 and 0.6021 into the calculator, the division turns out to be 3.0644. The difference in answers is due to round-off error.

Exercise 3.3

In Problems 1–6, rewrite the given exponential statement in the equivalent logarithmic form.

1. $10^{-3} = 0.001$

2. $10^6 = 1,000,000$

3. $10^{-0.6990} = \frac{1}{5}$

4. $10^{0.3010} = 2$

5. $e^{4.5539} = 95$

6. $e^{-1} = 0.3679$

In Problems 7–12, find the value of the given logarithm. Do not use a calculator.

7. $\log_{10}(0.000001)$

8. $\log_{10}(100)^2$

9. $\ln e^5$

10. $\ln \dfrac{1}{e^9}$

11. $\log_{10}\sqrt{10\sqrt{10}}$

12. $\ln(\ln e)$

Calculator Problems

In Problems 13–18, find the value of the given logarithm.

13. $\ln 0.062$

14. $\ln 845$

15. $\ln(\log_{10}6853)$

16. $\log_{10}5.446$

17. $\log_{10}\dfrac{735}{41}$

18. $\log_{10}(6.52)^3$

In Problems 19–24, solve for x.

19. $\log_{10}x = 2.2$

20. $\log_{10}x = -1.8$

21. $\log_{10}x = 0.3$

22. $\ln x = 0.71$

23. $\ln x = 4.36$

24. $\ln x = -5$

In Problems 25–30, find the value of the given logarithm.

25. $\log_2 5$

26. $\log_7 65$

27. $\log_{12}900$

28. $\log_6 18$

29. $\log_9 0.09$

30. $\log_{5.2}148$

31. Use a calculator to fill in the table for the function

$$f(h) = \frac{\ln(2 + h) - \ln 2}{h}.$$

Round your answers to five decimal places.

h	$f(h)$
0.1	
0.01	
0.001	
0.0001	
0.00001	

32. From inspection of Problem 31, conjecture the value of $\lim_{h \to 0} \dfrac{\ln(2 + h) - \ln 2}{h}$.

33. Use the fact that $y = \ln x$ is equivalent to $x = e^y$ to prove

a. $\ln x = \dfrac{\log_{10}x}{\log_{10}e}$

b. $\ln x \approx (2.3026)\log_{10}x$.

34. The following approximation is established in calculus:

$$\ln(1 + x) \approx x - \frac{x^2}{2} + \frac{x^3}{3} - \frac{x^4}{4}, \quad -1 < x \leq 1.$$

Use this result to approximate

a. $\ln 0.75$ **b.** $\ln 1.5$ **c.** $\ln 2$

Review Problems

---

35. Rewrite the statement $v = \log_u w$ as an exponential statement.

36. True or false: $\log_{1/b} x = -\log_b x$.

37. Simplify and write as one logarithm:

$$4 \ln x - \tfrac{1}{2}[\ln(x + 5) + \ln(x - 5)].$$

38. a. Graph the function $f(x) = e^x(e^x - 1)$.
   b. Use part a to graph $f(x) = e^{2x} - e^x + 2$.

3.4 Exponential and Logarithmic Equations

Exponential Equations

Recall from Example 1 of Section 3.1 that the exponential function

$$P(t) = 1000(\tfrac{3}{2})^t$$

described the population in a certain community. For $t = 1$, we saw that the population was $P(1) = 1500$. Suppose we now turn the problem around; that is, for a given value of P we find the corresponding value of t. For example, if $P = 2250$, then we must solve the **exponential equation**

$$2250 = 1000(\tfrac{3}{2})^t \quad \text{or} \quad (\tfrac{3}{2})^t = 2.25$$

for the unknown t. Rewriting this in the equivalent logarithmic form, we have

$$t = \log_{3/2} 2.25$$
$$= \log_{3/2} \tfrac{9}{4}$$
$$= \log_{3/2} (\tfrac{3}{2})^2$$
$$= 2 \log_{3/2} (\tfrac{3}{2})$$
$$= 2.$$

Example 1   Solve for x: $4^{x^2-1} = \frac{1}{2}$.

Solution   If we write the given equation in logarithmic form, we find

$$x^2 - 1 = \log_4 \tfrac{1}{2}$$
$$x^2 - 1 = -\tfrac{1}{2}$$
$$x^2 = \tfrac{1}{2}$$
$$x = \pm\sqrt{\tfrac{1}{2}}$$

Check: $4^{(\sqrt{1/2})^2-1} = 4^{-1/2} = \frac{1}{2}$ and $4^{(-\sqrt{1/2})^2-1} = 4^{-1/2} = \frac{1}{2}$.   ■

Example 2   Solve: $6^x = 27$.

Solution   Even though the given equation can be written immediately in the equivalent logarithmic form $x = \log_6 27$, this latter answer is of no particular help. Instead we proceed as in Example 3 of Section 3.3. By taking the common logarithm of both sides of the equation $6^x = 27$ and using the properties of logarithms, it follows that

$$\log_{10} 6^x = \log_{10} 27$$
$$x \log_{10} 6 = \log_{10} 27$$
$$x = \frac{\log_{10} 27}{\log_{10} 6}.$$

Computing and then dividing the logarithms yields $x \approx 1.8394$. In other words, $6^{1.8394} = 27$. Verify this using the $\boxed{y^x}$ key on a calculator.

Alternative Solution   We leave it to the reader to verify that

$$\ln 6^x = \ln 27$$
$$x = \frac{\ln 27}{\ln 6}$$

gives the same value for x.   ■

Example 3   Solve for x: $5^x - 5^{-x} = 2$.

Solution   We first multiply the equation by 5^x:

$$(5^x)^2 - 1 = 2(5^x).$$

We write the result as

$$(5^x)^2 - 2(5^x) - 1 = 0.$$

Since this last equation can be interpreted as a quadratic equation in 5^x, we use the quadratic formula. Solving for 5^x then gives

$$5^x = \frac{2 \pm \sqrt{4+4}}{2}$$

$$= 1 \pm \sqrt{2}.$$

But 5^x is always positive so we must reject the solution containing the negative square root, since $1 - \sqrt{2} < 0$. Hence

$$5^x = 1 + \sqrt{2}.$$

For variety let us take the natural logarithm of both sides of the last equation:

$$\ln 5^x = \ln(1 + \sqrt{2})$$

$$x \ln 5 = \ln(1 + \sqrt{2})$$

$$x = \frac{\ln(1 + \sqrt{2})}{\ln 5}$$

Computing and then dividing the logarithms gives

$$x \approx 0.5476.$$

∎

Since $f(x) = b^x$ is a one-to-one function, an equation

$$b^{x_1} = b^{x_2} \quad \text{implies} \quad x_1 = x_2. \tag{1}$$

Example 4   Solve for x: $(\frac{1}{3})^x = 9^{1-2x}$.

Solution   Write the given equation as

$$(3^{-1})^x = (3^2)^{(1-2x)}$$

$$3^{-x} = 3^{2-4x}.$$

From (1) it follows that the exponents of the same base must be equal:

$$-x = 2 - 4x$$

$$3x = 2$$

$$x = \tfrac{2}{3}.$$

∎

Logarithmic Equations

By using the equivalence of the statements $x = b^y$ and $y = \log_b x$, we can also solve certain equations involving logarithms.

Example 5   Solve for x: $\log_{10}(2x + 50) = 2$.

Solution   The equation can be written equivalently as

$$2x + 50 = 10^2$$

or $$2x + 50 = 100$$

$$2x = 50$$

$$x = 25.$$

The reader should check this solution.   ∎

Example 6   Solve for x: $\log_2 x + \log_2(x - 2) = 3$.

Solution   By rule I of Section 3.2 we can write the sum of logarithms as one logarithm:

$$\log_2 x(x - 2) = 3.$$

Hence we have $$x(x - 2) = 2^3$$

$$x^2 - 2x = 8$$

$$x^2 - 2x - 8 = 0$$

$$(x - 4)(x + 2) = 0.$$

We conclude that $x = 4$ or $x = -2$. However, $x = -2$ must be ruled out since the term $\log_2 x$ in the given equation is not defined for $x \leq 0$. Thus $x = 4$ is the only solution.

Check: $$\log_2 4 + \log_2 2 = \log_2 2^2 + \log_2 2$$

$$= 2\log_2 2 + \log_2 2$$

$$= 2 + 1 = 3.$$
      ∎

Example 7   Solve for x: $\log_3(7 - x) - \log_3(1 - x) = 1$.

Solution   By rule II of Section 3.2 we can write the given equation as

$$\log_3 \frac{7 - x}{1 - x} = 1$$

so that $$\frac{7 - x}{1 - x} = 3^1.$$

We multiply the latter equation by $1 - x$ and solve for x:

$$7 - x = 3(1 - x)$$

$$7 - x = 3 - 3x$$

$$2x = -4$$

$$x = -2.$$

Now at this point we should *not* conclude that $x = -2$ is not a solution of the given equation simply because it is negative. Indeed, $x = -2$ *is* a solution since

$$\log_3(7 - (-2)) - \log_3(1 - (-2)) = \log_3 9 - \log_3 3$$

$$= \log_3 \tfrac{9}{3}$$

$$= \log_3 3 = 1. \quad\blacksquare$$

Recall that $f(x) = \log_b x$ is also a one-to-one function. Thus,

$$\log_b x_1 = \log_b x_2 \quad \text{implies} \quad x_1 = x_2. \tag{2}$$

Example 8   Solve for x: $2 \log_2 x + 3 \log_2 2 = 3 \log_2 x - \log_2 \tfrac{1}{32}$.

Solution   First, we write each side of the equation as one logarithm:

$$\log_2 x^2 + \log_2 2^3 = \log_2 x^3 - \log_2 \tfrac{1}{32}$$

$$\log_2 2^3 x^2 = \log_2 \frac{x^3}{\frac{1}{32}}$$

$$\log_2 8x^2 = \log_2 32x^3.$$

It then follows from (2) that

$$8x^2 = 32x^3$$

$$8x^2(1 - 4x) = 0,$$

and so $x = 0$ or $x = \tfrac{1}{4}$. But $\log_2 x$ is not defined for $x = 0$; therefore the only possible solution is $x = \tfrac{1}{4}$. We leave it to the reader to check that this is a solution. $\quad\blacksquare$

Exercise 3.4

In Problems 1–24, solve each exponential equation.

1. $5^{x-2} = 1$

2. $3^x = 27^{x^2}$

3. $10^{-2x} = \frac{1}{10,000}$

4. $2^x + 2^{-x} = 2$

5. $2^x + 2^{-x} = 3$

6. $5 - 10^{-t} = 0$

7. $2^{x^2} = 8^{2x-3}$

8. $7^{-x} = 9$

9. $\tfrac{1}{4}(10^{-2x}) - 25(10^x) = 0$

10. $9^x = \dfrac{9^{2x-1}}{3^x}$

11. $4^{\log_2 x} = 9$

12. $5^{|x|-1} = 25$

13. $(\frac{1}{2})^{-x+2} = 8(2^{x-1})^3$

14. $6^{2t} + 6^t - 6 = 0$

15. $\frac{1}{3} = (2^{|x|-2} - 1)^{-1}$

16. $(2 - 2^{|x|})^2 = 4$

17. $3 \cdot 2^{2x} + 4^x = \frac{1}{8}$

18. $(\frac{2}{3})^x = 3.375$

19. $10^{x+1} = 21$

20. $13^{x^2-1} = 169$

21. $2^x \cdot 3^x = 36$

22. $9^{x/2} = 243$

23. $2^{x^2-6x} = \frac{1}{32}$

24. $(-7^x)^5 + 49 = 0$

In Problems 25–48, solve each logarithmic equation.

25. $2 \log_4 x = \log_4 25$

26. $\log_{10} x - \log_{10} 6 = 0$

27. $\log_2 x = 4$

28. $\log_5 |x| - 3 \log_5 3 = 0$

29. $\log_{10} \dfrac{1}{x^2} = 2$

30. $\log_{10} x = 1 + \log_{10} \sqrt{x}$

31. $\log_3 \sqrt{x^2 + 17} = 2$

32. $\log_2(x - 3) - \log_2(2x + 1) = -\log_2 4$

33. $\log_2(\log_3 x) = 2$

34. $\log_{10} 54 - \log_{10} 2 = 2 \log_{10} x - \log_{10} \sqrt{x}$

35. $\ln x = \ln 6 + \ln 10$

36. $\log_2 x + \log_2(10 - x) = 4$

37. $\log_6 2x - \log_6(x + 1) = 0$

38. $\log_5 |1 - x| = 1$

39. $\dfrac{\log_2 8^x}{\log_2 \frac{1}{4}} = \frac{1}{2}$

40. $\log_8 x + \log_8 x^2 = 1$

41. $\log_9 \sqrt{10x + 5} - \frac{1}{2} = \log_9 \sqrt{x + 1}$

42. $(\log_{10} x)^2 + \log_{10} x = 2$

43. $\log_{10} x^2 + \log_{10} x^3 + \log_{10} x^4 - \log_{10} x^5 = \log_{10} 16$

44. $\log_5 \left(\dfrac{1}{x} - \dfrac{1}{x + 1} \right) = -\log_5 30$

45. $\ln(4 + 2t) = \ln(5t - 7)$

46. $(\ln x)^2 = \ln x^2$

47. $x^{\ln x} = e^{16}$

48. $x^{\log_{10} x} = \dfrac{10,000}{x^3}$

In Problems 49 and 50, solve the simultaneous equations.

49. $5^{x-y} = 125$
$2y - x = 6$

50. $y = x^2 - x - 1$
$\log_2(x + y) = \log_2 x$

Review Problems

51. If $\log_b 5 = 2$ and $\log_b 10 = 6$, find $\log_b 50$, $\log_b \frac{1}{2}$, and $\log_b 100$.

52. Find the value of $\log_5(5 \cdot 5^2 \cdot 5^3)$.

53. Write $\ln 1 = 0$ as an equivalent exponential statement.

54. True or false: The values of $f(x) = (\frac{2}{3})^x$ decrease as x increases.

55. The graph of the function $P(t) = e^{a/b} e^{-ce^{-bt}}$ is called a **Gompertz curve** and is encountered in population studies and in actuarial predictions. Solve for t in terms of the other symbols.

3.5 Applications

In this section we are going to consider some applications of logarithmic and exponential functions.

pH of a Solution

The **pH**, or hydrogen potential, of a solution is defined by

$$pH = -\log_{10}[H^+] \tag{1}$$

where $[H^+]$ is the concentration of hydrogen ions in an aqueous solution in moles per liter. When $0 < pH < 7$ the solution is said to be *acid*; for $pH > 7$ the solution is *base* or *alkaline*; for $pH = 7$ the solution is *neutral* (for example, water). A strongly acid solution such as lemon juice has a pH in the range $pH \leq 3$. The average value of the pH of human urine is 6. Shellfish, such as mussels, die when the pH of the water is about 6. On the other hand, some varieties of lake trout can survive in water that has a pH level of about 4.5. Acid rain, caused in some parts of the country by coal-fired power plants, ranges in pH levels between 4 and 5.

Example 1 In a healthy person it is found that the concentration of hydrogen ions in blood is $[H^+] = 3.98 \times 10^{-8}$ moles/liter. Determine the pH of blood.

Solution From (1) we find that the pH of blood is given by

$$pH = -\log_{10}(3.98 \times 10^{-8})$$
$$= -(\log_{10}3.98 + \log_{10}10^{-8})$$
$$= -(\log_{10}3.98 - 8)$$
$$\approx -(0.5999 - 8)$$
$$\approx 7.4$$

Severe illness, or even death, can result when a person's blood pH falls outside the narrow limits $7.2 \leq pH \leq 7.6$. We note that values of pH are usually given to the nearest tenth of a unit.

Example 2 Determine the hydrogen-ion concentration $[H^+]$ of a solution with a pH of 5.2.

Solution Since $-\log_{10}[H^+] = \log_{10}\frac{1}{[H^+]}$ we have

$$\log_{10}\frac{1}{[H^+]} = 5.2 \quad \text{or} \quad \frac{1}{[H^+]} = 10^{5.2}.$$

With the aid of a calculator we find

$$\frac{1}{[H^+]} \approx 1.6 \times 10^5$$

$$[H^+] = \frac{1}{1.6 \times 10^5} \approx 6 \times 10^{-6} \text{ moles/liter.}$$ ∎

The Environment

The **intensity level** b of sound measured in decibels (dB)* is given by

$$b = 10 \log_{10} \frac{I}{I_0}, \tag{2}$$

where I is the intensity of the sound (measured in watts/cm^2) and I_0 is a reference intensity of 10^{-16} watt/cm^2 corresponding to the faintest sound that can be heard. When $I = I_0$ then (2) gives $b = 0$ dB. The intensity levels of frequently occurring sounds are given in the accompanying table.

Source	Intensity level (dB)
Threshold of hearing	0
Whisper	20
Normal talking	40–60
Some TV commercials	65
Smoke detector alarm	70
Jet airplane taking off	80–100
Threshold of pain	120

Example 3 Determine the intensity level of a sound having intensity 10^{-4} watt/cm^2.

Solution From (2) we see that the intensity level is

$$b = 10 \log_{10} \frac{10^{-4}}{10^{-16}} = 10 \log_{10} 10^{12}$$

$$= 120 \text{ dB.}$$ ∎

As indicated in the table, sound at an intensity level of around 120 dB can cause pain. Prolonged exposure to levels around 90 dB (easily produced by rock and roll groups) can cause temporary deafness.

---

* The decibel is (1/10) bell. This latter unit, named for Alexander Graham Bell (1847–1922), proved to be too large in practice.

Equation (2) can be used to obtain the intensity level b_2 of a sound at a distance d_2 from the source if one level b_1 is measured at a distance d_1 as follows. We utilize a principle from physics that states that the intensity of sound I is inversely proportional to the square of the distance d from the source:

$$I = \frac{k}{d^2}, \tag{3}$$

where k is the intensity at a unit distance from the source. Then substituting (3) into (2) and using rules II and III of logarithms gives

$$b = 10 \log_{10} \frac{k/d^2}{I_0}$$

$$= 10 \log_{10} \frac{k/I_0}{d^2}$$

$$= 10 \left(\log_{10} \frac{k}{I_0} - \log_{10} d^2 \right)$$

$$= 10 \left(\log_{10} \frac{k}{I_0} - 2 \log_{10} d \right).$$

Hence at distances d_2 and d_1, we find

$$b_2 = 10 \left(\log_{10} \frac{k}{I_0} - 2 \log_{10} d_2 \right) \tag{4}$$

and

$$b_1 = 10 \left(\log_{10} \frac{k}{I_0} - 2 \log_{10} d_1 \right). \tag{5}$$

Subtracting (5) from (4) then yields

$$b_2 - b_1 = 10(2 \log_{10} d_1 - 2 \log_{10} d_2)$$

$$= 20 \log_{10} \frac{d_1}{d_2}$$

or

$$\boldsymbol{b_2 = b_1 + 20 \log_{10} \frac{d_1}{d_2}.} \tag{6}$$

Airport Noise Reduction

In an experimental landing approach to an airport, called the two-segment approach, a plane starts a 6° glide path when it is 5.5 miles (measured horizontally) from the runway. The plane then switches to a 3° glide path when it is 1.5 miles out. The normal approach consists entirely of a 3° glide path. The purpose of the two-segment approach is of course noise reduction. See Figure 5.11 on page 293. At 5.5 miles from the field a plane P_1 using the normal landing approach has an altitude of 1522 feet, whereas a plane P_2 using the experimental glide path has an altitude of 2635 feet. Equation (3) implies that the sound intensity I of P_2 at ground level 5.5 miles from the airport is one-third that of P_1. (See Problem 26.) However, as the following example shows, this does not mean that the intensity level of P_2 is one-third the intensity level of P_1.

Example 4 At ground level the intensity level b_1 of plane P_1 5.5 miles out from the runway is found to be 80 dB. Determine the intensity level b_2 of plane P_2 at the same point.

Solution Identifying $d_1 = 1522$ and $d_2 = 2635$, it follows from (6) that

$$b_2 = 80 + 20 \log_{10} \frac{1522}{2635}$$

$$\approx 80 + 20 \log_{10} 0.58$$

$$\approx 80 - 4.732$$

$$\approx 75 \text{ dB.} \qquad \blacksquare$$

Earthquakes

The magnitude R of an earthquake of intensity I measured on the Richter scale is given by

$$R = \log_{10} \frac{I}{I_0} \qquad (7)$$

where I_0 is a reference intensity corresponding to $R = 0$.

Example 5 The San Francisco earthquake of 1906 had an estimated intensity of 8.25 on the Richter scale. How much greater was the intensity of the 1906 earthquake compared with the 1985 Mexico City earthquake of magnitude 7.8?

Solution Let I_{sf} denote the intensity of the 1906 San Francisco earthquake and let I_{mc} denote the intensity of the 1985 Mexico City earthquake. Then by (7) we have

$$\log_{10} \frac{I_{sf}}{I_0} = 8.25 \quad \text{and} \quad \log_{10} \frac{I_{mc}}{I_0} = 7.8.$$

These two equations are, in turn, equivalent to

$$\frac{I_{sf}}{I_0} = 10^{8.25} \quad \text{and} \quad \frac{I_{mc}}{I_0} = 10^{7.8}.$$

Since $8.25 = 0.45 + 7.8$, by the laws of exponents we can write

$$\frac{I_{sf}}{I_0} = 10^{8.25} = 10^{0.45}(10^{7.8}) = 10^{0.45}\left(\frac{I_{mc}}{I_0}\right).$$

Employing the $\boxed{10^x}$ key on a calculator (or the $\boxed{\text{INV}}$ $\boxed{\text{log}}$ keys) to compute $10^{0.45}$ it follows that

$$\frac{I_{sf}}{I_0} \approx (2.8)\frac{I_{mc}}{I_0} \quad \text{or} \quad I_{sf} \approx (2.8)I_{mc}.$$

In other words, the 1906 San Francisco earthquake was approximately 3 times as strong as the 1985 Mexico City earthquake. $\qquad \blacksquare$

The exponential function $y = e^x$ and the natural logarithm function $y = \ln x$ occur frequently in the analysis of problems involving population growth, radioactive decay, electrical circuits, temperature distributions, spread of diseases, growth of technology, and compounding of interest.

Continuous Compound Interest

When P dollars are deposited in a savings bank account with an annual rate of interest r compounded n times a year for t years, then the return S, or future value of P, is given by

$$S = P\left(1 + \frac{r}{n}\right)^{nt}. \tag{8}$$

Interest can be compounded semiannually ($n = 2$), quarterly ($n = 4$), daily ($n = 365$), and so on. If n is increased without bound ($n \to \infty$), then we say the interest is **compounded continuously**. Using calculus it can be shown that the return S on an investment of P dollars is given by

$$S = Pe^{rt} \tag{9}$$

when interest is compounded continuously for t years at a rate r.

Example 6

An amount of $5000 is deposited into a savings account paying 6% annual interest compounded continuously. What is the return after 10 years?

Solution

From (9) we have

$$S = 5000e^{(0.06)10}$$

$$= 5000e^{0.6}.$$

Using a calculator and the $\boxed{e^x}$ key (or $\boxed{\text{INV}}$ $\boxed{\ln}$) we find

$$S \approx 5000(1.8221)$$

$$\approx \$9110. \qquad\blacksquare$$

Population Growth

In biology it is often observed that the population $P(t)$ of small animals and bacteria can be predicted over short periods of time t by the function

$$P(t) = P_0 e^{kt} \tag{10}$$

where $k > 0$ is a constant and P_0 is the initial population.

Example 7

A culture initially has number of bacteria P_0. After t hours it is predicted that the number of bacteria will be $P(t) = P_0 e^{0.45t}$. Determine the number of hours in which the number of bacteria have tripled.

Solution When $P(t) = 3P_0$ we must find the value of t for which

$$3P_0 = P_0 e^{0.45t} \quad \text{or} \quad 3 = e^{0.45t}.$$

The equation at the right implies that

$$0.45t = \ln 3$$

$$t = \frac{\ln 3}{0.45} \approx 2.4 \text{ hours.} \qquad \blacksquare$$

Thomas R. Malthus (1776–1834), an English clergyman and economist, predicted that the world population would grow exponentially according to equation (10) and eventually surpass the ability of the world to produce the necessary food. The world population growth during the 18th and 19th centuries was remarkably close to the values given by equation (10) for appropriate choices of P_0 and k. Due to the growth of technology, however, the predicted food crisis did not occur.

Logistic Function When influences such as deaths due to predators or disease are taken into consideration, the population of certain species of animals is given by the **logistic function**

$$P(t) = \frac{aP_0 e^{at}}{bP_0 e^{at} + (a - bP_0)} \tag{11}$$

where a and b are constants. A population characterized by (11) exhibits limited, or bounded, growth.

Concentration of a Drug in the Blood Under some conditions the concentration C of a drug in the bloodstream at any time t after injection is given by

$$C(t) = \frac{a}{b} + \left(C_0 - \frac{a}{b}\right)e^{-bt} \tag{12}$$

where C_0 is the concentration of the drug at $t = 0$ and a and b are positive constants. Notice that as t increases the term e^{-bt} decreases. Hence, for large values of time the concentration C is close to the value a/b. This latter number is known as the **steady-state** concentration. (See Problem 47.)

Series Circuit As the next example shows, the number e appears in the analysis of the current in a **series circuit**.

Example 8 For a simple series circuit consisting of a constant voltage E, an inductance of L henries, and a resistance of R ohms (Figure 3.14), it can be shown that the current I at any time t is given by

$$I = \frac{E}{R}[1 - e^{-(R/L)t}]. \tag{13}$$

Solve for t in terms of I.

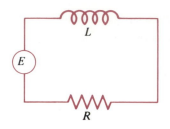

Figure 3.14

Solution   Using algebra, we find from (13)

$$\frac{IR}{E} = 1 - e^{-(R/L)t}$$

$$e^{-(R/L)t} = 1 - \frac{IR}{E}$$

$$-\left(\frac{R}{L}\right)t = \ln\left(1 - \frac{IR}{E}\right)$$

$$t = -\frac{L}{R}\ln\left(1 - \frac{IR}{E}\right).$$

Exercise 3.5

In Problems 1–6, determine the pH *of a solution with the given hydrogen-ion concentration* $[H^+]$.

1. $[H^+] = 10^{-6}$ **2.** $[H^+] = 4.0 \times 10^{-7}$ **3.** $[H^+] = 2.8 \times 10^{-8}$

4. $[H^+] = 7.6 \times 10^{-3}$ **5.** $[H^+] = 6.9 \times 10^{-7}$ **6.** $[H^+] = 5.1 \times 10^{-5}$

In Problems 7–12, determine the hydrogen-ion concentration $[H^+]$ *of a solution with the given* pH.

7. pH $= 3.3$ **8.** pH $= 4.6$ **9.** pH $= 6.6$

10. pH $= 7.3$ **11.** pH $= 8.1$ **12.** pH $= 5.9$

In Problems 13 and 14, determine how many more times acidic the first substance is compared to the second substance.

13. lemon juice, pH $= 2.3$ **14.** battery acid, pH $= 1$
      vinegar, pH $= 3.3$ lye, pH $= 13$

In Problems 15–20, determine the intensity level of a sound with the given intensity.

15. 10^{-7} **16.** 10^{-16} **17.** 10^{-10}

18. 5×10^{-6} **19.** 3×10^{-9} **20.** 2×10^{-6}

21. Show that the intensity level (2) can be expressed as $b = 160 + 10 \log_{10} I$.

22. The concentration of hydrogen ions in stomach fluid is given by $[H^+] = 9.3 \times 10^{-3}$. Determine the pH of the fluid.

23. A particular brand of beer has a pH of 4.4. Determine the concentration of hydrogen ions $[H^+]$.

24. At a distance of 10 meters the intensity level of a jet airplane engine is measured as 130 dB. What is the intensity level 500 meters from the engine?

25. The intensity level of normal conversation is 40 dB when measured 3 feet from the source. What is the intensity level 12 feet from the source?

26. Using equation (3), show that at ground level the sound intensity I of a plane P_2 at an altitude of 2635 feet is approximately one-third that of a plane P_1 at 1522 feet.

27. In 1979 an earthquake occurred in San Francisco of magnitude 5.95 on the Richter scale. Compared to this earthquake, how much greater was the intensity of the 1906 earthquake of magnitude 8.25? How much greater was the intensity of the 1906 San Francisco earthquake compared to the 1971 San Fernando Valley earthquake of magnitude 6.6?

28. Two of the most powerful earthquakes in recent times were the 1933 Japan earthquake of magnitude 8.9 and the 1964 Alaska earthquake of magnitude 8.5. How much greater was the intensity of the Japan earthquake compared to the Alaska earthquake? How much greater were the intensities of the Japan and Alaska earthquakes compared to the 1906 San Francisco earthquake? How much greater was the intensity of the 1933 Japan earthquake compared to the 1985 Mexico City earthquake of magnitude 7.8?

In Problems 29–32, find the future value S of the given amount invested at the indicated rate of interest, compounded continuously, for the specified time.

29. $1000, 7% per year, 5 years

30. $2500, 6% per year, 20 years

31. $8000, 10% per year, 2.5 years

32. $7500, 9% per year, 10 years

In Problems 33–36, find the present value $P = Se^{-rt}$ of the given amount invested at the indicated rate of interest, compounded continuously, for the specified time.

33. $10,000, 7% per year, 20 years

34. $25,000, 12% per year, 5 years

35. $4000, 5.5% per year, 10 years

36. $5000, 6.25% per year, 40 years

37. $10,000 is deposited in a savings account paying 10% annual interest compounded daily. Use continuous compounding of interest as an approximation to daily compounding to compute the time necessary to quadruple the investment.

38. Use formulas (8) and (9) of this section along with a calculator to complete the comparative table for the future values of $P = 1000 with $r = 0.06$ over 4 years.

Method of compounding	n	Future value S
annually	1	
semiannually	2	
quarterly	4	
monthly	12	
weekly	52	
daily	365	
hourly	8760	
continuously	$n \to \infty$	

39. The population of a community in t years is predicted by formula (10). Show that the time required to double the population is $t = (\ln 2)/k$. This is the so-called **law of Malthus**.

40. In physics the amount x of radioactive material remaining after t years is given by $x(t) = x_0 e^{-kt}$, where x_0 is the amount of material present at $t = 0$ and k is a positive constant. If $x(t) = x_0 e^{-0.007t}$, determine the time at which $x = 0.03 x_0$.

41. If x_0 represents an initial amount of highly radioactive radium, then the quantity remaining at any subsequent time t is given by $x(t) = x_0 e^{-0.000418t}$. The **half-life** of a radioactive substance, or the time necessary for one-half of the substance to decay, is a measure of the stability of the substance. Use the given formula to determine the half-life of radium.

42. Use $x(t) = x_0 e^{-kt}$ to show that the half-life of a radioactive substance is given by $t = (t_2 - t_1)\ln 2/\ln(x_1/x_2)$, when it is known that $x_1 = x(t_1)$ and $x_2 = x(t_2)$, $t_1 < t_2$.

43. According to the **Bouguer-Lambert law**, the intensity I (in lumens) of a vertical beam of light of intensity I_0 passing through a transparent substance decreases according to the exponential function $I(x) = I_0 e^{-kx}$, $k > 0$, where x is depth measured in meters. If the intensity of light 1 meter below the surface of water is 30% of I_0, what is the intensity 3 meters below the surface?

44. The intensity I (in lumens) of a light beam after passing through x centimeters of a medium having an absorption coefficient of 0.1 is given by $I(x) = 1000 e^{-0.1x}$.
a. How many centimeters of the material would reduce the illumination to 800 lumens?
b. Find the thickness of the medium that would reduce the illumination to 600 lumens.

45. For the logistic function given in (11), solve for t in terms of the other symbols.

46. The number of students on a 1000-student campus infected with a flu virus at time t (in days) is given by $x(t) = 1000/(1 + 999 e^{0.99t})$.
a. What are $x(4)$, $x(5)$, $x(6)$?
b. Determine the time at which $x(t) = 950$.

47. From equation (12) determine the time at which $C(t)$ is one-half the steady-state concentration.

48. When a series of deposits P are made at equally spaced intervals of time, the savings plan is called an **annuity**. If r is the annual rate of interest compounded continuously, then after m deposits the amount accrued is given by

$$S = P + Pe^r + Pe^{2r} + \cdots + Pe^{(m-1)r}.$$

Determine the value of the annuity in 10 years if $P = \$2000$ and the annual rate of interest is 12%.

49. The formula

$$x(t) = c_1 e^{m_1 t} + c_2 e^{m_2 t}, \quad c_1 c_2 < 0, \quad m_1 < 0, \quad m_2 < 0,$$

represents the vertical displacement of a weight attached to the end of a spring when the entire system is immersed in a heavy fluid. If $c_1 = \frac{3}{2}$, $c_2 = -2$, $m_1 = -1$, $m_2 = -2$, find the value of the time t for which $x(t) = 0$.

50. According to **Newton's law of cooling**, the temperature T of a roast removed from an oven into a room at temperature $75°F$ is $T = 75 + 275 e^{-0.218t}$, where t is measured in minutes. What is the initial temperature of the roast (that is, at $t = 0$)? What is the temperature of the roast after 15 minutes? Find the time when the temperature of the roast is $110°F$.

Review Problems

51. Use the natural logarithm to solve the equation $y = \dfrac{e^x + e^{-x}}{2}$ for x in terms of y.

52. Solve for x: $e^{\ln(2x+1)} = 5$.

53. Solve for x: $\ln(2x-3) + \ln(x+1) = 2\ln x$.

54. Use a calculator to find the value of $\ln(\log_{10}(\log_{10} 7290))$.

55. If $\log_b 8 = 2.0794$, find the value of $\log_b \sqrt{8b^5} + \log_b \dfrac{64}{b^3}$.

56. Solve the inequality $2 \le \log_2 x \le 4$ and express the solution in interval form. [*Hint:* Use the fact that $f(x) = \log_2 x$ is increasing on $(0, \infty)$.]

Chapter Review

An **exponential function** is any function of the form

$$f(x) = b^x, \qquad b > 0, b \ne 1.$$

For $b > 1$, the functional values increase as x increases; for $0 < b < 1$, the functional values decrease as x increases.

A **logarithm** is just another name for **exponent**. The statement

$$y = \log_b x \quad \text{is equivalent to} \quad x = b^y.$$

There are six basic **properties** of logarithms:

$$b^{\log_b N} = N$$

$$\log_b b = 1$$

$$\log_b 1 = 0$$

$$\log_b MN = \log_b M + \log_b N$$

$$\log_b \frac{M}{N} = \log_b M - \log_b N$$

$$\log_b N^c = c \log_b N$$

The **logarithmic function** $y = \log_b x$ and the exponential function $y = b^x$ are inverse functions of each other. In the most general case there is no restriction on the base b other than $b \ne 1$ and $b > 0$; however, it is usually assumed that $b > 1$.

Logarithms with the base 10 are called **common logarithms**. In many applications the irrational number

$$e = 2.71828182\ldots$$

is used as a logarithmic base; $\log_e x = \ln x$ is called the **natural logarithm** of x.

Chapter 3 Review Exercises

In Problems 1–14, answer true or false.

1. The exponential function $f(x) = 2^{-x}$ is the same as $f(x) = (\frac{1}{2})^x$. _____

2. The range of the exponential function $f(x) = b^x$ is the set of real numbers $y \geq 0$. _____

3. $e^x e^{2x} e^{3x} = (e^x)^6$. _____

4. The graph of $f(x) = 9^{-x}$ goes down as x increases. _____

5. $3^{\log_3 5^2} = 25$. _____

6. $\log_b(M + N) = \log_b M + \log_b N$. _____

7. $\dfrac{\log_b M}{\log_b N} = \log_b M - \log_b N$. _____

8. The domain of $f(x) = \log_4 x$ is the set of real numbers $x \geq 0$. _____

9. $x = -6$ is a solution of $\log_{18}(3x + x^2) = 1$. _____

10. If $b^r = b^s$, $b > 0$, then $r = s$. _____

11. $x = -2$ is a vertical asymptote for the graph of $f(x) = \log_{10}(x + 2)$. _____

12. The graphs of $y = e^x$ and $y = \ln x$ are symmetric with respect to the line $y = x$. _____

13. If $5^{-1} = 0.2$ then $-1 = \log_5 2 - \log_5 10$. _____

14. The function $f(x) = \log_{10}|x|$ is one-to-one. _____

In Problems 15–24, fill in the blank.

15. If $9^{1/2} = 3$ then $\frac{1}{2} = \log_{\underline{\ \ }}$ _____.

16. If $\log_{10} 2 = 0.3010$ then $2 =$ _____.

17. $\log_{10}(40 + 60) =$ _____.

18. If $\log_4 16 = x^2$ then $x =$ _____.

19. If $\log_b 25 = 2$ then $b =$ _____.

20. If $\log_5 N = -3$ then $N =$ _____.

21. $\dfrac{\log_{10} 64}{\log_{10} 8} =$ _____.

22. $\log_7 N = 0$ implies $N =$ _____.

23. $5^{3\log_5 10} =$ _____.

24. $\log_9 \frac{1}{27} =$ _____.

25. If $f(x) = -4^{x+2}$, find $f(-2)$, $f(-\frac{3}{2})$, $f(-1)$, $f(0)$, $f(\frac{1}{2})$, and $f(2)$.

26. If $f(x) = \log_3 x$, find $f(\frac{1}{3})$, $f(9)$, $f(27)$, $f(\frac{1}{81})$, and $f(\sqrt{243})$.

In Problems 27–32, graph the given function.

27. $f(x) = 3^{-|x|}$

28. $f(x) = 1 + \frac{1}{2} \cdot 4^x$

29. $f(x) = 1 + e^{-x}$

30. $f(x) = (\frac{2}{3})^x$

31. $f(x) = 4 \cdot 2^x$

32. $f(x) = -5^{x+1}$

33. Find the domain of the given function.

a. $f(x) = \dfrac{2^{x-1}}{4^{x+1}}$

b. $f(x) = \dfrac{1}{2^x - 1}$

c. $f(x) = 2^{1/x}$

d. $f(x) = 3^{\sqrt{x+1}}$

e. $f(x) = 4^{-2x}$

f. $f(x) = e^x - e^{-x}$

34. Find the domain of the given function.

a. $f(x) = \log_2(x - 1)$

b. $f(x) = \log_5|x + 2|$

c. $f(x) = \log_{10}x^4$

35. Simplify and write as one logarithm:

$$\log_3 18 - 2\log_3 6 + \log_3 9 + \log_3 \tfrac{1}{3} - \log_3 81 + 3\log_3 4.$$

36. Simplify: $\log_2 \dfrac{(4)(256)}{(16)^2(8)^3}$.

In Problems 37–40, graph the given function.

37. $f(x) = \log_{10}\dfrac{1}{x}$

38. $f(x) = \log_{10}x^2$

39. $f(x) = \log_{10}|x + 10|$

40. $f(x) = 2 + \log_{10}x$

In Problems 41–46, solve for x.

41. $\log_4(8 + 2x) = \tfrac{1}{2}$

42. $\log_5(\log_3 2x) = 0$

43. $2^{2x-3} = 128$

44. $10^{x^2} = 100^{4(x-2)}$

45. $5^x = 3 \cdot 5^{-x}$

46. $4\log_2 x - \log_2(x + 2) = \log_2 x^2$

47. Find the value of $\log_7 19$.

48. Find the pH of a solution for which $[\mathrm{H}^+] = 3.8 \times 10^{-7}$.

49. A radioactive substance decays by radiation. After t hours the amount remaining is given by $x(t) = x_0 e^{-0.025t}$ where x_0 is the initial amount. Determine the time when 75% of the substance has decayed.

50. The absolute magnitudes M_A and M_B of two stars A and B are related to their absolute luminosities L_A and L_B by $M_B - M_A = 2.5\log_{10}(L_A/L_B)$. Calculate the absolute magnitude of the star Betelgeuse if it is known that the absolute magnitude of the sun is 4.7 and that the absolute luminosity of Betelgeuse is 10^5 times the absolute luminosity of the sun.

In Problems 51 and 52, find the slope of the line L in the given figure.

51. $f(x) = 3^{-x-1}$

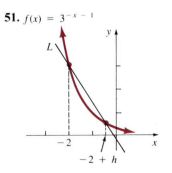

Figure 3.15

52.

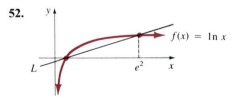

$f(x) = \ln x$

Figure 3.16

In Problems 53 and 54, find an equation of the indicated form for the exponential graph given in the figure.

53.

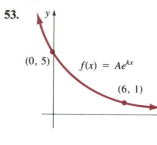

$(0, 5)$ $f(x) = Ae^{kx}$

$(6, 1)$

Figure 3.17

54.

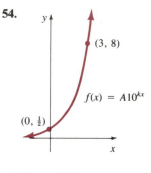

$(3, 8)$

$f(x) = A10^{kx}$

$(0, \frac{1}{2})$

Figure 3.18

55. The graph of $f(x) = a^x + 1$ is shown in Figure 3.19. Find a.

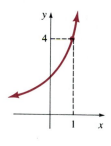

Figure 3.19

56. The graph of $f(x) = a + \log_3 x$ is shown in Figure 3.20. Find a.

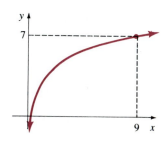

Figure 3.20

4 Trigonometric Functions

4.1 The Measurement of Angles

In this section we shall discuss two units of angular measure. In subsequent sections an important class of of functions, the trigonometric functions, will be considered. Historically, these functions were defined using angles in right triangles.

Angle

An **angle** is formed by two half-lines which have a common endpoint, called the **vertex**. For the study of trigonometry it is convenient to designate one half-line the **initial side** of the angle and the other the **terminal side**. Alternatively, we may consider the angle as having been formed by a rotation from the initial side to the terminal side. See Figure 4.1(a).

(a)   (b)

Figure 4.1

Standard Position

As shown in Figure 4.1(b), we can place the angle in a Cartesian coordinate plane with its vertex at the origin and initial side coinciding with the positive x axis. Such an angle is said to be in **standard position**. For the remainder of this section all angles are considered to be in standard position.

Degrees

Two units of measure commonly used for angles are degrees and radians. The first of these is based on the assignment of 360 **degrees** (written 360°) to the angle formed by one complete counterclockwise rotation. Other angles are then measured in terms of a 360° angle. If the rotation is counterclockwise, the measure will be *positive*; if clockwise, the measure will be *negative*. For example, if an angle is obtained by one-fourth of a complete *counterclockwise* rotation, it will be

$$\tfrac{1}{4}(360°) = 90°.$$

If the angle is obtained by one-fourth of a complete *clockwise* rotation, it will be −90°. [*Note:* An angle of 1° is formed by $\frac{1}{360}$ of a complete counterclockwise rotation.] In Figure 4.2 we have sketched angles of 90°, 180°, 270°, and −90°.

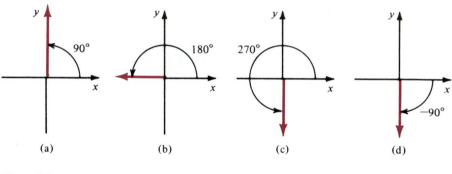

(a)   (b)   (c)   (d)

Figure 4.2

Example 1   Sketch an angle of 120°.

Solution   Since $120° = \tfrac{1}{3}(360°)$, this angle corresponds to $\tfrac{1}{3}$ of a complete counterclockwise rotation. This is shown in Figure 4.3.

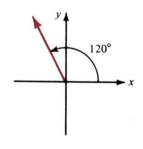

Figure 4.3   ∎

Example 2   Locate the terminal side of an angle of 960°.

Solution   Dividing 960 by 360, we obtain a quotient of 2 and a remainder of 240; that is,

$$960 = 2(360) + 240.$$

Thus, this angle is formed by making two complete counter-clockwise rotations before sweeping out

$$\frac{240}{360} = \frac{2}{3}$$

of another rotation. This is illustrated in Figure 4.4.

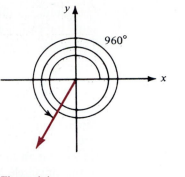

Figure 4.4 ∎

Minutes and Seconds

Traditionally, fractions of degrees have been represented using **minutes** and **seconds**. That is, 1 degree = 60 minutes. This is written

$$1° = 60'.$$

Similarly, 1 minute = 60 seconds,

$$1' = 60''.^*$$

Thus one-half of a 15° angle is 7°30'. (In this text we shall avoid using seconds.) Since most scientific calculators use decimal notation to represent fractions of degrees, the reader should be able to convert between minutes and decimals.

Example 3 Convert 42.23° to degrees and minutes.

Solution Since 0.23° represents $\frac{23}{100}$ of 1°, we may write

$$0.23° = (0.23)60'$$
$$= 13.8' \approx 14'.$$

Thus $42.23° \approx 42°14'$ ∎

Example 4 Convert 17°47' to decimal notation.

Solution Write 47' as

$$\frac{47}{60} \times 1°$$

and divide. Then, since $\frac{47}{60} \approx 0.78$, we have

$$17°47' \approx 17.78°.$$ ∎

---

* The use of the number 60 as a base dates back to the Babylonians. Another example of the use of this base in our culture is in the measurement of time (1 hour = 60 minutes).

Radians

In calculus, the most convenient unit of measure for an angle is the radian. We will see some indication of this in Section 4.6, where we graph the trigonometric functions. The radian measure of angle is based on the length of an arc on the **unit circle**

$$x^2 + y^2 = 1.$$

As seen in Figure 4.5(a) an angle in standard position can be viewed as having been formed by a half-line rotating from the positive x axis to the terminal side. Thus we can measure the angle by the distance t traversed along the circumference of the unit circle by the half-line. This unit of angular measure is called a **radian**. In Figure 4.5(b) we see that an angle of one radian subtends an arc whose length is equal to the radius of the unit circle. We refer to the angle shown in Figure 4.5(a) as an angle of t radians, or more simply as the angle t.

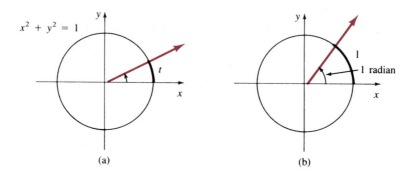

(a)   (b)

Figure 4.5

We have the same convention as before: an angle formed by a counterclockwise rotation is considered positive, whereas an angle formed by a clockwise rotation is negative. The radian measure of an angle formed by a single complete counterclockwise rotation is equal to the circumference of the unit circle. Thus, this is an angle of 2π radians, since the circumference of a circle of radius r is $2\pi r$ and $r = 1$ for a unit circle. In Figure 4.6 we illustrate angles of $\pi/2$, $-\pi/2$, π, and 3π radians, respectively.

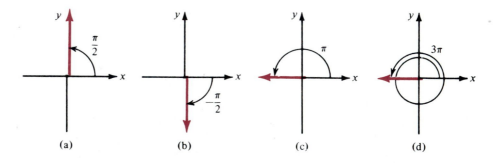

(a)   (b)   (c)   (d)

Figure 4.6

Note that the angle of $\pi/2$ radians shown in (a) is obtained by one fourth of a complete counterclockwise rotation; that is,

$$\frac{1}{4}(2\pi \text{ radians}) = \frac{\pi}{2} \text{ radians}.$$

The angle shown in (b), obtained by one fourth of a complete clockwise rotation, is $-\pi/2$ radians.

Coterminal Angles

From Figure 4.6(c) and (d) we see that the terminal side of 3π radians coincides with the terminal side of an angle of π radians. When two angles in standard position have the same terminal sides, we say they are **coterminal**. For example, the angles $\theta, \theta + 2\pi$, and $\theta - 2\pi$ shown in Figure 4.7 are coterminal. In fact, the addition of any integer multiple of 2π radians to a given angle results in a coterminal angle. Conversely, any two coterminal angles differ by an integer multiple of 2π radians.

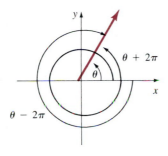

Figure 4.7

Example 5

Find the angle between 0 and 2π that is coterminal with the angle $-55\pi/6$.

Solution

We can repeatedly add 2π to $-35\pi/6$ until we obtain an angle between 0 and 2π:

$$-\frac{35\pi}{6} + 2\pi + 2\pi + 2\pi = \frac{\pi}{6}.$$

Thus angles of $\pi/6$ and $-35\pi/6$ radians are coterminal. ∎

Conversion Formulas

Many scientific calculators have a key which converts between degree and radian measure. There are also simple formulas which enable us to do this by hand. The degree measure of the angle corresponding to one complete counterclockwise rotation is $360°$, while the radian measure of the same angle is 2π radians. Thus we have

$$360° = 2\pi \text{ radians,*}$$

or

$$180° = \pi \text{ radians.} \tag{1}$$

From (1) we obtain

$$1° = \frac{\pi}{180}\text{radians} \tag{2}$$

$$1 \text{ radian} = \left(\frac{180}{\pi}\right)° \tag{3}$$

Equation (2) enables us to convert from degrees to radians, whereas equation (3) allows us to convert from radians to degrees.

Example 6   Convert $20°$ to radians.

Solution   From (2) we see that there are $\pi/180$ radians in 1 degree. Therefore in 20 degrees there are $20 \times \pi/180$ radians. Thus

$$20° = 20\left(\frac{\pi}{180}\right)\text{radians} = \frac{\pi}{9}\text{radians.} \qquad \blacksquare$$

Example 7   Convert $7\pi/6$ radians to degrees.

Solution   From (3) the number of degrees in 1 radian is $180/\pi$. Thus in $7\pi/6$ radians there are $7\pi/6 \times 180/\pi$ degrees so that

$$\frac{7\pi}{6}\text{radians} = \frac{7\pi}{6}\left(\frac{180}{\pi}\right)° = 210°. \qquad \blacksquare$$

Example 8   Convert 2 radians to degrees.

Solution   From (3) we have

$$2 \text{ radians} = 2\left(\frac{180}{\pi}\right)°$$

$$= \left(\frac{360}{\pi}\right)° \approx 114.6°. \qquad \blacksquare$$

* This does not mean that $360 = 2\pi$, just as 12 inches = 1 foot does not mean that $12 = 1$. Rather, it says that two measures of the one angle are equivalent.

Example 9   Convert $153°40'$ to radians.

Solution   We first write $153°40'$ in decimal form:

$$153°40' = (153 + \tfrac{40}{60})°$$
$$= (153 + \tfrac{2}{3})° \approx 153.67°.$$

Then we convert from degrees to radians using equation (2):

$$153.67° = (153.67)\left(\frac{\pi}{180}\right) \text{radians}$$

$$\approx 2.68 \text{ radians.}$$

■

Exercise 4.1

In Problems 1–16, draw the given angle in standard position.

1. $60°$	**2.** $-120°$	**3.** $135°$	**4.** $150°$
5. $1140°$	**6.** $-315°$	**7.** $-240°$	**8.** $-210°$
9. $\dfrac{\pi}{3}$	**10.** $\dfrac{5\pi}{4}$	**11.** $\dfrac{7\pi}{6}$	**12.** $-\dfrac{2\pi}{3}$
13. $-\dfrac{\pi}{6}$	**14.** -3π	**15.** $\dfrac{5\pi}{2}$	**16.** $\dfrac{3\pi}{4}$

In Problems 17–24, express the given angle in decimal notation.

17. $30°40'$	**18.** $20°15'$	**19.** $10°39'$	**20.** $43°7'$
21. $-5°10'$	**22.** $10°25'$	**23.** $50°1'$	**24.** $-115°50'$

In Problems 25–32, express the given angle in terms of degrees and minutes (rounded to the nearest minute).

25. $17.3°$	**26.** $12.21°$	**27.** $10.78°$	**28.** $15.45°$
29. $-30.81°$	**30.** $-110.5°$	**31.** $5.9°$	**32.** $83.56°$

In Problems 33–36, find the angle between $0°$ and $360°$ that is coterminal with the given angle.

33. $875°$	**34.** $400°$	**35.** $-610°$	**36.** $-150°$

In Problems 37–40, find the angle between 0 and 2π radians that is coterminal with the given angle.

37. $-\dfrac{\pi}{4}$	**38.** $\dfrac{17\pi}{2}$	**39.** 5.3π	**40.** $-\dfrac{9\pi}{5}$

In Problems 41–56, convert from degrees to radians.

41. $45°$	**42.** $30°$	**43.** $270°$	**44.** $60°$

45. 1°	**46.** 0°	**47.** 131°40′	**48.** −120°
49. −230°	**50.** 52°	**51.** 540°	**52.** −47.2°
53. 112°25′	**54.** 71°6′	**55.** 127.5°	**56.** 32.7°

In Problems 57–68, convert from radians to degrees.

57. $\dfrac{2\pi}{3}$	**58.** $\dfrac{\pi}{12}$	**59.** $\dfrac{\pi}{6}$	**60.** 7π
61. $\dfrac{5\pi}{4}$	**62.** $\dfrac{19\pi}{2}$	**63.** 3.1	**64.** 0.76
65. 12	**66.** −1.6	**67.** −2.5	**68.** 7.2

4.2 The Sine and Cosine Functions

As we have indicated in Section 4.1, originally the trigonometric functions were defined using angles in a right triangle. However, we shall take a more modern approach that employs the unit circle. There are numerous applications of trigonometry in surveying, navigation, biology, physics, and engineering. Many examples of these will be discussed in Chapter 5. For the present our primary concern is the development and theory of the trigonometric functions.

Sine and Cosine

To each real number t there corresponds an angle of t radians in standard position. As shown in Figure 4.8, we denote the point of intersection of the terminal side of the angle t with the unit circle by $P_t(x, y)$. The abscissa x is called the **cosine** of t and the ordinate y is called the **sine** of t. This is written

$$x = \cos t,$$
$$y = \sin t. \tag{1}$$

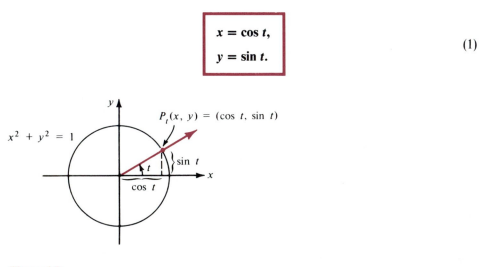

Figure 4.8

Since to each real number t there corresponds a unique point $P_t(\cos t, \sin t)$, we have two functions, the **cosine function**, denoted by $\cos t$, and the **sine function**, denoted by $\sin t$, each with domain the set R of real numbers.

Since the point $P_t(x, y)$ is on the unit circle, its coordinates satisfy

$$x^2 + y^2 = 1.$$

Substituting (1) into this equation gives an important relationship between the sine and the cosine:

$$(\cos t)^2 + (\sin t)^2 = 1.$$

This identity is usually written as

$$\cos^2 t + \sin^2 t = 1. \qquad (2)$$

From the fact that P_t lies on the unit circle it also follows that

$$-1 \le x \le 1 \quad \text{and} \quad -1 \le y \le 1,$$

or

$$|\cos t| \le 1 \quad \text{and} \quad |\sin t| \le 1, \qquad (3)$$

and that the range of each function is the interval $[-1, 1]$.

Since $\cos t$ and $\sin t$ are coordinates of a point on the terminal side of the angle t, these values are positive or negative depending on the quadrant in which the terminal side of t is located.

Example 1 If $t = 5\pi/3$, then the corresponding angle of $5\pi/3$ radians (300°) has its terminal side in the fourth quadrant. The x coordinate of a point in the fourth quadrant is positive and the y coordinate is negative. Therefore $\cos(5\pi/3) > 0$ and $\sin(5\pi/3) < 0$. See Figure 4.9.

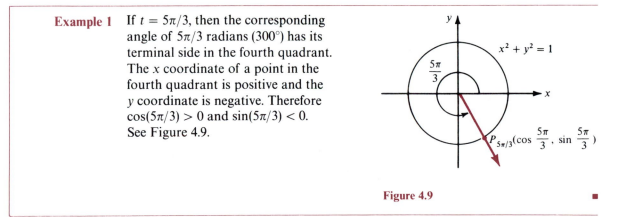

Figure 4.9

Figure 4.10 displays the signs of the cosine and sine functions in each of the four quadrants.

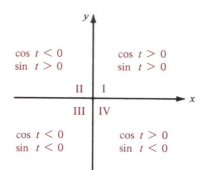

Figure 4.10

Example 2   Given that $\cos t = \frac{1}{3}$ and that the terminal side of the angle of t radians is in the fourth quadrant, find $\sin t$.

Solution   Substituting $\cos t = \frac{1}{3}$ into $\cos^2 t + \sin^2 t = 1$ gives

$$\frac{1}{9} + \sin^2 t = 1$$
$$\sin^2 t = \frac{8}{9}.$$

Since $\sin t$ is negative for a fourth quadrant angle of t radians, we take the negative square root:

$$\sin t = -\sqrt{\frac{8}{9}} = -\frac{2\sqrt{2}}{3}.$$

■

Values of the Sine and Cosine

Inspection of Figure 4.11 yields the following numerical values of $\cos t$ and $\sin t$ for $t = 0,\ \pi/2,\ \pi,\ 3\pi/2,$ and 2π.

$\cos 0 = 1$	$\sin 0 = 0$
$\cos \dfrac{\pi}{2} = 0$	$\sin \dfrac{\pi}{2} = 1$
$\cos \pi = -1$	$\sin \pi = 0$
$\cos \dfrac{3\pi}{2} = 0$	$\sin \dfrac{3\pi}{2} = -1$
$\cos 2\pi = 1$	$\sin 2\pi = 0$

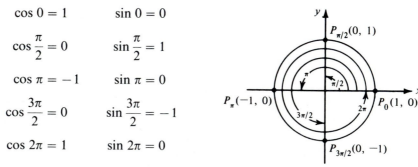

Figure 4.11

Using results from plane geometry, we can determine the values of the sine and cosine for $t = \pi/6$, $t = \pi/4$, and $t = \pi/3$. For $t = \pi/4$ we draw an angle of $\pi/4$ radians (45°) and form a right triangle by dropping a perpendicular from $P_{\pi/4}$ to the x axis as shown in Figure 4.12. Since the sum of the angles in any triangle is 180°, the third angle of this triangle is also 45°, and hence the triangle is isosceles. Therefore the coordinates of $P_{\pi/4}$ are equal; that is,

$$\cos \frac{\pi}{4} = \sin \frac{\pi}{4}.$$

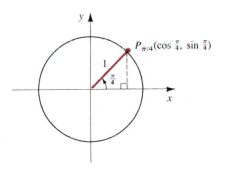

Figure 4.12

It follows from (2) that

$$\cos^2 \frac{\pi}{4} + \cos^2 \frac{\pi}{4} = 1 \qquad ,$$

$$2 \cos^2 \frac{\pi}{4} = 1$$

$$\cos^2 \frac{\pi}{4} = \frac{1}{2}$$

$$\cos \frac{\pi}{4} = \frac{1}{\sqrt{2}} = \frac{\sqrt{2}}{2},$$

and so
$$\sin \frac{\pi}{4} = \frac{\sqrt{2}}{2}.$$

(We consider only the positive square root, since $\cos(\pi/4)$ is the x coordinate of a point in the first quadrant.)

To find the values of $\sin(\pi/6)$ and $\cos(\pi/6)$, we construct angles of $\pi/6$ radians (30°) in the first and fourth quadrants as shown in Figure 4.13(a). Drawing perpendicular line segments from $P_{\pi/6}$ and Q to the x axis, we obtain two right triangles, each with hypotenuse of length 1 and angles of 30°, 60°, and 90°. Hence these are congruent triangles. It follows that we have a triangle $QOP_{\pi/6}$ that is equilateral, with each side having length 1. Since $\sin(\pi/6)$ is equal to half the length of the vertical

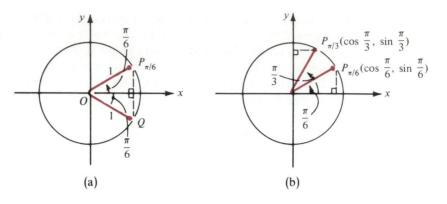

Figure 4.13

side of $QOP_{\pi/6}$, we have
$$\sin\frac{\pi}{6} = \frac{1}{2}.$$

We use this result and equation (2) to find the value of $\cos(\pi/6)$:

$$\cos^2\frac{\pi}{6} + \left(\frac{1}{2}\right)^2 = 1$$

$$\cos^2\frac{\pi}{6} = \frac{3}{4}$$

$$\cos\frac{\pi}{6} = \frac{\sqrt{3}}{2}.$$

To find the values $\cos(\pi/3)$ and $\sin(\pi/3)$, we have constructed two congruent right triangles in Figure 4.13(b). (See Problem 35 for the details of the construction.) It follows from the congruence of these triangles that

$$\cos\frac{\pi}{3} = \sin\frac{\pi}{6} = \frac{1}{2} \quad \text{and} \quad \sin\frac{\pi}{3} = \cos\frac{\pi}{6} = \frac{\sqrt{3}}{2}.$$

The table below summarizes the values of the cosine and sine functions that we have determined so far.

t	0	$\dfrac{\pi}{6}$	$\dfrac{\pi}{4}$	$\dfrac{\pi}{3}$	$\dfrac{\pi}{2}$	π	$\dfrac{3\pi}{2}$
$\cos t$	1	$\dfrac{\sqrt{3}}{2}$	$\dfrac{\sqrt{2}}{2}$	$\dfrac{1}{2}$	0	-1	0
$\sin t$	0	$\dfrac{1}{2}$	$\dfrac{\sqrt{2}}{2}$	$\dfrac{\sqrt{3}}{2}$	1	0	-1

Reference Angle

If t is the radian measure of an angle in quadrants II, III, or IV, $\cos t$ and $\sin t$ can always be determined by employing a related angle in the first quadrant. For example, suppose we want to find $\cos(7\pi/6)$ and $\sin(7\pi/6)$. As Figure 4.14(a) shows, the angle of $7\pi/6$ radians extends $\pi/6$ radians (from the horizontal) into the third quadrant. Therefore in Figure 4.14(b), by constructing a **reference angle** t' of $\pi/6$ radians ($30°$) in the first quadrant, we obtain a point $P_{t'}$ whose coordinates are equal *in absolute value* to the respective coordinates of P_t. Using the fact that the coordinates of a point in the third quadrant are negative, we have

$$\cos\frac{7\pi}{6} = -\cos\frac{\pi}{6} = -\frac{\sqrt{3}}{2} \quad \text{and} \quad \sin\frac{7\pi}{6} = -\sin\frac{\pi}{6} = -\frac{1}{2}.$$

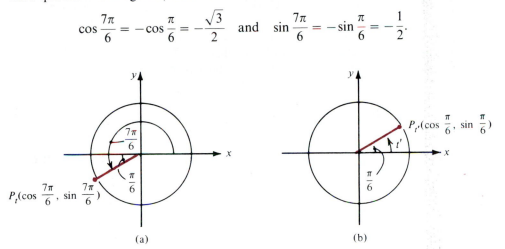

Figure 4.14

In general for any angle t (with P_t not on the y axis) *the measure of the reference angle t' in the first quadrant is equal to that of the acute angle formed by the terminal side of t and the x axis.* Figure 4.15 illustrates this for angles with terminal sides in each of the three quadrants II, III, and IV, respectively.

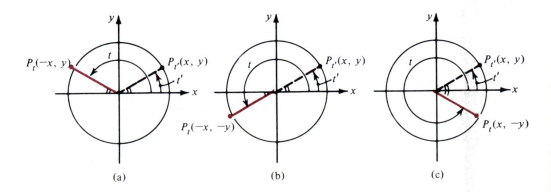

Figure 4.15

If t' is the reference angle of t then

$$\sin t = \sin t' \quad \text{or} \quad \sin t = -\sin t'$$

and
$$\cos t = \cos t' \quad \text{or} \quad \cos t = -\cos t',$$

where the positive or negative value is chosen according to the quadrant in which t lies.

Example 3 Find $\cos(5\pi/3)$ and $\sin(5\pi/3)$.

Solution From Figure 4.16 we see that the terminal side of the angle of $5\pi/3$ radians makes an acute angle of $\pi/3$ radians with the positive x axis. Thus the reference angle for $5\pi/3$ is $\pi/3$. We know that $\cos(\pi/3) = 1/2$ and $\sin(\pi/3) = \sqrt{3}/2$. Using the fact that the abscissa of a point in the fourth quadrant is positive and the ordinate is negative, we have

$$\cos \frac{5\pi}{3} = \frac{1}{2} \quad \text{and} \quad \sin \frac{5\pi}{3} = -\frac{\sqrt{3}}{2}.$$

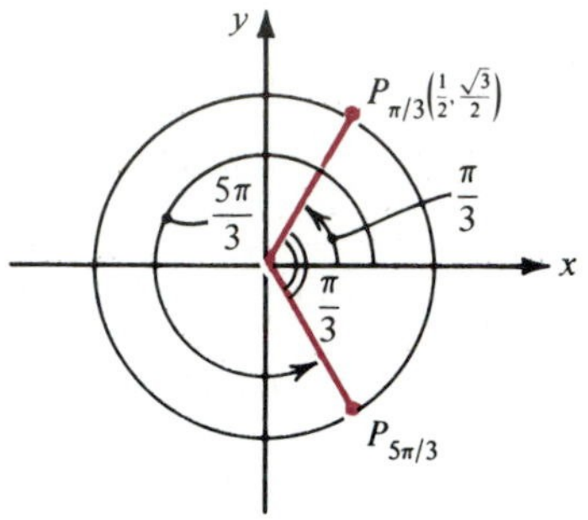

Figure 4.16

Periodicity

In Section 4.1 we saw that for any real number t, the angles t and $t \pm 2\pi$ are coterminal. Thus they determine the same point (x, y) on the unit circle. Hence

$$\boxed{\begin{aligned} \cos t &= \cos(t \pm 2\pi), \\ \sin t &= \sin(t \pm 2\pi). \end{aligned}} \tag{4}$$

In other words, the sine and cosine functions repeat their values every 2π units. It follows that for any integer n

$$\boxed{\begin{aligned} \cos t &= \cos(t + 2n\pi), \\ \sin t &= \sin(t + 2n\pi). \end{aligned}} \tag{5}$$

In general a nonconstant function f is said to be **periodic** (or p-periodic) if there is a positive number p such that

$$f(t) = f(t + p) \tag{6}$$

for every t in the domain of f. If p is the smallest positive number for which (6) is true, then p is called the **period** of the function f. Thus the properties in (4) imply that the cosine and sine functions are periodic. To see that the period of $\sin t$ is actually 2π, we note that there is only one point P_t on the unit circle with y coordinate 1, namely $(0, 1)$. That is,

$$\sin t = 1 \quad \text{only for} \quad t = \frac{\pi}{2}, \frac{\pi}{2} \pm 2\pi, \frac{\pi}{2} \pm 4\pi, \text{ and so on.}$$

Thus the smallest possible positive value of p is 2π, and the sine function is 2π-periodic. The verification that the cosine function has period 2π is left as an exercise. See Problem 31.

Example 4   Evaluate $\cos(9\pi/4)$.

Solution   We can write

$$\frac{9\pi}{4} = 2\pi + \frac{\pi}{4},$$

so that by (4)

$$\cos\frac{9\pi}{4} = \cos\left(\frac{\pi}{4} + 2\pi\right) = \cos\frac{\pi}{4} = \frac{\sqrt{2}}{2}. \qquad \blacksquare$$

Example 5   Evaluate $\sin(16\pi/3)$.

Solution   Since

$$\frac{16\pi}{3} = 5\pi + \frac{\pi}{3} = 4\pi + \frac{4\pi}{3},$$

it follows from (5) that

$$\sin\frac{16\pi}{3} = \sin\left(\frac{4\pi}{3} + 4\pi\right) = \sin\frac{4\pi}{3}.$$

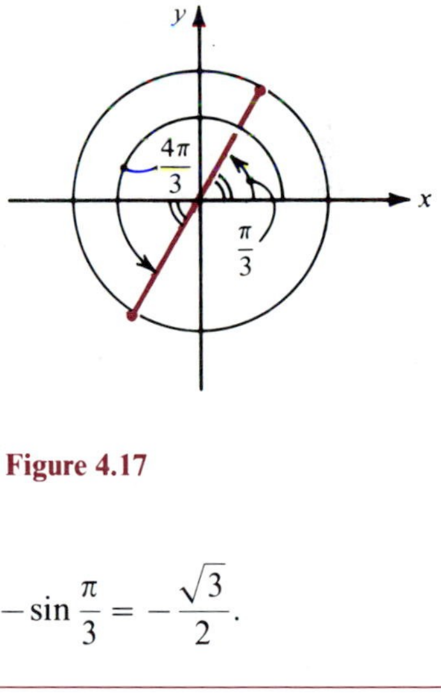

Now the first quadrant reference angle for $4\pi/3$ is $\pi/3$, as shown in Figure 4.17. Therefore we obtain

Figure 4.17

$$\sin\frac{16\pi}{3} = \sin\frac{4\pi}{3} = -\sin\frac{\pi}{3} = -\frac{\sqrt{3}}{2}. \qquad \blacksquare$$

Additional Properties

As shown in Figure 4.18, for an angle of t radians with terminal side in the first quadrant, the angle $-t$ lies in the fourth quadrant, and the two triangles formed with the x axis are congruent. Thus

$$\boxed{\begin{aligned} \cos(-t) &= \cos t, \\ \sin(-t) &= -\sin t. \end{aligned}}$$

(7)

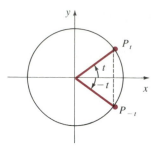

Figure 4.18

In fact, equations (7) hold for any angle t, regardless of the location of its terminal side. See Problem 32.

Example 6   Evaluate $\sin(-\pi/6)$.

Solution   Applying (7), we have that

$$\sin\left(-\frac{\pi}{6}\right) = -\sin\frac{\pi}{6} = -\frac{1}{2}.$$

Of course, this problem can also be solved by means of a reference angle. ∎

The following additional properties may be verified by arguments similar to the one above for equations (7). See Problems 33–34.

$$\cos\left(\frac{\pi}{2} - t\right) = \sin t$$
$$\sin\left(\frac{\pi}{2} - t\right) = \cos t$$

(8)

$$\cos(t + \pi) = -\cos t$$
$$\sin(t + \pi) = -\sin t$$

(9)

Example 7   The reader may have observed that $\cos(\pi/3) = \sin(\pi/6)$. This is a special case of the second equation in (8) since $\sin(\pi/6) = \sin(\pi/2 - \pi/3)$. ∎

Exercise 4.2

1. Complete the following table.

t	$\cos t$	$\sin t$
0	1	0
$\dfrac{\pi}{6}$	$\dfrac{\sqrt{3}}{2}$	$\dfrac{1}{2}$
$\dfrac{\pi}{4}$	$\dfrac{\sqrt{2}}{2}$	$\dfrac{\sqrt{2}}{2}$
$\dfrac{\pi}{3}$	$\dfrac{1}{2}$	$\dfrac{\sqrt{3}}{2}$
$\dfrac{\pi}{2}$	0	1
$\dfrac{2\pi}{3}$		
$\dfrac{3\pi}{4}$		
$\dfrac{5\pi}{6}$		
π	-1	0
$\dfrac{7\pi}{6}$	$-\dfrac{\sqrt{3}}{2}$	$-\dfrac{1}{2}$
$\dfrac{5\pi}{4}$		
$\dfrac{4\pi}{3}$		
$\dfrac{3\pi}{2}$	0	-1
$\dfrac{5\pi}{3}$	$\dfrac{1}{2}$	$-\dfrac{\sqrt{3}}{2}$
$\dfrac{7\pi}{4}$		
$\dfrac{11\pi}{6}$		
2π	1	0

2. Given that $\sin t = \frac{1}{4}$ and that the terminal side of the angle of t radians is in the second quadrant, find $\cos t$.

3. Given that $\cos t = -\frac{2}{5}$ and that the terminal side of the angle of t radians is in the second quadrant, find $\sin t$.

4. Given that $\cos t = \frac{1}{3}$ and that the terminal side of the angle of t radians is in the fourth quadrant, find $\sin t$.

5. Given that $\sin t = -\frac{2}{3}$ and that the terminal side of the angle of t radians is in the third quadrant, find $\cos t$.

6. If $\cos t = \frac{3}{10}$, find all possible values of $\sin t$.

7. If $\sin t = -\frac{2}{7}$, find all possible values of $\cos t$.

8. If $\sin t = 0.4$, find all possible values of $\cos t$.

9. If $\cos t = -0.2$, find all possible values of $\sin t$.

10. If $2 \sin t - \cos t = 0$, find all possible values of $\sin t$ and $\cos t$.

11. If $3 \sin t - 2 \cos t = 0$, find all possible values of $\sin t$ and $\cos t$.

12. If $\cos t / \sin t = -2$, and the terminal side of the angle of t radians is in the second quadrant, find the values of $\sin t$ and $\cos t$.

13. If $\sin t / \cos t = 3$, and the terminal side of the angle of t radians is in the third quadrant, find the values of $\sin t$ and $\cos t$.

In Problems 14–25, find the indicated value.

14. $\cos \dfrac{7\pi}{3}$

15. $\cos 5\pi$

16. $\sin\left(-\dfrac{9\pi}{2}\right)$

17. $\sin\left(-\dfrac{7\pi}{6}\right)$

18. $\cos\left(-\dfrac{5\pi}{2}\right)$

19. $\cos \dfrac{23\pi}{4}$

20. $\sin \dfrac{17\pi}{6}$

21. $\sin 9\pi$

22. $\cos \dfrac{13\pi}{6}$

23. $\cos\left(-\dfrac{10\pi}{3}\right)$

24. $\sin \dfrac{11\pi}{4}$

25. $\sin\left(-\dfrac{4\pi}{3}\right)$

26. Determine all possible values of the angle t if $\sin t = \frac{1}{2}$.

27. Determine all possible values of t if $\cos t = -\frac{1}{2}$.

28. Determine all possible values of t if $\cos t = \sin t$.

29. If $t = n\pi$, where $n = 0, \pm 1, \pm 2, \ldots$, what are the values of $\cos t$ and $\sin t$?

30. If $t = (2n + 1)\pi/2$, where $n = 0, \pm 1, \pm 2, \ldots$, what are the values of $\cos t$ and $\sin t$?

31. Show that $\cos t$ is 2π-periodic.

32. Verify formulas (7) for an angle of t radians with terminal side in quadrants II, III, and IV.

33. Verify formulas (8).

34. Verify formulas (9).

35. In Figure 4.13(b) we constructed two right triangles by drawing a line from $P_{\pi/6}$ perpendicular to the x axis and a line from $P_{\pi/3}$ perpendicular to the y axis. State why the two right triangles are congruent.

36. A particle moves in the Cartesian plane according to

$$x = 2 \cos t \qquad y = 4 \sin^2 t,$$

where t is time. Eliminate t in these equations to obtain an equation of motion in terms of only x and y.

Review Problems

37. Find the angle between 0 and 2π radians that is coterminal with an angle of $27\pi/4$ radians.

38. Find the angle between 0 and 2π radians that is coterminal with an angle of 7.4 radians.

39. Convert $405°$ to radians and find the sine and cosine of this angle.

40. Convert $-570°$ to radians and find the sine and cosine of this angle.

4.3 Other Trigonometric Functions

Four additional trigonometric functions—**tangent**, **cotangent**, **secant**, and **cosecant** (abbreviated **tan**, **cot**, **sec**, and **csc**, respectively)—are defined in terms of the sine and cosine functions as follows:

$$\tan t = \frac{\sin t}{\cos t} \tag{1}$$

$$\cot t = \frac{\cos t}{\sin t} \tag{2}$$

$$\sec t = \frac{1}{\cos t} \tag{3}$$

$$\csc t = \frac{1}{\sin t}. \tag{4}$$

We note that

$$\cot t = \frac{\cos t}{\sin t} = \frac{1}{\dfrac{\sin t}{\cos t}} = \frac{1}{\tan t}.$$

From (1) and (3) we see that the domains of the tangent and the secant functions consist of all real numbers t such that $\cos t \neq 0$. That is,

$$\left\{ t \,\middle|\, t \neq \frac{\pi}{2} + n\pi \quad \text{for any integer } n \right\}.$$

Similarly, it follows from (2) and (4) that the cotangent and cosecant functions are defined for all real numbers t such that $\sin t \neq 0$; that is, for

$$\{ t \,|\, t \neq n\pi \quad \text{for any integer } n \}.$$

Since $|\sin t| \leq 1$ and $|\cos t| \leq 1$, we have

$$|\csc t| = \left| \frac{1}{\sin t} \right| = \frac{1}{|\sin t|} \geq \frac{1}{1} = 1$$

and

$$|\sec t| = \left| \frac{1}{\cos t} \right| = \frac{1}{|\cos t|} \geq \frac{1}{1} = 1.$$

The values of the tangent and the cotangent functions are unrestricted. Figure 4.19(a) illustrates the signs of the tangent, cotangent, secant, and cosecant in each of the four quadrants. This is easily verified using the signs of the sine and cosine functions that were displayed in Figure 4.10. As an aid in remembering the signs of the various trigonometric functions in each quadrant, we display in Figure 4.19(b) only the functions which are positive in each quadrant.

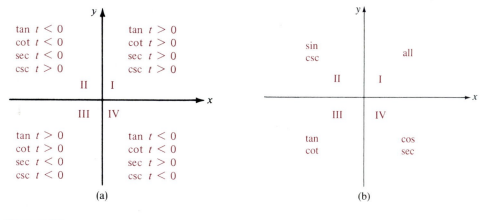

(a)   (b)

Figure 4.19

Example 1   Evaluate $\tan t$, $\cot t$, $\sec t$, and $\csc t$ for $t = -\pi/3$.

Solution   From Figure 4.20 we see that the angle $-\pi/3$ radians is in quadrant IV and has reference angle $\pi/3$. Thus

$$\sin\left(-\frac{\pi}{3} \right) = -\sin \frac{\pi}{3} = -\frac{\sqrt{3}}{2}$$

and

$$\cos\left(-\frac{\pi}{3}\right) = \cos\frac{\pi}{3} = \frac{1}{2}.$$

Hence we have

$$\tan\left(-\frac{\pi}{3}\right) = \frac{-\sqrt{3}/2}{1/2} = -\sqrt{3},$$

$$\cot\left(-\frac{\pi}{3}\right) = \frac{1}{\tan\left(-\frac{\pi}{3}\right)} = \frac{-1}{\sqrt{3}} = -\frac{\sqrt{3}}{3},$$

$$\sec\left(-\frac{\pi}{3}\right) = \frac{1}{\frac{1}{2}} = 2,$$

$$\csc\left(-\frac{\pi}{3}\right) = \frac{1}{-\sqrt{3}/2} = \frac{-2}{\sqrt{3}} = -\frac{2\sqrt{3}}{3}.$$

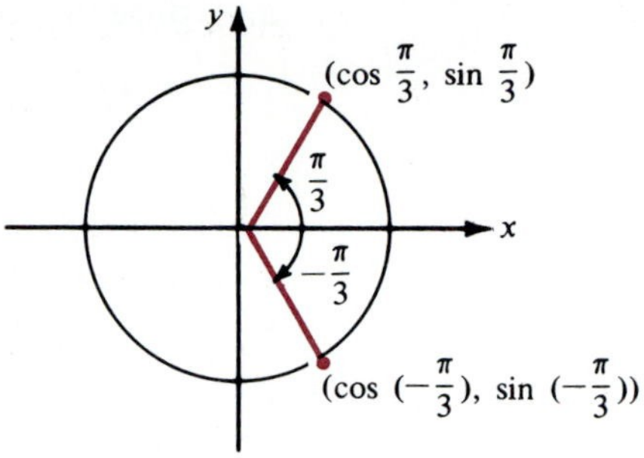

Figure 4.20

Two Identities

The tangent is related to the secant by a useful identity. If we divide

$$\cos^2 t + \sin^2 t = 1 \tag{5}$$

by $\cos^2 t$, we have

$$\frac{\cos^2 t}{\cos^2 t} + \frac{\sin^2 t}{\cos^2 t} = \frac{1}{\cos^2 t},$$

or

$$\boxed{1 + \tan^2 t = \sec^2 t.} \tag{6}$$

Similarly, it can be shown that

$$\boxed{1 + \cot^2 t = \csc^2 t.} \tag{7}$$

See Problem 55.

Example 2 Given that $\csc t = -5$ and the terminal side of the angle t lies in the fourth quadrant, determine $\tan t$ and $\cot t$.

Solution We first compute $\cot t$. From equation (7) it follows that

$$\cot^2 t = \csc^2 t - 1,$$

and, since the cotangent is negative in the fourth quadrant,

$$\begin{aligned} \cot t &= -\sqrt{\csc^2 t - 1} \\ &= -\sqrt{(-5)^2 - 1} \\ &= -\sqrt{24} \\ &= -2\sqrt{6}. \end{aligned}$$

Since $\cot t = 1/\tan t$, we find

$$\begin{aligned} \tan t = \frac{1}{\cot t} &= \frac{1}{-2\sqrt{6}} \\ &= -\frac{\sqrt{6}}{12}. \end{aligned}$$

$\blacksquare$

Periodicity

From formulas (9) in Section 4.2 we have that

$$\tan(t + \pi) = \frac{\sin(t + \pi)}{\cos(t + \pi)} = \frac{-\sin t}{-\cos t} = \tan t.$$

Thus the tangent function is periodic with period less than or equal to π. By examining the values of t for which $\tan t = 0$, we can see that the period is, in fact, π. Since the cotangent is the reciprocal of the tangent, it follows that the cotangent is π-periodic. Similarly, the secant and cosecant, as reciprocals of the cosine and sine, respectively, are 2π-periodic.

Trigonometric Functions of Angles

In certain applications it is sometimes more convenient to view the trigonometric functions as functions of angles measured in degrees. This is easily accomplished. For instance, if θ is an angle measured in degrees, we define the cosine of θ to be the cosine of the number t, where t is the radian measure of θ. Thus, for example,

$$\cos 45° = \cos\left(\frac{\pi}{4} \text{ radians}\right) = \cos\frac{\pi}{4} = \frac{\sqrt{2}}{2}.$$

A similar definition can be made for each of the other trigonometric functions. Thus,

$$\sin 45° = \sin\left(\frac{\pi}{4} \text{ radians}\right) = \sin\frac{\pi}{4} = \frac{\sqrt{2}}{2},$$

$$\tan 45° = \tan\left(\frac{\pi}{4} \text{ radians}\right) = \tan\frac{\pi}{4} = \frac{\sin\frac{\pi}{4}}{\cos\frac{\pi}{4}} = 1,$$

and so on. The table below gives the values of $\cos\theta$, $\sin\theta$, and $\tan\theta$ for some standard angles.

θ(degrees)	θ(radians)	$\cos \theta$	$\sin \theta$	$\tan \theta$
$0°$	0	1	0	0
$30°$	$\dfrac{\pi}{6}$	$\dfrac{\sqrt{3}}{2}$	$\dfrac{1}{2}$	$\dfrac{\sqrt{3}}{3}$
$45°$	$\dfrac{\pi}{4}$	$\dfrac{\sqrt{2}}{2}$	$\dfrac{\sqrt{2}}{2}$	1
$60°$	$\dfrac{\pi}{3}$	$\dfrac{1}{2}$	$\dfrac{\sqrt{3}}{2}$	$\sqrt{3}$
$90°$	$\dfrac{\pi}{2}$	0	1	$-$
$180°$	π	-1	0	0
$270°$	$\dfrac{3\pi}{2}$	0	-1	$-$

Example 3   Find $\sin 210°$, $\cos 210°$, $\tan 210°$, $\cot 210°$, $\sec 210°$, and $\csc 210°$.

Solution   From Figure 4.21 we see that the angle of $30°$ is the reference angle for $210°$. Therefore we have

$$\sin 210° = -\sin 30° = -\frac{1}{2},$$

$$\cos 210° = -\cos 30° = -\frac{\sqrt{3}}{2}$$

$$\tan 210° = \frac{\sin 210°}{\cos 210°} = \frac{-1/2}{-\sqrt{3}/2} = \frac{\sqrt{3}}{3}$$

$$\cot 210° = \frac{\cos 210°}{\sin 210°} = \frac{-\sqrt{3}/2}{-1/2} = \sqrt{3}$$

$$\sec 210° = \frac{1}{\cos 210°} = \frac{1}{-\sqrt{3}/2} = -\frac{2\sqrt{3}}{3}$$

$$\csc 210° = \frac{1}{\sin 210°} = \frac{1}{-1/2} = -2.$$

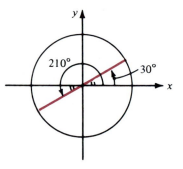

Figure 4.21

Use of a Calculator

To compute the values of trigonometric functions of angles with reference angles other than 0, $\pi/6$, $\pi/4$, and $\pi/3$, we can use Table IV and V or a calculator. A discussion on the use of trigonometric tables is given in Appendix A-4. Most scientific calculators have keys for the sine, cosine, and tangent functions. By computing the reciprocals of these, we can obtain the values of the other three trigonometric functions. In general, the calculator will need to be set to either degree or radian mode, depending on the mode of the input angle. When in degree mode it is usually the case that fractions of a degree are represented in decimal form rather than in minutes and seconds.

Example 4   Use a calculator to evaluate

a. $\sin(\pi/8)$ **b.** $\tan 40°20'$ **c.** $\sec 17°$.

Solution   **a.** To calculate $\sin \pi/8$, first select radian mode. Next, compute $\pi/8$ by first entering $\pi \approx 3.1416$ (many calculators have a π key) and then dividing by 8. Now press the $\boxed{\text{SIN}}$ key to obtain

$$\sin(\pi/8) \approx \sin 0.3927 \approx 0.3827.$$

b. To calculate $\tan 40°20'$ we first convert the angle to decimal form:

$$40°20' = 40\tfrac{20}{60}° \approx 40.3333°.$$

Then

$$\tan 40°20' \approx \tan 40.3333° \approx 0.8491.$$

Note that if we use $40°20' \approx 40.33°$, we obtain

$$\tan 40°20' \approx \tan 40.33° \approx 0.8490.$$

The fact that the two calculated values of $\tan 40°20'$ differ indicates that it is important to use as many decimal places as the calculator will accept when entering an angle.

c. To calculate the secant of $17°$ we use the fact that $\sec t = 1/\cos t$:

$$\sec 17° = \frac{1}{\cos 17°} \approx \frac{1}{0.9563} \approx 1.0457. \qquad \blacksquare$$

The results obtained when computing trigonometric function values on a calculator may differ slightly depending on the calculator used. This will be most likely to occur for values of the tangent, cotangent, secant, and cosecant functions which are very large.

Exercise 4.3

1. Complete the following table.

t	$\tan t$	$\cot t$	$\sec t$	$\csc t$
0				
$\dfrac{\pi}{6}$				
$\dfrac{\pi}{4}$	1	1	$\sqrt{2}$	$\sqrt{2}$
$\dfrac{\pi}{3}$				
$\dfrac{\pi}{2}$	—	0		
$\dfrac{2\pi}{3}$				
$\dfrac{3\pi}{4}$				
$\dfrac{5\pi}{6}$				
π				
$\dfrac{7\pi}{6}$				
$\dfrac{5\pi}{4}$	1	1		
$\dfrac{4\pi}{3}$				
$\dfrac{3\pi}{2}$				
$\dfrac{5\pi}{3}$				
$\dfrac{7\pi}{4}$				
$\dfrac{11\pi}{6}$				
2π				

In Problems 2–21, find the indicated value without the use of a calculator.

2. $\sec\dfrac{10\pi}{3}$ **3.** $\cot\dfrac{13\pi}{6}$ **4.** $\csc\left(-\dfrac{3\pi}{2}\right)$ **5.** $\tan\dfrac{9\pi}{2}$

6. $\sec 7\pi$ **7.** $\csc\left(-\dfrac{\pi}{6}\right)$ **8.** $\cot\left(-\dfrac{13\pi}{3}\right)$ **9.** $\tan\dfrac{23\pi}{4}$

10. $\tan\left(-\dfrac{5\pi}{6}\right)$ **11.** $\sec\dfrac{10\pi}{3}$ **12.** $\cot\dfrac{17\pi}{6}$ **13.** $\csc 5\pi$

14. $\sec\dfrac{29\pi}{4}$ **15.** $\cot\left(-\dfrac{5\pi}{4}\right)$ **16.** $\cos(-45°)$ **17.** $\sin 150°$

18. $\tan 405°$ **19.** $\sec(-120°)$ **20.** $\cot(-720°)$ **21.** $\csc 495°$

Calculator Problems

In Problems 22–37, use a calculator to find the indicated value. Give at least eight digits in your answer.

22. $\cos(\pi/5)$ **23.** $\sin(2\pi/9)$ **24.** $\tan(3\pi/5)$ **25.** $\sec(4\pi/5)$

26. $\cot(2\pi/7)$ **27.** $\csc(\pi/7)$ **28.** $\tan 3.14$ **29.** $\sin 0.62$

30. $\csc 1.3$ **31.** $\sec 1.57$ **32.** $\sin 29°$ **33.** $\cos(-112°)$

34. $\sec 3°$ **35.** $\tan 50°30'$ **36.** $\cos 25°10'$ **37.** $\sec 12°40'$

38. Given that $\cot t = 2$ and that the terminal side of the angle t lies in third quadrant, find $\csc t$.

39. Given that $\tan t = -2$ and that the terminal side of the angle t lies in the second quadrant, find $\sec t$.

40. Given that $\sec t = -3$, and that the terminal side of the angle t lies in the second quadrant, find $\tan t$.

41. If $\cot t = \frac{3}{4}$, find all possible values of $\csc t$.

42. If $\sec t = -5$, find all possible values of $\cos t$ and $\sin t$.

43. Given that $\cos t = \frac{1}{10}$ and that the terminal side of the angle t lies in the fourth quadrant, find $\sin t$, $\tan t$, $\cot t$, $\sec t$, and $\csc t$.

44. If $3\cos t = \sin t$, find all values of $\tan t$, $\cot t$, $\sec t$, and $\csc t$.

45. If $2\tan t + \sec t = 0$, find all values of $\sin t$, $\cos t$, $\cot t$, and $\csc t$.

46. If $\csc t = \cot t$ and $0 \le t \le \pi$, find $\sin t$, $\cos t$, and $\tan t$.

47. Determine all possible angles t for which $\tan t = \sqrt{3}$.

48. Determine all possible angles t for which $\sec t = \sqrt{2}$.

49. If $t = (2n+1)\pi/2$, where $n = 0, \pm 1, \pm 2, \ldots$, what are the values of $\tan t$, $\cot t$, $\sec t$ and $\csc t$?

50. If $t = n\pi$, where $n = 0, \pm 1, \pm 2, \ldots$, what are the values of $\tan t$, $\cot t$, $\sec t$, and $\csc t$?

51. As t varies from 0 to $\pi/2$, how do the values of $\tan t$ vary?

52. As t varies from 0 to π, how do the values of $\cot t$ vary?

53. As t varies from 0 to $\pi/2$, how do the values of $\sec t$ vary?

54. As t varies from 0 to π, how do the values of $\csc t$ vary?

55. Derive formula (7).

In Problems 56–59, eliminate the variable t in the equations to obtain an equation in terms of only x and y.

56. $x = 3 \tan t, y = 2 \sec^2 t$

57. $x = 4 \cot^2 t, y = \csc t$

58. $x = 2 \tan t, y = \sec t$

59. $x = \cot t, y = 5 \csc t$

Calculator Problems

60. The involute function, $\text{inv}(x) = \tan x - x$, is used in the design of gear teeth. (If x is given in degrees, it must be converted to radians before evaluating the function.) Find
 a. inv (0) **b.** inv ($\pi/6$) **c.** inv ($20°$)

61. The function $\text{sinc}(u) = (\sin u)/u$ arises frequently in the study of optics. (We require that u be in radians.)
 a. Evaluate sinc(1), sinc($\pi/6$), and sinc($\pi/4$).
 b. Evaluate sinc(u) when $u = 10^{-n}$, for n an integer from 1 to 5. Give at least eight digits in your answer.

62. A block slides down an inclined plane with constant speed if $\tan \theta = \mu$, where θ is the angle of inclination of the plane and μ is the coefficient of sliding friction between the block and plane. Suppose it is determined experimentally that a brass block will slide down a steel plane with a constant speed when the angle of inclination of the plane is $23°45'$. Find the coefficient of sliding friction for brass on steel.

Review Problems

63. Convert from radians to degrees.
 a. $13\pi/4$ **b.** 2.51

64. Find the angle between 0 and 2π that is coterminal with an angle of 7.5 radians.

65. Given that $\cos t = -\frac{3}{5}$ and that the terminal side of the angle of t radians is in the third quadrant, find sin t.

66. If $\sin^2 t - \cos^2 t + \frac{3}{4} = 0$ and the terminal side of the angle of t radians is in the fourth quadrant, find the values of sin t and cos t.

4.4 Special Formulas

4.4.1 Addition Formulas

Addition formulas for the sine and cosine In practice it is extremely useful to have formulas that can reduce $\cos(u - v)$, $\cos(u + v)$, $\sin(u - v)$, and $\sin(u + v)$ to expressions involving cos u, cos v, sin u, and sin v. In order to derive the formula for $\cos(u - v)$, we let u and v be angles as shown in Figure 4.22(a). If we place the angle $u - v$ in standard position as shown in Figure 4.22(b), then the distance d from R to S equals the distance from P to Q shown in Figure 4.22(a). Equating the squares of these distances,

$$[d(P, Q)]^2 = [d(R, S)]^2,$$

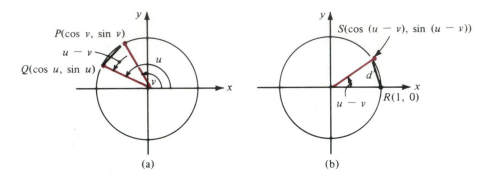

Figure 4.22

and using the distance formula gives

$$(\cos u - \cos v)^2 + (\sin u - \sin v)^2 = (\cos(u-v) - 1)^2 + \sin^2(u-v),$$

or

$$\cos^2 u - 2\cos u \cos v + \cos^2 v + \sin^2 u - 2\sin u \sin v + \sin^2 v$$
$$= \cos^2(u-v) - 2\cos(u-v) + 1 + \sin^2(u-v).$$

Since $\cos^2 u + \sin^2 u = 1$, $\cos^2 v + \sin^2 v = 1$ and $\cos^2(u-v) + \sin^2(u-v) = 1$, the preceding equation simplifies to

$$2 - 2\cos u \cos v - 2\sin u \sin v = 2 - 2\cos(u-v),$$

or
$$\cos(u - v) = \cos u \cos v + \sin u \sin v. \tag{1}$$

To obtain an analogous formula for $\cos(u + v)$, we write

$$\cos(u + v) = \cos(u - (-v))$$
$$= \cos u \cos(-v) + \sin u \sin(-v).$$

Since

$$\cos(-v) = \cos v \quad \text{and} \quad \sin(-v) = -\sin v,$$

it follows that

$$\cos(u + v) = \cos u \cos v - \sin u \sin v. \tag{2}$$

In a similar manner, also using formula (1), we can show that

$$\sin(u - v) = \sin u \cos v - \cos u \sin v \tag{3}$$

and

$$\sin(u + v) = \sin u \cos v + \cos u \sin v. \qquad (4)$$

See Problems 45 and 46.

Formulas (1), (2), (3), and (4) are called the **addition formulas** for the sine and cosine. These formulas can be used to derive further formulas which are useful in calculus. See Problems 41–44. In addition, the formulas can be used to find *exact* values of the sine and cosine functions of angles that can be represented as sums or differences of $\pi/3$, $\pi/4$, $\pi/6$, $2\pi/3$, and so on. For example, we can compute precisely the sine and cosine of angles such as

$$\frac{\pi}{12} = \frac{\pi}{3} - \frac{\pi}{4} \quad \text{or} \quad 15° = 60° - 45°,$$

and

$$\frac{7\pi}{12} = \frac{\pi}{3} + \frac{\pi}{4} \quad \text{or} \quad 105° = 60° + 45°.$$

Recall that calculators and trigonometric tables provide only decimal approximations to these values.

Example 1 Evaluate $\cos \dfrac{7\pi}{12}$.

Solution We can write

$$\cos \frac{7\pi}{12} = \cos\left(\frac{\pi}{3} + \frac{\pi}{4}\right);$$

hence by (2) it follows that

$$\cos\left(\frac{\pi}{3} + \frac{\pi}{4}\right) = \cos \frac{\pi}{3} \cos \frac{\pi}{4} - \sin \frac{\pi}{3} \sin \frac{\pi}{4}$$

$$= \frac{1}{2} \frac{\sqrt{2}}{2} - \frac{\sqrt{3}}{2} \frac{\sqrt{2}}{2}$$

$$= \frac{\sqrt{2}}{4}(1 - \sqrt{3}).$$

Note that $\cos(7\pi/12) < 0$ as expected. ∎

Example 2 Evaluate $\sin \dfrac{7\pi}{12}$.

Solution From (4) we have

$$\sin\frac{7\pi}{12} = \sin\left(\frac{\pi}{3} + \frac{\pi}{4}\right)$$

$$= \sin\frac{\pi}{3}\cos\frac{\pi}{4} + \cos\frac{\pi}{3}\sin\frac{\pi}{4}$$

$$= \frac{\sqrt{3}}{2}\frac{\sqrt{2}}{2} + \frac{1}{2}\frac{\sqrt{2}}{2}$$

$$= \frac{\sqrt{2}}{4}(1 + \sqrt{3}). \tag{5}$$

Alternatively, we can obtain the value of $\sin(7\pi/12)$ from

$$\cos^2\frac{7\pi}{12} + \sin^2\frac{7\pi}{12} = 1.$$

Using the value of $\cos(7\pi/12)$ from Example 1, it follows that

$$\sin\frac{7\pi}{12} = \sqrt{1 - \cos^2\frac{7\pi}{12}}$$

$$= \sqrt{1 - \left[\frac{\sqrt{2}}{4}(1 - \sqrt{3})\right]^2}$$

$$= \sqrt{1 - \frac{1}{8}(1 - 2\sqrt{3} + 3)}$$

$$= \sqrt{\frac{4 + 2\sqrt{3}}{8}}.$$

Although this number does not look like the result given in (5), the values are the same since

$$\sqrt{\frac{4 + 2\sqrt{3}}{8}} = \sqrt{\frac{2}{16}(1 + 2\sqrt{3} + 3)}$$

$$= \sqrt{\frac{2}{16}(1 + \sqrt{3})^2}$$

$$= \frac{\sqrt{2}}{4}(1 + \sqrt{3}). \qquad\blacksquare$$

Addition formulas for the tangent If we divide equation (4) by equation (2) and simplify, we obtain the following addition formula for the tangent:

$$\tan(u + v) = \frac{\tan u + \tan v}{1 - \tan u \tan v}. \tag{6}$$

Similarly,

$$\boxed{\tan(u - v) = \frac{\tan u - \tan v}{1 + \tan u \tan v}.}$$ (7)

The derivations of (6) and (7) are not difficult and are left as exercises. See Problems 47 and 48.

Example 4 Evaluate $\tan(\pi/12)$.

Solution Writing $\pi/12 = \pi/4 - \pi/6$ we obtain from (7)

$$\tan \frac{\pi}{12} = \tan\left(\frac{\pi}{4} - \frac{\pi}{6}\right)$$

$$= \frac{\tan \dfrac{\pi}{4} - \tan \dfrac{\pi}{6}}{1 + \tan \dfrac{\pi}{4} \tan \dfrac{\pi}{6}}$$

$$= \frac{1 - \dfrac{1}{\sqrt{3}}}{1 + 1 \cdot \dfrac{1}{\sqrt{3}}}$$

$$= \frac{\sqrt{3} - 1}{\sqrt{3} + 1}$$

$$= \frac{\sqrt{3} - 1}{\sqrt{3} + 1} \cdot \frac{\sqrt{3} - 1}{\sqrt{3} - 1} \qquad \text{(rationalizing the denominator)}$$

$$= \frac{(\sqrt{3} - 1)^2}{2}$$

$$= \frac{4 - 2\sqrt{3}}{2} = 2 - \sqrt{3}.$$

∎

Example 5 Verify that $\tan(u + \pi) = \tan u$.

Solution From equation (6) with $v = \pi$, we have

$$\tan(u + \pi) = \frac{\tan u + \tan \pi}{1 - \tan u \tan \pi} = \tan u,$$

since $\tan \pi = 0$.

∎

While an addition formula for the cotangent function could be similarly derived, it is customary to use formula (11) and take the reciprocal of the result. Corresponding remarks can be made regarding addition formulas for the secant and cosecant functions.

The formulas that have been derived in this section can be used to derive many additional properties of the trigonometric functions. See Problems 29–44.

**4.4.2
Double- and
Half-Angle
Formulas**

The formulas for the cosine and sine of the sum of two angles can be used to derive four additional formulas which are very useful in calculus.

Double-angle formulas   If we set $v = u$ in formula (2), then, since $\cos(u + u) = \cos 2u$, we find

$$\boxed{\cos 2u = \cos^2 u - \sin^2 u.} \tag{8}$$

Similarly, putting $v = u$ in (4) gives

$$\sin(u + u) = \sin u \cos u + \cos u \sin u,$$

or

$$\boxed{\sin 2u = 2 \sin u \cos u.} \tag{9}$$

Equations (8) and (9) are known as the **double-angle formulas**.

Example 6   If $\sin t = -\frac{1}{4}$ and $\pi < t < 3\pi/2$, find $\cos 2t$ and $\sin 2t$.

Solution   First, we compute $\cos t$ from

$$\cos^2 t + \sin^2 t = 1$$

$$\cos^2 t = 1 - \sin^2 t$$

$$\cos t = -\sqrt{1 - \sin^2 t} \qquad (t \text{ is a third quadrant angle})$$

$$= -\sqrt{1 - (-\tfrac{1}{4})^2} = -\frac{\sqrt{15}}{4}.$$

Now from (8) and (9) we obtain

$$\cos 2t = \left(-\frac{\sqrt{15}}{4}\right)^2 - \left(-\frac{1}{4}\right)^2 = \frac{15}{16} - \frac{1}{16} = \frac{14}{16} = \frac{7}{8},$$

and

$$\sin 2t = 2\left(-\frac{1}{4}\right)\left(-\frac{\sqrt{15}}{4}\right) = \frac{\sqrt{15}}{8}. \qquad \blacksquare$$

Half-angle formulas Equation (8) is the source of two **half-angle formulas**. Replacing $\cos^2 u$ by $1 - \sin^2 u$ gives

$$\cos 2u = (1 - \sin^2 u) - \sin^2 u$$

$$= 1 - 2\sin^2 u,$$

or

$$2\sin^2 u = 1 - \cos 2u$$

$$\sin^2 u = \tfrac{1}{2}(1 - \cos 2u)$$

If we let $2u = t$, then

$$\sin^2\frac{t}{2} = \frac{1}{2}(1 - \cos t). \tag{10}$$

Similarly, it can be shown that

$$\cos^2\frac{t}{2} = \frac{1}{2}(1 + \cos t). \tag{11}$$

See Problem 49.

To find the sine or cosine of $t/2$ we take the square root of the right-hand side of equation (10) or (11). The positive or negative square root is chosen, depending on the quadrant in which $t/2$ lies.

Example 7 Evaluate $\sin(5\pi/8)$ and $\cos(5\pi/8)$.

Solution With $t = 5\pi/4$, formulas (10) and (11) yield

$$\sin^2\frac{5\pi}{8} = \frac{1}{2}\left[1 - \cos\frac{5\pi}{4}\right]$$

$$= \frac{1}{2}\left[1 - \left(-\frac{\sqrt{2}}{2}\right)\right]$$

$$= \frac{1}{2}\left[1 + \frac{\sqrt{2}}{2}\right],$$

and

$$\cos^2\frac{5\pi}{8} = \frac{1}{2}\left[1 + \cos\frac{5\pi}{4}\right]$$

$$= \frac{1}{2}\left[1 + \left(-\frac{\sqrt{2}}{2}\right)\right]$$

$$= \frac{1}{2}\left[1 - \frac{\sqrt{2}}{2}\right].$$

Since $5\pi/8$ is a second quadrant angle, it follows that

$$\sin\frac{5\pi}{8} = \sqrt{\frac{1}{2}\left(1 + \frac{\sqrt{2}}{2}\right)} = \frac{1}{2}\sqrt{2 + \sqrt{2}}$$

and

$$\cos\frac{5\pi}{8} = -\frac{1}{2}\sqrt{2 - \sqrt{2}},$$

respectively. ∎

Exercise 4.4

In Problems 1–24, use an appropriate addition or half-angle formula to find the exact value of the given expression.

1. $\cos\dfrac{\pi}{12}$ **2.** $\sin\dfrac{\pi}{12}$ **3.** $\sin\dfrac{3\pi}{8}$ **4.** $\cos\dfrac{3\pi}{8}$

5. $\sin\dfrac{5\pi}{12}$ **6.** $\cos\dfrac{13\pi}{8}$ **7.** $\tan\dfrac{5\pi}{12}$ **8.** $\cos\dfrac{\pi}{8}$

9. $\sin\dfrac{\pi}{8}$ **10.** $\tan\dfrac{11\pi}{12}$ **11.** $\sin\dfrac{11\pi}{12}$ **12.** $\tan\dfrac{7\pi}{12}$

13. $\cos\left(-\dfrac{9\pi}{8}\right)$ **14.** $\tan\dfrac{17\pi}{12}$ **15.** $\sin\left(-\dfrac{7\pi}{8}\right)$ **16.** $\cos\left(-\dfrac{5\pi}{12}\right)$

17. $\cos 165°$ **18.** $\sin 165°$ **19.** $\tan 165°$ **20.** $\cos 195°$

21. $\sin 195°$ **22.** $\tan 195°$ **23.** $\cos 75°$ **24.** $\sin 345°$

25. Given that $\sin t = \sqrt{2}/3$ and t is a second quadrant angle, compute $\sin(t/2)$, $\cos(t/2)$, $\sin 2t$, and $\cos 2t$.

26. Given that $\cos t = \sqrt{3}/5$ and t is a fourth quadrant angle, compute $\sin(t/2)$, $\cos(t/2)$, $\sin 2t$, and $\cos 2t$.

27. Given that $\sin t = -1/3$ and t is a third quadrant angle, compute $\sin(t/2)$, $\cos(t/2)$, $\sin 2t$, and $\cos 2t$.

28. Using the results of Problem 1, compute the value of $\sin(\pi/12)$ without the aid of an addition formula. Verify that your answer is the same as that obtained in Problem 2.

In Problems 29–44, verify the given formula.

29. $\sin\left(t + \dfrac{\pi}{2}\right) = \cos t$ **30.** $\cos\left(t + \dfrac{\pi}{2}\right) = -\sin t$

31. $\cos(t + \pi) = -\cos t$ **32.** $\sin(t + \pi) = -\sin t$

33. $\cos(\pi - t) = -\cos t$ **34.** $\sin(\pi - t) = \sin t$

35. $\sin(t + \pi/4) = (\sqrt{2}/2)(\sin t + \cos t)$ **36.** $\cos(t + \pi/4) = (\sqrt{2}/2)(\cos t - \sin t)$

37. $\cot(t + \pi) = \cot t$

38. $\sin\left(t - \dfrac{3\pi}{2}\right) = \cos t$

39. $\sin 3t = 3 \sin t - 4 \sin^3 t$

40. $\cos 3t = 4 \cos^3 t - 3 \cos t$

41. $\sin u \sin v = \frac{1}{2}[\cos(u - v) - \cos(u + v)]$

42. $\cos u \cos v = \frac{1}{2}[\cos(u - v) + \cos(u + v)]$

43. $\sin u \cos v = \frac{1}{2}[\sin(u + v) + \sin(u - v)]$

44. $\cos u \sin v = \frac{1}{2}[\sin(u + v) - \sin(u - v)]$

45. Show that $\cos\left(\dfrac{\pi}{2} - t\right) = \sin t$. Use this result to derive formula (4) by observing that

$$\sin(u + v) = \cos\left[\frac{\pi}{2} - (u + v)\right] = \cos\left[\left(\frac{\pi}{2} - u\right) - v\right]$$

46. Use formula (4) to derive formula (3).

47. Derive formula (6)   **48.** Derive formula (7).   **49.** Derive formula (11).

50. Show that $\tan 2t = \dfrac{2 \tan t}{1 - \tan^2 t}$

51. Show that $\tan \dfrac{t}{2} = \dfrac{1 - \cos t}{\sin t} = \dfrac{\sin t}{1 + \cos t}$.

In Problems 52–57, use Problem 51 to find the exact value of the given expression.

52. $\tan \dfrac{\pi}{12}$

53. $\tan \dfrac{\pi}{8}$

54. $\tan \dfrac{3\pi}{8}$

55. $\cot \dfrac{5\pi}{8}$

56. $\cot\left(-\dfrac{3\pi}{8}\right)$

57. $\cot\left(-\dfrac{\pi}{12}\right)$

In Problems 58–60 use an appropriate addition or half-angle formula to find the exact value of the given expression.

58. $\sec \dfrac{11\pi}{12}$

59. $\csc \dfrac{7\pi}{8}$

60. $\sec \dfrac{11\pi}{8}$

Review Problems

61. Find the exact value of the given expression.

a. $\cos\left(-\dfrac{2\pi}{3}\right)$

b. $\tan \dfrac{17\pi}{6}$

c. $\csc \dfrac{9\pi}{2}$

62. Given that $\csc t = 4$ and that the terminal side of the angle t lies in the second quadrant, find $\cot t$.

63. Eliminate the variable t in the equations $x = 3 \sin t$ and $y = 2 \cos t$ to obtain an equation in terms of only x and y.

64. Use a calculator to evaluate $f(1.3)$ when $f(t) = \sec t + \tan t$.

4.5 Identities

An equation such as

$$2(x - 1) = 2x - 2,$$

which is valid for all real numbers x, is called an **identity**. Also, an equation such as

$$\frac{x^2 - 4x}{x} = x - 4$$

is called an identity since it is valid for all real numbers for which both sides of the equation are defined, in this case, all $x \neq 0$.

Example 1 The trigonometric equation

$$\frac{\sin t}{\tan t} = \cos t$$

is an identity, since

$$\frac{\sin t}{\tan t} = \frac{\sin t}{\frac{\sin t}{\cos t}}$$

$$= \sin t \left(\frac{\cos t}{\sin t} \right)$$

$$= \cos t$$

for all real numbers t for which $\tan t$ is defined and $\tan t \neq 0$. ∎

To show that a trigonometric equation is an identity, we use the basic definitions and identities that were given in Sections 4.2 and 4.3:

$$\frac{\sin t}{\cos t} = \tan t \qquad \frac{\cos t}{\sin t} = \cot t$$

$$\frac{1}{\sin t} = \csc t \qquad \frac{1}{\cos t} = \sec t$$

$$\cos^2 t + \sin^2 t = 1$$

$$1 + \cos^2 t = \csc^2 t$$

$$1 + \tan^2 t = \sec^2 t$$

Example 2 Show that

$$\sec^2 t + \csc^2 t = \sec^2 t \csc^2 t$$

is an identity.

Solution   We show that the left-hand side of the equation is equivalent to the right-hand side.

$$\sec^2 t + \csc^2 t = \frac{1}{\cos^2 t} + \frac{1}{\sin^2 t}$$

$$= \frac{1}{\cos^2 t}\left(\frac{\sin^2 t}{\sin^2 t}\right) + \frac{1}{\sin^2 t}\left(\frac{\cos^2 t}{\cos^2 t}\right)$$

$$= \frac{\sin^2 t + \cos^2 t}{\cos^2 t \sin^2 t}$$

$$= \frac{1}{\cos^2 t \sin^2 t}$$

$$= \left(\frac{1}{\cos t}\right)^2 \left(\frac{1}{\sin t}\right)^2$$

$$= \sec^2 t \csc^2 t \qquad \blacksquare$$

Implicit in the above example is the assumption that the identity is valid only for those values of t for which both sides of the identity are defined. Note in this case that we must require $t \neq k\pi$ and $t \neq \pi/2 + k\pi$, where k is an integer. Hereafter we shall not belabor the restrictions on t.

Example 3   Show that

$$\sin t \cos t = \frac{1}{\tan t + \cot t}$$

is an identity.

Solution   We show that the right-hand side of the equation is equivalent to the left-hand side.

$$\frac{1}{\tan t + \cot t} = \frac{1}{\dfrac{\sin t}{\cos t} + \dfrac{\cos t}{\sin t}}$$

$$= \frac{1}{\dfrac{\sin^2 t + \cos^2 t}{\sin t \cos t}}$$

$$= \frac{\sin t \cos t}{\sin^2 t + \cos^2 t}$$

$$= \sin t \cos t. \qquad \blacksquare$$

Example 4   Show that

$$\sin t + \sin t \cot^2 t = \cos t \csc t \sec t$$

is an identity.

Solution   In this case we reduce both sides of the equation to the same expression:

$$\sin t + \sin t \cot^2 t = \sin t(1 + \cot^2 t)$$

$$= \sin t(\csc^2 t)$$

$$= \frac{1}{\csc t}\csc^2 t$$

$$= \csc t;$$

$$\cos t \csc t \sec t = \cos t \csc t \frac{1}{\cos t}$$

$$= \csc t.$$

Since both sides of the given equation are equivalent to $\csc t$, they are equivalent to each other. Therefore the equation is an identity.   ∎

Example 5   Show that

$$\tan 2t = \frac{2\tan t}{2 - \sec^2 t}$$

is an identity.

Solution   We use the addition formula for the tangent given on page 246 to obtain a double-angle formula for the tangent.

$$\tan 2t = \tan(t + t) = \frac{2\tan t}{1 - \tan^2 t}.$$

We now show that the left-hand side of the given equation is equal to the right-hand side.

$$\tan 2t = \frac{2\tan t}{1 - \tan^2 t}$$

$$= \frac{2\tan t}{1 - (\sec^2 t - 1)}$$

$$= \frac{2\tan t}{2 - \sec^2 t}.$$ ∎

In order to verify a trigonometric identity, we are required to show that the given expressions are equivalent. Notice that in the preceding four examples we worked *independently* with the expressions on each side of the equation to show that they were equivalent. That is, we did *not* perform the same algebraic operations on *both* sides of the equation simultaneously. This is standard practice in verifying trigonometric identities.

There is no general method for showing that a trigonometric equation is an identity. We list below a few techniques that may be useful.

1. Simplify the more complicated side of the equation first.
2. Find common denominators for sums or differences of fractions.
3. Express all trigonometric functions in terms of sines and cosines and then simplify.

Exercise 4.5

In Problems 1–32, show that the given equation is an identity.

1. $\dfrac{\sin t}{\csc t} = 1 - \dfrac{\cos t}{\sec t}$

2. $\dfrac{1 + \sin t}{\cos t} = \sec t + \tan t$

3. $1 - \cos^4 t = (2 - \sin^2 t)\sin^2 t$

4. $\dfrac{1 + \tan t}{\tan t} = \cot t + \sec^2 t - \tan^2 t$

5. $1 - 2\sin^2 t = 2\cos^2 t - 1$

6. $\tan^2 t - \sin^2 t = \tan^2 t \sin^2 t$

7. $\dfrac{\sec t - \csc t}{\sec t + \csc t} = \dfrac{\tan t - 1}{\tan t + 1}$

8. $\dfrac{\sin t + \tan t}{1 + \cos t} = \tan t$

9. $\dfrac{\sec^4 t - \tan^4 t}{1 + 2\tan^2 t} = 1$

10. $\dfrac{1 + \sin t}{\cos t} + \dfrac{\cos t}{1 + \sin t} = 2\sec t$

11. $\sin^2 t \cot^2 t + \cos^2 t \tan^2 t = 1$

12. $\dfrac{\sin t + \tan t}{\cot t + \csc t} = \sin^2 t \sec t$

13. $\sec t - \dfrac{\cos t}{1 + \sin t} = \tan t$

14. $\dfrac{1}{\sec t - \tan t} = \sec t + \tan t$

15. $\dfrac{\tan^2 t}{1 + \cos t} = \dfrac{\sec t - 1}{\cos t}$

16. $\dfrac{\tan^2 t - 1}{\sin t + \cos t} = \dfrac{\sin t - \cos t}{\cos^2 t}$

17. $(\csc t - \cot t)^2 = \dfrac{1 - \cos t}{1 + \cos t}$

18. $\cos t - \sin t + \csc t = \dfrac{\sin t + \cos t}{\tan t}$

19. $1 + \dfrac{1}{\cos t} = \dfrac{\tan^2 t}{\sec t - 1}$

20. $\dfrac{\tan t + \cot t}{\cos^2 t} - \sin t \sec^3 t = \sec t \csc t$

21. $\dfrac{\cot t - \tan t}{\cot t + \tan t} = 1 - 2\sin^2 t$

22. $\dfrac{1 + \sec t}{\sin t + \tan t} = \csc t$

23. $2\cos^2 t - 1 = \dfrac{1 - \tan^2 t}{1 + \tan^2 t}$

24. $\dfrac{\sin t}{1 + \cos t} = 2\csc t$

25. $\dfrac{\cos t}{\sin t} + \dfrac{\sin t}{\cos t} = \dfrac{\sec t}{\sin t}$

26. $\dfrac{1 + \tan^2 t}{1 + \cot^2 t} = \tan^2 t$

27. $\dfrac{1 - \cos 2t}{\sin 2t} = \tan t$

28. $\dfrac{2\tan t}{1 + \tan^2 t} = \sin 2t$

29. $\dfrac{1 - \tan^2 t}{\cos 2t} = \dfrac{2\tan t}{\sin 2t}$

30. $\cot 2t = \dfrac{\csc^2 t - 2}{2\cot t}$

31. $\dfrac{\sin^2 2t}{(1 + \cos 2t)^2} = \sec^2 t - 1$

32. $\tan 4t = \dfrac{4 \tan t - 4 \tan^3 t}{1 - 6 \tan^2 t + \tan^4 t}$

In Problems 33 and 34, show that the given equation is an identity in the variables u and v.

33. $\sin^2 u - \sin^2 v = \sin(u + v)\sin(u - v)$

34. $\cos^2 u - \cos^2 v = -\sin(u + v)\sin(u - v)$

35. Under appropriate circumstances, the equation of motion of a vibrating string stretched between two points on the x axis is

$$y = A \sin(\omega t - kx) - A \sin(\omega t + kx)$$

where t is time and A, ω, and k are constants. Show that y can be equivalently represented in the form

$$y = -2A \cos \omega t \sin kx.$$

36. Show that the following equations are not identities by finding one value of t for which the equation is not valid.
   a. $\sin^2 t - \tan^2 t = \sin^2 t \tan^2 t$ **b.** $\sin(\csc t) = 1$

Review Problems

37. If $\tan t = \sec t$, find all values of $\sin t$, $\cos t$, and $\tan t$.

38. Find the exact value of $\cos \dfrac{21\pi}{6}$.

39. Find the exact value of $\sin \dfrac{13\pi}{12}$.

40. Show that $\sin 5t = 5 \sin t - 20 \sin^3 t + 16 \sin^5 t$.

41. Show that $\sin 4t = 8 \cos^3 t \sin t - 4 \cos t \sin t$.

42. Find functions $A(h)$ and $B(h)$ such that

$$\frac{\sin(x + h) - \sin x}{h} = A(h)\sin x + B(h)\cos x.$$

4.6 Graphs of the Trigonometric Functions

One of the best ways to further our understanding of the trigonometric functions is to examine their graphs.

We obtain an idea of the graph of the sine function, $f(t) = \sin t$, in Figure 4.23(b) by considering various positions of the terminal side of the angle t on the unit circle as

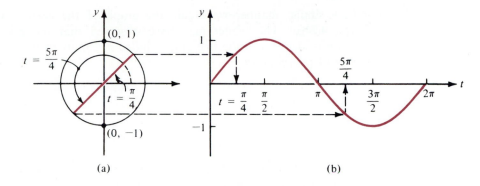

(a)   (b)

Figure 4.23

shown in Figure 4.23(a). As t varies from 0 to $\pi/2$, the value $f(t) = \sin t$ increases from 0 to its maximum value 1. But, as the angle t varies from $\pi/2$ to $3\pi/2$, the value $\sin t$ decreases from 1 to its minimum value -1. We note that $\sin t$ changes from positive to negative at $t = \pi$. For t between $3\pi/2$ and 2π we see that the corresponding values of $\sin t$ increase from -1 to 0.

Using the values of the sine for $0 \le t \le 2\pi$ obtained in Problem 1, Exercise 4.2, we have plotted points on the graph of $y = \sin t$ and joined them with a smooth curve in Figure 4.24(a). Since $\sin(t + 2\pi) = \sin t$, the graph of $y = \sin t$ for $2\pi \le t \le 4\pi$ is the same as the graph for $0 \le t \le 2\pi$. Using the periodicity of the sine function, we can extend the graph in either direction as shown in Figure 4.24(a).

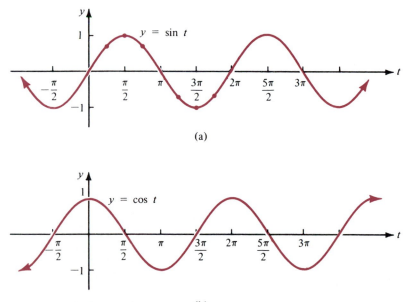

(a)

(b)

Figure 4.24

In a similar manner we obtain the graph of the cosine function shown in Figure 4.24(b). The reader may have observed that the graph of the cosine function is identical to the sine graph shifted $\pi/2$ units to the left. This is a consequence of the property

$$\cos t = \sin\left(\frac{\pi}{2} + t\right)$$

from Problem 29 of Section 4.4. For a further discussion of shifted graphs see Section 4.7.

We leave it to the reader to verify the graphs of the remaining trigonometric functions given in Figure 4.25. See Problems 13–16.

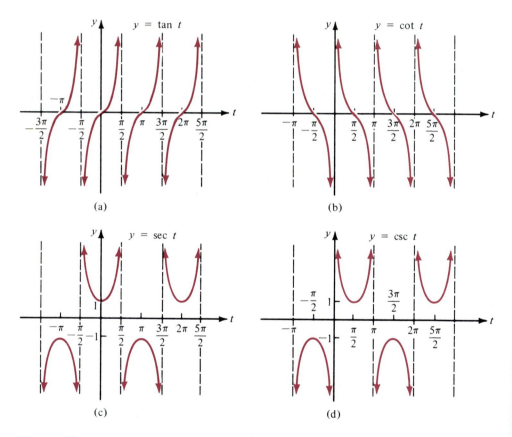

Figure 4.25

**Addition of
y Coordinates**

Recall from Section 2.5 that the graph of the sum of two functions $f(t) + g(t)$ can be obtained by graphing $f(t)$ and $g(t)$ on the same set of axes and then adding the ordinates.

Example 1   Graph $f(t) = 1 + \sin t$.

$f(t) = 1 + \sin t$

Solution   The given function can be viewed as the sum of the two functions $g(t) = 1$ and $h(t) = \sin t$; that is, $f(t) = g(t) + h(t)$. The addition of the constant function $g(t)$ to $h(t)$ merely shifts the graph of $h(t) = \sin t$ up one unit. The resulting graph is shown in Figure 4.26.

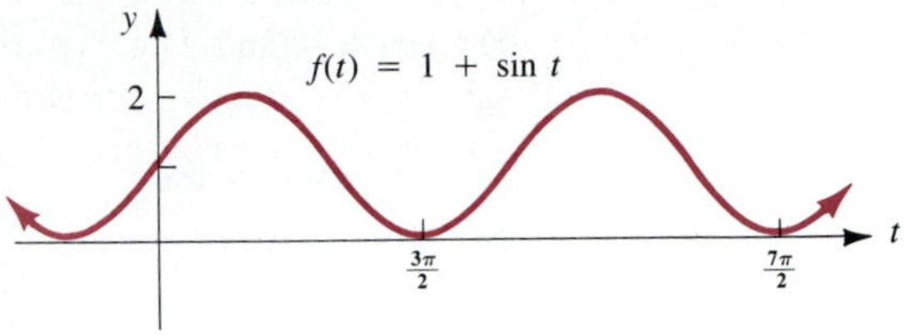

Figure 4.26

Example 2   Graph $f(t) = \sin t + \cos t$.

$f(t) = \sin t + \cos t$

Solution   In Figure 4.27 we have drawn the graphs of $g(t) = \sin t$ and $h(t) = \cos t$ with dashed curves. To obtain a point on the graph of $f(t) = \sin t + \cos t$, we choose an abscissa t, and add the ordinates $\sin t$ and $\cos t$, as indicated in Figure 4.27. After a sufficient number of points are located, we join them with a smooth curve.

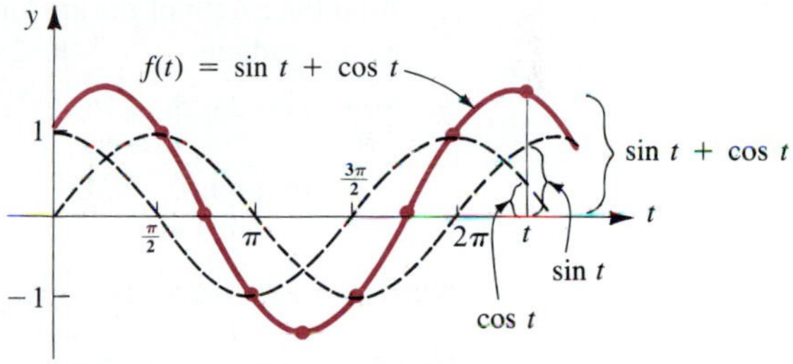

Figure 4.27

Exercise 4.6

In Problems 1–12, graph the given function.

1. $f(t) = -\sin t$

2. $f(t) = -\cos t$

3. $f(t) = 1 + \cos t$

4. $f(t) = -1 + \sin t$

5. $f(t) = 2 - \sin t$

6. $f(t) = -(3 + \cos t)$

7. $f(t) = \cos t - \sin t$

8. $f(t) = \sin t + \cos t + 1$

9. $f(t) = t + \sin t$

10. $f(t) = \sin t - \cos t$

11. $f(t) = 2 \sin t$

12. $f(t) = 2 \sin t + \cos t$

13. Use the values of $\tan t$ obtained in Problem 1, Exercise 4.3, and the properties $\tan(-t) = -\tan t$ and $\tan(t + \pi) = \tan t$, to graph $f(t) = \tan t$.

14. Use Problem 13 and the property $\tan(\pi/2 - t) = \cot t$ to graph $f(t) = \cot t$.

15. Use the graph of the sine function in Figure 4.24(a) and the fact that $\csc t = 1/\sin t$ to graph $f(t) = \csc t$.

16. Use Problem 15 and the property $\csc(t + \pi/2) = \sec t$ to graph $f(t) = \sec t$.

In Problems 17–28, graph the given function.

17. $f(t) = -\tan t$ **18.** $f(t) = 2 + \sec t$ **19.** $f(t) = 2 \cot t$

20. $f(t) = 2 \sec t$ **21.** $f(t) = \cot t + 1$ **22.** $f(t) = 2 \csc t$

23. $f(t) = \sin t + \csc t$ **24.** $f(t) = t - \tan t$ **25.** $f(t) = 1 - \csc t$

26. $f(t) = -\sec t$ **27.** $f(t) = t + \csc t$ **28.** $f(t) = \cos t + \sec t$

Review Problems

29. On a circle of radius r the length of the arc subtended by a central angle of t radians is tr. Find the length of the arc for a circle of radius 3 which is subtended by the given angle.
     a. $\pi/3$ radians   **b.** 2 radians   **c.** $30°$ **d.** $180°$

30. Show that $\cos 5t = 16 \cos^5 t - 20 \cos^3 t + 5 \cos t$.

31. Show that $\tan(t + \pi/4) = \dfrac{1 + \tan t}{1 - \tan t}$.

In Problems 32 and 33, show that the given equation is an identity.

32. $\csc t - \sin t = \cot t \cos t$

33. $\dfrac{1 + \csc t}{\sec t} = \cos t + \cot t$

4.7 Additional Trigonometric Graphs

In this section we shall study the properties and graphs of the functions

$$f(t) = a \sin(bt + c) \tag{1}$$

and

$$g(t) = a \cos(bt + c), \tag{2}$$

where a, b, and c are real numbers.

     We first consider the function $F(t) = a \sin t$, which is a special case of (1) with $b = 1$ and $c = 0$. For example, as shown in Figure 4.28(a), we obtain the graph of $f(t) = 2 \sin t$ by doubling each ordinate on the graph of $g(t) = \sin t$. Note that the maximum and minimum values of $f(t) = 2 \sin t$ occur at the same abscissas as the maximum and minimum values of $g(t) = \sin t$, respectively. However, as illustrated in Figure 4.28(b), this situation is reversed for $h(t) = -2 \sin t$; that is, a minimum value occurs when $g(t) = \sin t$ has a maximum value, and conversely. We also observe that the graph of $h(t) = -2 \sin t$ is the reflection in the t axis of the graph

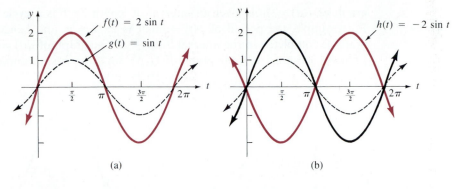

Figure 4.28

of $f(t) = 2 \sin t$. In general the graph of $F(t) = a \sin t$ can be obtained by multiplying the ordinates of the graph of $g(t) = \sin t$ by the number a. Similar remarks can be made for $G(t) = a \cos t$.

Amplitude

It follows from the preceding discussion that the maximum distance from the t axis of any point on the graph of $F(t) = a \sin t$ or $G(t) = a \cos t$ is $|a|$. This number is called the **amplitude** of each of these functions.

Example 1   Compare the graphs of $f(t) = 2 \cos t$ and $g(t) = \frac{1}{2} \cos t$.

Solution   The given functions have amplitudes 2 and $\frac{1}{2}$ respectively. Limiting our attention to the interval $0 \le t \le 2\pi$, we know that the cosine attains its maximum at $t = 0$ and its minimum at $t = \pi$. Thus $2 \cos 0 = 2$ and $2 \cos \pi = -2$ are the maximum and minimum values, respectively, of $f(t) = 2 \cos t$. For $g(t) = \frac{1}{2} \cos t$ we find the maximum and minimum values to be $\frac{1}{2} \cos 0 = \frac{1}{2}$ and $\frac{1}{2} \cos \pi = -\frac{1}{2}$, respectively. The graphs are shown on the same axes in Figure 4.29.

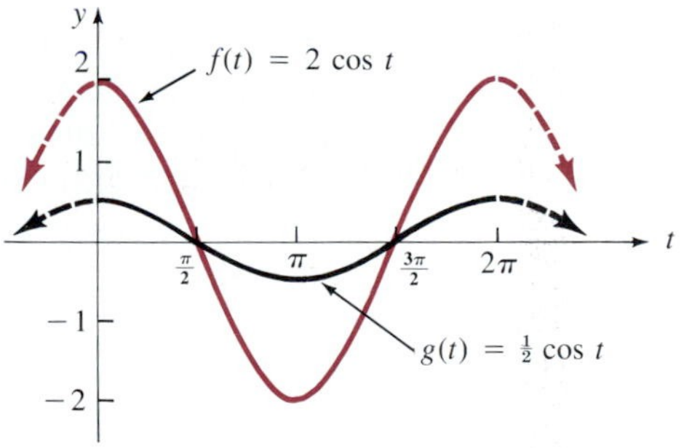

Figure 4.29

Period and Cycle

We now consider the graph of $F(t) = \sin bt$ for $b > 0$. Recall that $f(t) = \sin t$ has period 2π. Thus, starting at $t = 0$, $F(t) = \sin bt$ will repeat its values beginning at $bt = 2\pi$, or $t = 2\pi/b$. It follows that $F(t) = \sin bt$ has **period** $2\pi/b$; this means the graph will repeat itself every $2\pi/b$ units. For this reason we say that the graph of $F(t) = \sin bt$ over an interval of length $2\pi/b$ is a **cycle** of the sine curve. For example, the period of $g(t) = \sin 2t$ is $2\pi/2 = \pi$, and hence one cycle of the graph is completed in the interval $0 \le t \le \pi$. The graph of $g(t) = \sin 2t$ is shown in Figure 4.30 along with the graph of $f(t) = \sin t$ (the dashed curve) for comparison.

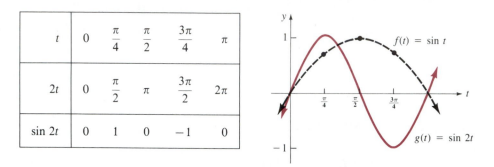

t	0	$\dfrac{\pi}{4}$	$\dfrac{\pi}{2}$	$\dfrac{3\pi}{4}$	π
$2t$	0	$\dfrac{\pi}{2}$	π	$\dfrac{3\pi}{2}$	2π
$\sin 2t$	0	1	0	-1	0

Figure 4.30

Similar remarks hold for the cosine function. That is, for $b > 0$ the function $G(t) = \cos bt$ has period $2\pi/b$, and one cycle of the graph is completed in the interval $0 \le t \le \pi$.

Example 2   Find the period of $f(t) = \cos 4t$ and graph.

Solution   The period is $2\pi/4 = \pi/2$. Hence one cycle of the graph is completed in any interval of length $\pi/2$. To graph the function, we draw one cycle of the cosine curve on the interval $0 \le t \le \pi/2$. Then, as shown in Figure 4.31, we use periodicity to extend the graph over the entire t axis.

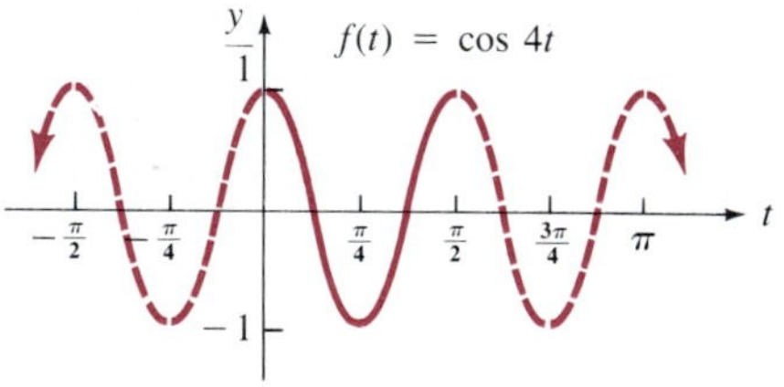

Figure 4.31

Combining the results of the preceding discussions, we find that the graphs of $F(t) = a \sin bt$ and $G(t) = a \cos bt$, $b > 0$, each have

$$\text{amplitude} = |a|$$

$$\text{period} = \frac{2\pi}{b}.$$

If $b < 0$ in either $F(t) = a \sin bt$ or $G(t) = a \cos bt$, we use the property

$$\sin(-t) = -\sin t \tag{3}$$

or

$$\cos(-t) = \cos t \tag{4}$$

to rewrite the expression in the standard form. This is illustrated in the following example.

Example 3   Find the amplitude and period of $f(t) = \sin(-\frac{1}{2}t)$. Graph

Solution   Since we want $b > 0$, we use (3) to write

$$f(t) = \sin(-\tfrac{1}{2}t) = -\sin \tfrac{1}{2}t.$$

It follows that the amplitude is $|a| = |-1| = 1$, and the period is $2\pi/\frac{1}{2} = 4\pi$. Hence the graph of the given function completes one cycle on the interval $0 \le t \le 4\pi$. In Figure 4.32 the solid curve is the graph of $f(t) = -\sin \frac{1}{2} t$ and is a reflection in the t axis of the dashed graph $g(t) = \sin \frac{1}{2}t$.

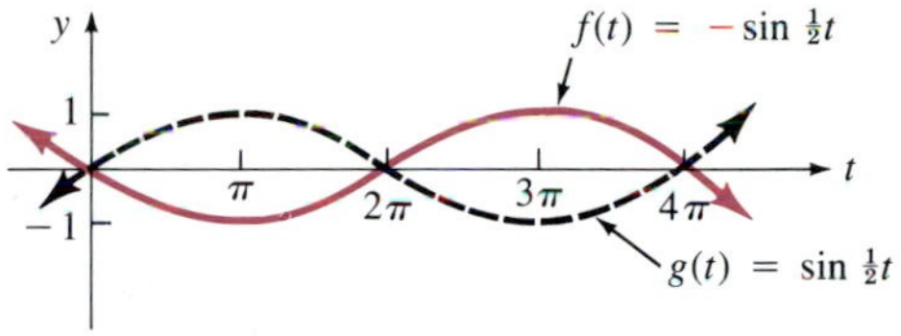

Figure 4.32

Example 4   Graph $f(t) = \frac{5}{2} \sin 2\pi t$.

Solution   The amplitude is $\frac{5}{2}$ and the period is $2\pi/2\pi = 1$. Thus the function completes one cycle on the interval $0 \le t \le 1$. Since the sine attains its maximum at $\pi/2$, we solve $2\pi t = \pi/2$ for t to see that the maximum ordinate $y = \frac{5}{2}$ occurs at $t = \frac{1}{4}$. Similarly, solving $2\pi t = 3\pi/2$ for t, we see that the minimum ordinate $y = -\frac{5}{2}$ occurs at $t = \frac{3}{4}$. In Figure 4.33 one cycle of the graph is shown as a solid curve.

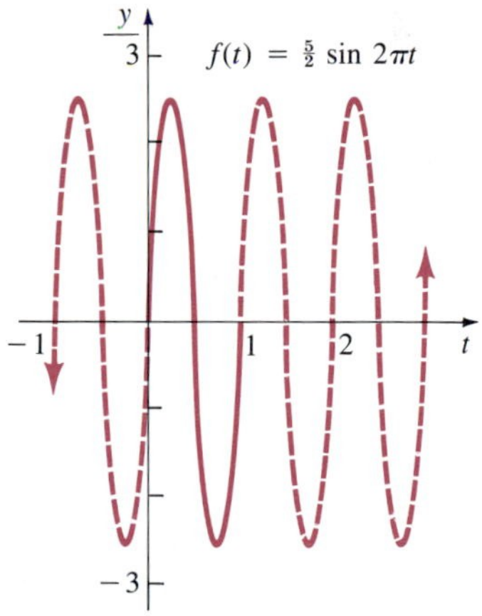

Figure 4.33

We note that there is no amplitude associated with $f(t) = \tan t$, $g(t) = \cot t$, $h(t) = \sec t$, or $k(t) = \csc t$, since their graphs are unbounded. However, $H(t) = \sec bt$ and $K(t) = \csc bt$, $b > 0$, each have the period $2\pi/b$, while $F(t) = \tan bt$ and $G(t) = \cot bt$, $b > 0$, each have the period π/b. See Problems 29–32.

Phase Shift

We now consider the graph of (1), $f(t) = a \sin(bt + c)$, for $b > 0$. Since the values of $\sin(bt + c)$ range from -1 to 1, it follows that the amplitude of (1) is $|a|$. As $bt + c$ varies from 0 to 2π, the graph will complete one cycle. Hence by solving $bt + c = 0$ and $bt + c = 2\pi$, we find that one cycle is completed as t varies from $-c/b$ to $(2\pi - c)/b$. Therefore $f(t) = a \sin(bt + c)$ has period

$$\frac{2\pi - c}{b} - \left(-\frac{c}{b}\right) = \frac{2\pi}{b}.$$

Writing $a \sin(bt + c) = a \sin[b(t + c/b)]$, we see that the graph of $f(t) = a \sin(bt + c)$ can be obtained by *shifting* the graph of $f(t) = a \sin bt$ horizontally a distance of $|c|/b$. If $c < 0$, the shift is to the right, whereas if $c > 0$, the shift is to the left. (See Section 2.3.) The number $|c|/b$ is called the **phase shift** of the graph of (1).

Example 5 Compare the graphs of $f(t) = 3 \sin 2t$ and $g(t) = 3 \sin(2t - \pi/3)$.

Solution The amplitude of the first function is $|a| = 3$, and the period is $2\pi/2 = \pi$. To find the t intercepts t_n, we solve

$$\sin 2t = 0,$$

which yields

$$2t = n\pi, \qquad n = 0, \pm 1, \pm 2, \ldots$$

or

$$t_n = \frac{n\pi}{2}, \qquad n = 0, \pm 1, \pm 2, \ldots$$

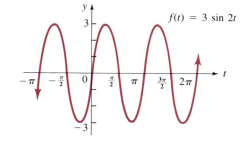

Figure 4.34

The graph is given in Figure 4.34.

Now the amplitude and period of $g(t) = 3 \sin(2t - \pi/3)$ are identical to those of the first function, but the phase shift is $|c|/b = |-\pi/3|/2 = \pi/6$. Since $c = -\pi/3 < 0$, the shift is to the right. The graph of $g(t) = 3 \sin(2t - \pi/3)$ is given by the colored curve in Figure 4.35.

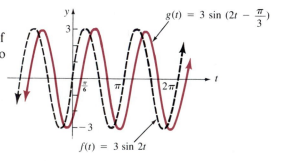

Figure 4.35

An analysis similar to the one above for $f(t) = a \sin(bt + c)$ can be made for the graph of $g(t) = a \cos(bt + c)$, with the result that for both graphs

$$
\begin{aligned}
\text{amplitude} &= |a|, \\[4pt]
\text{period} &= \frac{2\pi}{b}, \\[4pt]
\text{phase shift} &= \frac{|c|}{b}.
\end{aligned}
$$

The graphs have the same form as those of $F(t) = a \sin bt$ and $G(t) = a \cos bt$, *except* they are shifted to the right or to the left $|c|/b$ units depending on whether $c < 0$ or $c > 0$, respectively.

Example 6   Determine the amplitude, period, phase shift and direction of shift for

a. $f(t) = 5 \cos\left(5t - \dfrac{3\pi}{2}\right)$ **b.** $g(t) = 10 \sin\left(-2t - \dfrac{\pi}{6}\right)$.

Solution   **a.** We make the identifications $a = 5$, $b = 5$, and $c = -3\pi/2$. Thus the amplitude is $|a| = 5$, the period is $2\pi/b = 2\pi/5$, and the phase shift is $|c|/b = |-3\pi/2|/5 = 3\pi/10$. Since $c = -3\pi/2 < 0$, the graph of $f(t) = 5 \cos(5t - 3\pi/2)$ is the graph of $F(t) = 5 \cos 5t$ shifted $3\pi/10$ units to the right.

b. Since we require that $b > 0$, we first use (3) to write

$$
g(t) = 10 \sin\left(-2t - \frac{\pi}{6}\right) = -10 \sin\left(2t + \frac{\pi}{6}\right).
$$

Now with $a = -10$, $b = 2$, and $c = \pi/6$, we find that the amplitude is $|a| = 10$, the period is $2\pi/2 = \pi$, and the phase shift is $(\pi/6)/2 = \pi/12$. Since $c = \pi/6 > 0$, the graph of $G(t) = -10 \sin 2t$ is shifted $\pi/12$ units to the left.   ∎

Example 7   Graph $f(t) = -2 \cos\left(t + \dfrac{\pi}{3}\right)$.

Solution   Since $b = 1$ and $c = \pi/3$, the phase shift is $\pi/3$. Thus the graph shown in Figure 4.36 is obtained by shifting the graph of $F(t) = -2 \cos t$ (the dashed curve) $\pi/3$ units to the left.

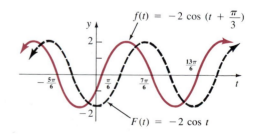

Figure 4.36   ∎

Exercise 4.7

In Problems 1–12, graph the given function. State the amplitude and period. Give the y intercept and the first three positive t intercepts, if they exist.

1. $f(t) = 4 \cos t$

2. $f(t) = -2 \cos t$

3. $f(t) = -\frac{1}{2} \sin t$

4. $f(t) = \frac{3}{2} \sin \pi t$

5. $f(t) = 2 \cos 2t$

6. $f(t) = \cos \frac{1}{4} t$

7. $f(t) = 4 \sin(-t)$.

8. $f(t) = 2 \sin 4t$

9. $f(t) = 5 \cos 2\pi t$

10. $f(t) = \cos(-\frac{1}{2}t)$

11. $f(t) = \sin \frac{2}{3} t$

12. $f(t) = -3 \sin(-2t)$

In Problems 13–24, graph the given function. State the amplitude, period, and phase shift.

13. $f(t) = \sin\left(t - \frac{\pi}{6}\right)$

14. $f(t) = \sin\left(3t - \frac{\pi}{4}\right)$

15. $f(t) = \cos\left(t + \frac{\pi}{4}\right)$

16. $f(t) = -2 \cos\left(2t - \frac{\pi}{6}\right)$

17. $f(t) = 4 \cos\left(2t - \frac{3\pi}{2}\right)$

18. $f(t) = 3 \sin\left(2t + \frac{\pi}{4}\right)$

19. $f(t) = 3 \sin\left(\frac{t}{2} - \frac{\pi}{3}\right)$

20. $f(t) = -\cos\left(\frac{t}{2} - \pi\right)$

21. $f(t) = 5 \cos\left(\frac{2t}{3} - \frac{\pi}{12}\right)$

22. $f(t) = 2 \sin\left(-t + \frac{\pi}{8}\right)$

23. $f(t) = 2 \cos\left(-4t - \frac{4\pi}{3}\right)$

24. $f(t) = 5 \sin(\pi t - 1)$

In Problems 25–28, state the amplitude and period of the given function.

25. $f(t) = 2 \sin 2t \cos 2t$

26. $f(t) = 5 \cos^2 4t - 5 \sin^2 4t$

27. $f(t) = 3 \sin^2 5t - 3 \cos^2 5t$

28. $f(t) = 4 \sin 6t \cos 6t$

29. Prove that the period of $f(t) = \tan bt$ for $b > 0$ is $\dfrac{\pi}{b}$.

30. Prove that the period of $g(t) = \cot bt$ for $b > 0$ is $\dfrac{\pi}{b}$.

31. Prove that the period of $h(t) = \sec bt$ for $b > 0$ is $\dfrac{2\pi}{b}$.

32. Prove that the period of $k(t) = \csc bt$ for $b > 0$ is $\dfrac{2\pi}{b}$.

In Problems 33–40, graph the given function. State the period. Give the y intercept and the first three positive t intercepts, if they exist.

33. $f(t) = \tan \pi t$

34. $f(t) = \cot \frac{\pi}{2} t$

35. $f(t) = 3 \sec \frac{1}{2} t$

36. $f(t) = -\csc 2t$

37. $f(t) = 2 \csc 3t$

38. $f(t) = 2 \tan 0.7t$

39. $f(t) = 4 \cot 1.5t$ **40.** $f(t) = \sec(-2t)$

In Problems 41 and 42, write the given function in the form $G(t) = a \cos(bt + c), a > 0.$

41. $g(t) = -4 \sin 3t$ **42.** $g(t) = 2 \sin 2t$

In Problems 43 and 44, write the given function in the form $F(t) = a \sin(bt + c), a > 0.$

43. $f(t) = 2 \cos 5t$ **44.** $f(t) = 3 \cos(-4t)$

In Problems 45–48, determine if the given function is periodic. Explain.

45. $f(t) = t + \cos t$ **46.** $f(t) = t \sin t$

47. $f(t) = \sin^2 3t$ **48.** $f(t) = 1 - \sec^2 t$

In Problems 49–51, find the period of the given function.

49. $f(t) = \sin \frac{1}{2}t + \sin 2t$

50. $f(t) = \cos \pi t - 2 \cos 3\pi t$

51. $f(t) = 3 \tan \frac{1}{2}t + 2 \tan 3t$

52. The general equation for simple harmonic motion is

$$y = A \sin(\omega t + \phi).$$

This equation can be written in the equivalent form

$$y = B \sin \omega t + C \cos \omega t.$$

Find B and C in terms of A and ϕ.

In Problems 53 and 54, find a shifted sine function with the given graph.

53.

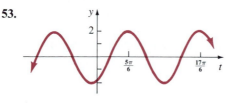

Figure 4.37

54.

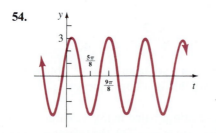

Figure 4.38

Review Problems

55. Find how many radians the hour hand of a clock travels in the given amount of time.
    a. 40 minutes   **b.** 2 hours, 25 minutes

56. Show $\tan(t + \pi/2) = -\cot t$.

In Problems 57 and 58, show that the given equation is an identity.

57. $\dfrac{1 + \cos^2 t \, \csc^2 t}{1 + \cot^2 t} = 1$ **58.** $\dfrac{1 - \cos t}{\sin^2 t} = \dfrac{1 + 2\cos t}{2\cos^2 t + 3\cos t + 1}$

In Problems 59 and 60, graph the given function.

59. $f(t) = 1 - \sin(t - \pi/4)$ **60.** $f(t) = 2 + 3\cos(2t + \pi/6)$

4.8 Trigonometric Equations

In the study of advanced mathematics, science, and engineering, equations involving the trigonometric functions are often encountered. For example,

$$\sin x = \frac{\sqrt{2}}{2} \tag{1}$$

and

$$4\sin^2 x - 8\sin x + 3 = 0 \tag{2}$$

are trigonometric equations. We consider the problem of finding all real numbers x satisfying $\sin x = \sqrt{2}/2$ by examining the graph of $f(x) = \sin x$. As Figure 4.39 indicates, there exists an infinite number of solutions

$$\ldots, \; -\frac{7\pi}{4}, \frac{\pi}{4}, \frac{9\pi}{4}, \frac{17\pi}{4}, \ldots$$

and $\tag{3}$

$$\ldots, \; -\frac{5\pi}{4}, \frac{3\pi}{4}, \frac{11\pi}{4}, \frac{19\pi}{4}, \ldots$$

Note that in each list in (3) every solution may be obtained by adding 2π to the preceding solution. Of course, this is a consequence of the periodicity of the sine function.

    In practice, to obtain the solutions listed in (3), we employ a unit circle rather than the graph of $f(x) = \sin x$. Since $\sin x = \sqrt{2}/2$, the reference angle for x is $\pi/4$ radians. The fact that the value $\sin x = \sqrt{2}/2$ is positive implies that the angle of x radians can

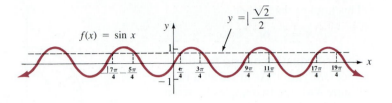

Figure 4.39

be in either the first or the second quadrant. Thus, as illustrated in Figure 4.40, the only solutions between 0 and 2π are

$$x = \frac{\pi}{4} \quad \text{and} \quad x = \frac{3\pi}{4}.$$

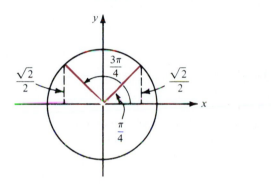

Figure 4.40

All other solutions may be obtained by adding integer multiples of 2π to these solutions. Hence the solution set of (1) is

$$\left\{ x \,\middle|\, x = \frac{\pi}{4} + 2n\pi \quad \text{or} \quad x = \frac{3\pi}{4} + 2n\pi, \quad n = 0, \pm 1, \pm 2, \ldots \right\}.$$

When faced with a more complicated equation, such as (2), the basic approach is to solve for sin x, cos x, or tan x by using methods similar to those employed in solving algebraic equations. Then the values of x are determined using the unit circle. We illustrate this technique in the following examples.

Example 1 Find all solutions of $4 \sin^2 x - 8 \sin x + 3 = 0$.

Solution We first observe that this is a quadratic equation in sin x, and that it factors as

$$(2 \sin x - 3)(2 \sin x - 1) = 0.$$

This implies that either

$$\sin x = \tfrac{3}{2} \quad \text{or} \quad \sin x = \tfrac{1}{2}.$$

The first equation has no solution since $|\sin x| \le 1$. As illustrated in Figure 4.41, the two angles between 0 and 2π with sine equal to $\tfrac{1}{2}$ are

$$x = \frac{\pi}{6} \quad \text{and} \quad x = \frac{5\pi}{6}.$$

Therefore, by the periodicity of the sine function, the solution set is

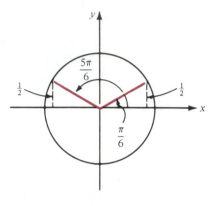

Figure 4.41

$$\left\{ x \,\middle|\, x = \frac{\pi}{6} + 2n\pi \quad \text{or} \quad x = \frac{5\pi}{6} + 2n\pi, \qquad n = 0, \pm 1, \pm 2, \dots \right\}. \qquad \blacksquare$$

Example 2   Solve $\sin x = \cos x$. $\qquad\qquad$ (4)

Solution   Dividing both sides of the equation by $\cos x$ gives

$$\tan x = 1. \qquad\qquad (5)$$

This equation is equivalent to (4) provided that $\cos x \neq 0$.
   We observe that if $\cos x = 0$, then

$$x = \frac{\pi}{2} + 2n\pi \quad \text{or} \quad x = \frac{3\pi}{2} + 2n\pi$$

for any integer n. Since

$$\sin\!\left(\frac{\pi}{2} + 2n\pi\right) \neq 0$$

and

$$\sin\!\left(\frac{3\pi}{2} + 2n\pi\right) \neq 0,$$

these values of x do not satisfy the original equation. Thus we shall find *all* the solutions to (4) by solving equation (5).
   Now $\tan x = 1$ implies that the reference angle for x is $\pi/4$ radians. Since $\tan x = 1 > 0$, the angle of x radians can lie in either the first or the third quadrant, as shown in Figure 4.42. Thus the solution

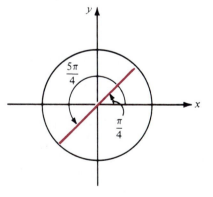

Figure 4.42

set is

$$\left\{x \mid x = \frac{\pi}{4} + 2n\pi \quad \text{or} \quad x = \frac{5\pi}{4} + 2n\pi, n = 0, \pm 1, \pm 2, \dots\right\},$$

which is equivalent to

$$\left\{x \mid x = \frac{\pi}{4} + n\pi, \quad n = 0, \pm 1, \pm 2, \dots\right\}.$$

∎

Example 3 Solve $2 \sin x \cos^2 x = -\dfrac{\sqrt{3}}{2} \cos x$.

Solution We write the equation as

$$2 \sin x \cos^2 x + \frac{\sqrt{3}}{2} \cos x = 0$$

and factor:

$$\cos x \left(2 \sin x \cos x + \frac{\sqrt{3}}{2} \right) = 0.$$

Thus either

$$\cos x = 0 \quad \text{or} \quad 2 \sin x \cos x + \frac{\sqrt{3}}{2} = 0.$$

Now the cosine is zero for all odd multiples of $\pi/2$, that is,

$$x = (2n + 1)\frac{\pi}{2} = \frac{\pi}{2} + n\pi, \quad n = 0, \pm 1, \pm 2, \dots.$$

In the second equation we use formula (9) of Section 4.4 to replace $2 \sin x \cos x$ by $\sin 2x$ and obtain

$$\sin 2x + \frac{\sqrt{3}}{2} = 0 \quad \text{or} \quad \sin 2x = -\frac{\sqrt{3}}{2}.$$

Thus the reference angle of $2x$ is $\pi/3$. Since the sine is negative, the angle $2x$ must be in either the third or the fourth quadrant. As Figure 4.43 indicates, either

$$2x = \frac{4\pi}{3} + 2n\pi$$

or

$$2x = \frac{5\pi}{3} + 2n\pi,$$

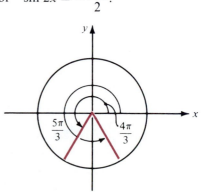

Figure 4.43

and hence

$$x = \frac{2\pi}{3} + n\pi \quad \text{or} \quad x = \frac{5\pi}{6} + n\pi.$$

Therefore the solution set is

$$\left\{ x \,\middle|\, x = \frac{\pi}{2} + n\pi, \qquad x = \frac{2\pi}{3} + n\pi, \quad \text{or} \quad x = \frac{5\pi}{6} + n\pi, \qquad n = 0, \pm 1, \pm 2, \ldots \right\}.$$

∎

Example 4   Solve $3 \cos^2 x - \cos 2x = 1$.

Solution   From formula (8) of Section 4.4 we have that

$$\cos 2x = \cos^2 x - \sin^2 x.$$

Thus the given equation is equivalent to

$$3 \cos^2 x - (\cos^2 x - \sin^2 x) = 1,$$

or

$$2 \cos^2 x + \sin^2 x = 1.$$

Replacing $\sin^2 x$ by $1 - \cos^2 x$, we have

$$2 \cos^2 x + 1 - \cos^2 x = 1$$
$$\cos^2 x = 0$$
$$\cos x = 0.$$

Therefore the solution set is

$$\left\{ x \,\middle|\, x = (2n + 1)\frac{\pi}{2}, \qquad n = 0, \pm 1, \pm 2, \ldots \right\}.$$

∎

In the next section we will consider trigonometric equations with solutions which do not have reference angles of 0, $\pi/6$, $\pi/4$, $\pi/3$, or $\pi/2$.

Exercise 4.8

In Problems 1–20, find all solutions of the given equation.

1. $\tan x = 0$

2. $\cot x + 1 = 0$

3. $2 \cos x + \sqrt{2} = 0$

4. $\sqrt{3} \sin x = \cos x$

5. $\cos^2 x - 1 = 0$

6. $2 \sin^2 x + (2 - \sqrt{3})\sin x - \sqrt{3} = 0$

7. $2 \cos^2 x - 3 \cos x - 2 = 0$

8. $\tan^2 x + (\sqrt{3} - 1)\tan x - \sqrt{3} = 0$

9. $3 \sec^2 x = \sec x$

10. $2 \sin^2 x - \sin x - 1 = 0$

11. $\sin 2x + \sin x = 0$

12. $\cos 2x + \sin^2 x = 1$

13. $\cos 2x = \sin x$

14. $\sin 2x + 2 \sin x - 2 \cos x = 2$

15. $\sec x \sin^2 x = \tan x$

16. $\dfrac{1 + \cos x}{\cos x} = 2$

17. $\sqrt{\dfrac{1 + 2 \sin x}{2}} = 1$

18. $\sin x + \sqrt{\sin x} = 0$

19. $\cos x - \sqrt{\cos x} = 0$

20. $\cos x \sqrt{1 + \tan^2 x} = 1$

In Problems 21–30, find the solutions of the given equation in the interval $[0, 2\pi)$.

21. $1 + \sin x = \cos x$

22. $\tan x = \cot x$

23. $2 \sin x = \tan x$

24. $\sec 2x \csc x = 0$

25. $\sin x - 2 \cos^2 x = 4$

26. $\cot x = \csc^2 x - 1$

27. $2 \sec x - \sec^3 x = 0$

28. $2 \sin^2 x - \cos x = 2$

29. $\tan^2 x \cos x + 2 \tan^2 x = 3 \cos x + 6$

30. $2 \sin 2x + 6 \sin x + 2 \cos x + 3 = 0$

In Problems 31–34, show that the given equation has no solutions.

31. $\sin x + \cos x = 3$ 32. $\cos x + \cos^2 2x = -1$ 33. $\sec^2 x + \csc^2 x = 1$

34. $\tan^2 x - \sec^2 x = 1$

35. Determine by graphing whether the equation $\tan x = x$ has any solutions.

36. Consider a ray of light passing from one medium (such as air) into another medium (such as water). Let ϕ be the angle of incidence and θ the angle of refraction. As shown in Figure 4.44, these angles are measured from a normal line. According to Snell's law, there is a constant c, depending on the two mediums, such that

$$\frac{\sin \phi}{\sin \theta} = c.$$

Suppose that for light passing from air into water $c = 1.33$. Find ϕ and θ so that the angle of incidence will be twice the angle of refraction.

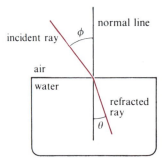

Figure 4.44

Review Problems

37. For a circle of radius r, the area of the sector of the circle subtended by a central angle of t radians is $r^2t/2$. Find the area of the sector of a circle of radius 3 subtended by the given angle.

    a. $\pi/4$ radians   **b.** $5\pi/6$ radians

    c. $15°$ **d.** $225°30'$

38. Show that $\cos^2 u - \sin^2 v = \cos(u + v)\cos(u - v)$ is an identity in u and v.

39. Show that $\sin^2 t + \sin^2(\pi/2 - t) = 1$.

40. Graph $f(t) = 2\sin(t + \pi/3)$. Find the amplitude, period, and phase shift.

41. Find the period of $f(t) = \sin \frac{1}{2}t - \cos 4t$.

42. Find the equation of the line through the points on the graph of $f(x) = 2\sin 3x$ with x coordinates $\pi/9$ and $\pi/6$.

4.9 Inverse Trigonometric Functions

From Section 2.6 we know that a function has an inverse if and only if it is one-to-one. Inspection of the graphs of the various trigonometric functions clearly shows that these functions are not one-to-one. However, by suitably restricting each of their domains, the resulting functions are one-to-one.

Arcsine

From Figure 4.45 we see that the function $f(x) = \sin x$ is one-to-one on the closed interval $[-\pi/2, \pi/2]$, since on this interval any horizontal line intersects the graph at most once. Hence the sine function restricted to this particular interval has an inverse. We denote this inverse by

$$\sin^{-1}x \quad \text{or} \quad \arcsin x.$$

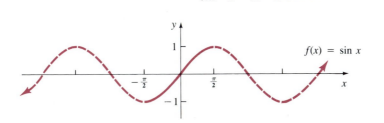

Figure 4.45

These symbols are read "inverse sine of x" and "arc sine of x," respectively. We define

$$y = \arcsin x \quad \text{if and only if} \quad x = \sin y,$$

where $-\pi/2 \le y \le \pi/2$. In other words, *the arcsine of the number x is that number y whose sine is x.*

Examples 1 Find **a.** arcsin $\frac{1}{2}$ **b.** $\sin^{-1}(-\frac{1}{2})$ **c.** $\sin^{-1}(-1)$

Solutions **a.** If we let $y = \arcsin \frac{1}{2}$, then $-\pi/2 \le y \le \pi/2$ and $\sin y = \frac{1}{2}$. It follows that $y = \pi/6$.
b. If we let $y = \sin^{-1}(-\frac{1}{2})$, then $\sin y = -\frac{1}{2}$. Since we must choose y such that $-\pi/2 \le y \le \pi/2$, we find that $y = -\pi/6$.
c. Letting $y = \sin^{-1}(-1)$, we have that $\sin y = -1$ and $-\pi/2 \le y \le \pi/2$. Hence $y = -\pi/2$. ∎

Note that the "-1" in $\sin^{-1} x$ is *not* an exponent; rather, it denotes an inverse function; that is,

$$(\sin x)^{-1} = \frac{1}{\sin x} \ne \sin^{-1} x.$$

Recall from Section 2.6 that the graph of an inverse function is the reflection of the graph of the given function in the line $y = x$. This technique is used in Figure 4.46(a) to obtain the graph of $g(x) = \arcsin x$. From Figure 4.46(b) we see that the domain of arcsin x is $[-1, 1]$ and the range is $[-\pi/2, \pi/2]$.

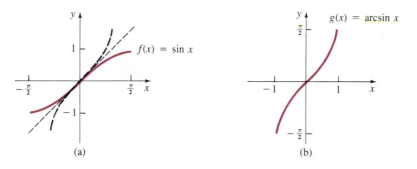

(a)   (b)

Figure 4.46

Example 2 Find $\tan(\sin^{-1} \frac{1}{4})$.

Solution We must find the tangent of the angle of t radians with sine equal to $\frac{1}{4}$; that is, tan t where $t = \sin^{-1} \frac{1}{4}$. The angle t is shown in Figure 4.47. Since

$$\tan t = \frac{\sin t}{\cos t} = \frac{\frac{1}{4}}{\cos t}, \tag{1}$$

we wish to determine cos t. Using $\sin t = \frac{1}{4}$ and the identity $\cos^2 t + \sin^2 t = 1$ we see that

$$\cos^2 t + (\tfrac{1}{4})^2 = 1,$$

or

$$\cos t = \frac{\sqrt{15}}{4}.$$

Substituting this value into (1), we have

$$\tan t = \frac{\frac{1}{4}}{\frac{\sqrt{15}}{4}} = \frac{\sqrt{15}}{15},$$

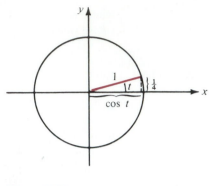

Figure 4.47

and thus

$$\tan(\sin^{-1}\tfrac{1}{4}) = \tan t = \frac{\sqrt{15}}{15}. \qquad \blacksquare$$

Arccosine

To define the inverse of $\cos x$, the domain of the cosine function is restricted to the closed interval $[0, \pi]$. We define

$$y = \textbf{arccos } x \quad \text{if and only if} \quad x = \cos y.$$

The graphs of $f(x) = \cos x$ and $g(x) = \arccos x$ are shown in Figure 4.48. The domain of $\arccos x$ is $[-1, 1]$ and the range is $[0, \pi]$.

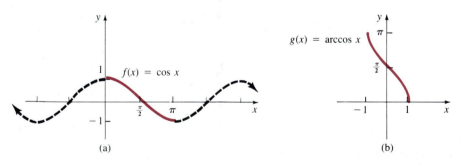

(a)   (b)

Figure 4.48

Example 3   Find   **a.** $\arccos \dfrac{\sqrt{2}}{2}$ **b.** $\cos^{-1}\left(-\dfrac{\sqrt{3}}{2}\right)$

Solution   **a.** If we let $y = \arccos(\sqrt{2}/2)$, then $\cos y = \sqrt{2}/2$, and $0 \leq y \leq \pi$. Thus $y = \pi/4$.
b. Letting $y = \cos^{-1}(-\sqrt{3}/2)$, we have that $\cos y = -\sqrt{3}/2$, and we must choose y such that $0 \leq y \leq \pi$. Therefore $y = 5\pi/6$. ∎

Example 4   Find $\sin(\cos^{-1}x)$.

Solution   In Figure 4.49 we show an angle of t radians with cosine equal to x. To find $\sin(\cos^{-1}x)$ $= \sin t$, we use $\cos t = x$ and the identity $\cos^2 t + \sin^2 t = 1$. Then

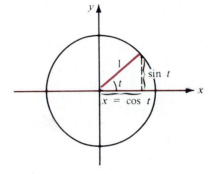

$$x^2 + \sin^2 t = 1$$

$$\sin^2 t = 1 - x^2$$

$$\sin t = \sqrt{1 - x^2}.$$

We use the *positive* square root of $1 - x^2$ since the range of $\cos^{-1}x$ is $[0, \pi]$, and the sine of an angle in the first or second quadrant is positive.

Figure 4.49

∎

Arctangent   If we restrict the domain of $\tan x$ to the open interval $(-\pi/2, \pi/2)$, then the resulting function is one-to-one and consequently has an inverse. We define

$$\boxed{y = \textbf{arctan } x \quad \text{if and only if} \quad x = \textbf{tan } y,}$$

where $-\pi/2 < y < \pi/2$. The graphs of $f(x) = \tan x$ and $g(x) = \arctan x$ are given in Figure 4.50. The domain of $\arctan x$ is R and the range is $(-\pi/2, \pi/2)$.

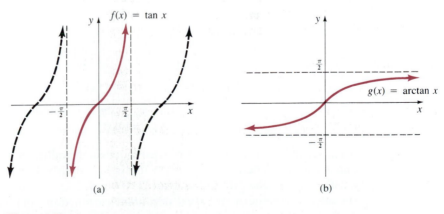

(a)   (b)

Figure 4.50

Example 5   Find $\tan^{-1}(-1)$.

Solution   If $\tan^{-1}(-1) = y$, then $\tan y = -1$, where $-\pi/2 < y < \pi/2$. It follows that $\tan^{-1}(-1) = y = -\pi/4$. ∎

Example 6   Find $\sin(\tan^{-1}(-\frac{5}{3}))$.

Solution   If we let $t = \tan^{-1}(-\frac{5}{3})$, then $\tan t = -\frac{5}{3}$. We use the identity $1 + \tan^2 t = \sec^2 t$ to find $\sec t$:

$$1 + (-\tfrac{5}{3})^2 = \sec^2 t$$

$$\sec t = \sqrt{\tfrac{25}{9} + 1}$$

$$= \sqrt{\tfrac{34}{9}}$$

$$= \tfrac{1}{3}\sqrt{34}.$$

We take the positive square root, since the range of $\tan^{-1} t$ is $(-\pi/2, \pi/2)$, and the secant of an angle in the first or fourth quadrant is positive. Now we find $\cos t$:

$$\cos t = \frac{1}{\sec t}$$

$$= \frac{1}{\tfrac{1}{3}\sqrt{34}}$$

$$= \frac{3}{\sqrt{34}}.$$

Finally we can use the identity

$$\tan t = \frac{\sin t}{\cos t}$$

to compute $\sin(\tan^{-1}(-\frac{5}{3}))$. We find that

$$\sin t = \tan t \cos t$$

$$= -\frac{5}{3}\left(\frac{3}{\sqrt{34}}\right).$$

$$= -\frac{5\sqrt{34}}{34}$$ ∎

Use of a Calculator

Most scientific calculators can be used to calculate values of the arcsine, arccosine, and arctangent functions. Frequently this is accomplished by first pressing an $\boxed{\text{INV}}$ key followed by the $\boxed{\text{SIN}}$, $\boxed{\text{COS}}$, or $\boxed{\text{TAN}}$ key. If the calculator is in degree mode the calculated value of an inverse trigonometric function will normally be an angle in degrees, with fractions of a degree in decimal form. In radian mode the calculated value will be an angle measured in radians.

Example 7 Use a calculator to evaluate

$$\textbf{a. } \sin^{-1}0.5 \qquad \textbf{b. } \cos^{-1}(-0.7) \qquad \textbf{c. } \tan^{-1}(-1)$$

Solution **a.** To calculate $\sin^{-1}0.5$ we first enter 0.5 and then press $\boxed{\text{INV}}$ followed by $\boxed{\text{SIN}}$. If the calculator is in degree mode the result is $30°$. In radian mode the result is 0.5236, which is $\pi/6$ correct to 4 decimal places.

b. Since -0.7 is negative, we expect $\cos^{-1}(-0.7)$ to be a second quadrant angle. We obtain

$$\cos^{-1}(-0.7) \approx 134.4270°$$

or

$$\cos^{-1}(-0.7) \approx 2.3462 \text{ radians.}$$

c. Since -1 is negative, we expect $\tan^{-1}(-1)$ to be a fourth quadrant angle. We obtain

$$\tan^{-1}(-1) = -45°$$

or

$$\tan^{-1}(-1) \approx -0.7854 \text{ radians.} \qquad \blacksquare$$

The functions $\cot x$, $\sec x$, and $\csc x$ when restricted to the domains $(0, \pi)$, $[0, \pi/2) \cup (\pi/2, \pi]$, and $[-\pi/2, 0) \cup (0, \pi/2]$, respectively, have inverses. See Problems 50–52. Any calculator that computes arcsine, arccosine, and arctangent can be used to calculate arccotangent, arcsecant, and arccosecant. See Problems 53–55.

Trigonometric Equations

Many trigonometric equations have solutions involving angles with reference angles not equal to 0, $\pi/6$, $\pi/4$, $\pi/3$, or $\pi/2$. In this case we can frequently express a solution in terms of an inverse trigonometric function and then use a calculator to find the decimal form of the solution set.

Example 8 Use a calculator to find all solutions of $3 \sin^2 x - 2 \sin x - 1 = 0$ in $[0, 2\pi)$.

Solution Factoring the equation we obtain

$$(\sin x - 1)(3 \sin x + 1) = 0.$$

This implies that

$$\sin x - 1 = 0 \quad \text{or} \quad 3 \sin x + 1 = 0.$$

From the first equation we have $\sin x = 1$. The angle between 0 and 2π with sine equal to 1 is $\pi/2 \approx 1.5708$. From the second equation we have $\sin x = -\frac{1}{3}$. Using a calculator set to radian mode we obtain

$$x = \sin^{-1}(-\tfrac{1}{3}) \approx -0.3398.$$

The corresponding reference angle is 0.3398 radians, and angles between 0 and 2π with $\sin x = -1/3$ are

$$\pi + 0.3398 \approx 3.4814 \quad \text{and} \quad 2\pi - 0.3398 \approx 5.9433.$$

Thus, the solutions of the equation in $[0, 2\pi)$ are 1.5708, 3.4814, and 5.9433. ∎

Exercise 4.9

In Problems 1–16, find the indicated value without the aid of a calculator.

1. $\sin^{-1}0$
2. $\tan^{-1}\sqrt{3}$
3. $\cos^{-1}(-1)$
4. $\sin^{-1}\dfrac{\sqrt{3}}{2}$

5. $\sec^{-1}2$
6. $\cot^{-1}(-\sqrt{3})$
7. $\csc^{-1}\sqrt{2}$
8. $\cos^{-1}\dfrac{\sqrt{3}}{2}$

9. $\tan^{-1}1$
10. $\csc^{-1}(-1)$
11. $\cot^{-1}\left(-\dfrac{\sqrt{3}}{3}\right)$
12. $\sec^{-1}(-\sqrt{2})$

13. $\sin^{-1}\left(-\dfrac{\sqrt{2}}{2}\right)$
14. $\tan^{-1}0$
15. $\csc^{-1}2$
16. $\sec^{-1}\dfrac{2\sqrt{3}}{3}$

In Problems 17–32, find the indicated value without the aid of a calculator.

17. $\tan(\cos^{-1}(-\frac{2}{3}))$
18. $\cos(\sin^{-1}\frac{1}{3})$
19. $\sin(\sec^{-1}2)$
20. $\cot(\cot^{-1}(-3))$

21. $\csc(\sin^{-1}\frac{3}{5})$
22. $\cos(\cos^{-1}\frac{4}{5})$
23. $\tan(\sin^{-1}\frac{2}{5})$
24. $\sec(\tan^{-1}4)$

25. $\sin(\csc^{-1}3)$
26. $\sec(\cos^{-1}\frac{4}{5})$

27. $\cos^{-1}\left(\cos\dfrac{\pi}{5}\right)$
28. $\tan^{-1}\left(\tan\dfrac{15\pi}{8}\right)$

29. $\sin^{-1}(\sin(-35°))$
30. $\sin^{-1}(\sin(210°))$

31. $\tan^{-1}(\tan 225°)$
32. $\cos^{-1}(\cos(-30°))$

Calculator Problems

In Problems 33–41, use a calculator to compute the value of each expression.

33. $\sin^{-1}0.7033$
34. $\cos^{-1}0.2675$
35. $\tan^{-1}5.798$

36. $\sin^{-1}(-0.419)$
37. $\tan^{-1}(-3.53)$
38. $\cos^{-1}(-0.015)$

39. $\sin(\cos^{-1}0.7317)$
40. $\tan(\sin^{-1}0.1296)$
41. $\cos(\tan^{-1}1.369)$

In Problems 42–49, write the given expression in terms of x without any trigonometric functions.

42. $\cos(\tan^{-1}x)$
43. $\sin(\tan^{-1}x)$
44. $\sec(\cos^{-1}x)$
45. $\tan(\sin^{-1}x)$

46. $\cos(\sin^{-1}x)$
47. $\cot(\sin^{-1}x)$
48. $\tan(\cot^{-1}x)$
49. $\cos(\sec^{-1}x)$

50. Graph $g(x) = \text{arccot } x$ using $f(x) = \cot x$ restricted to $(0, \pi)$. Give the domain and range of arccot x.

51. Graph $g(x) = \text{arcsec } x$ using $f(x) = \sec x$ restricted to $[0, \pi/2) \cup (\pi/2, \pi]$. Give the domain and range of arcsec x.

52. Graph $g(x) = \operatorname{arccsc} x$ using $f(x) = \csc x$ restricted to $[-\pi/2, 0) \cup (0, \pi/2]$. Give the domain and range of $\operatorname{arccsc} x$.

53. Using the fact that $\sec(\sec^{-1} x) = x$, show that

$$\sec^{-1} x = \cos^{-1}\left(\frac{1}{x}\right).$$

54. Derive the formula

$$\csc^{-1} x = \sin^{-1}\left(\frac{1}{x}\right).$$

55. Express $\cot^{-1} x$ in terms of $\tan^{-1}\left(\frac{1}{x}\right)$. [*Hint:* Consider separately the cases $x < 0$, $x = 0$, and $x > 0$. Note that the domains of the tangent and cotangent functions are restricted to different sets when determining the inverses of these functions.]

In Problems 56–64, use a calculator and the results of Problems 53–55 to compute the indicated value.

56. $\sec^{-1} 2.5$ 57. $\cot^{-1} 0.75$ 58. $\csc^{-1}(-1.3)$

59. $\cot^{-1}(-1.5)$ 60. $\cot^{-1}(-0.3)$ 61. $\sec^{-1}(-1.2)$

62. $\sin(\sec^{-1} 3.2)$ 63. $\csc(\cot^{-1}(-0.28))$ 64. $\sec(\csc^{-1}(-5.9))$

In Problems 65–72, find all solutions of the given equation in $[0, 2\pi)$.

65. $\sin x = 4 \cos x$ 66. $5 \sin x - \tan x = 0$

67. $2 \tan^2 x - \tan x - 1 = 0$ 68. $8 \cos^2 x - 2 \cos x - 1 = 0$

69. $(\sin x - 4)(\tan x + 2) = 0$ 70. $(5 \sin x + 3)(2 \csc x - 3) = 0$

71. $3 \sin 2x + 4 \sin x = 3 \cos x + 2$ 72. $\csc^2 x = 2 \cot x + 4$

In Problems 73–76, use the techniques of Section 2.3 to graph the given function.

73. $f(x) = \sin^{-1}(x - 1)$ 74. $f(x) = \pi + \sin^{-1}(x + 2)$

75. $f(x) = \cos^{-1}(x + 1)$ 76. $f(x) = 1 + \tan^{-1}(x + 2)$

Review Problems

77. Determine whether the given function is even, odd, or neither.
   a. $f(t) = \sin t$ b. $f(t) = \cos t$ c. $f(t) = \tan t$
   d. $f(t) = \tan^2 t$ e. $f(t) = \sin t \cos t$ f. $f(t) = \sin^2 t - \cos^2 t$

78. Find the exact value of $\tan \dfrac{\pi}{8}$.

79. Show that $\tan 3t = \dfrac{3 \tan t - \tan^3 t}{1 - 3 \tan^2 t}$.

80. Show that $\sin 2t - \tan t = \cos 2t \tan t$ is an identity.

81. Find the amplitude and period of $f(t) = 4 \cos \frac{1}{3} t$. Graph.

82. Solve $(2 \sin x - 1)(\cot x + 1) = 0$ without the aid of a calculator.

Chapter Review

An **angle** with its **vertex** at the origin and **initial side** coinciding with the positive x axis is in **standard position**. The angle is measured from its initial side to its **terminal side** in either **degrees** or **radians**. From $360° = 2\pi$ radians we obtain the conversion formulas

$$1° = \left(\frac{\pi}{180}\right) \text{ radians} \quad \text{and} \quad 1 \text{ radian} = \left(\frac{180}{\pi}\right)°.$$

For any real number t there exists a unique angle of t radians in standard position with terminal side intersecting the unit circle at $P_t(x, y)$. We define the **cosine** of t to be x and the **sine** of t to be y, and write

$$x = \cos t \quad \text{and} \quad y = \sin t.$$

These two functions satisfy

$$|\sin t| \le 1, \quad |\cos t| \le 1, \quad \text{and} \quad \cos^2 t + \sin^2 t = 1.$$

The sine and cosine are **periodic** with **period** 2π.

The **tangent, cotangent, secant**, and **cosecant** functions are defined in terms of the sine and cosine functions by

$$\tan x = \frac{\sin x}{\cos x} \qquad \cot x = \frac{\cos x}{\sin x}$$

$$\sec x = \frac{1}{\cos x} \qquad \csc x = \frac{1}{\sin x}$$

These functions satisfy

$$1 + \tan^2 t = \sec^2 t \quad \text{and} \quad 1 + \cot^2 t = \csc^2 t.$$

The tangent and cotangent are π-periodic, while the secant and cosecant are 2π-periodic.

The **addition formulas** for the sine and cosine functions are

$$\sin(u + v) = \sin u \cos v + \cos u \sin v,$$

$$\sin(u - v) = \sin u \cos v - \cos u \sin v,$$

$$\cos(u + v) = \cos u \cos v - \sin u \sin v,$$

$$\cos(u - v) = \cos u \cos v + \sin u \sin v.$$

The **double-angle formulas** are

$$\sin 2u = 2 \sin u \cos u,$$

$$\cos 2u = \cos^2 u - \sin^2 u.$$

The **half-angle formulas** are

$$\sin^2 \frac{t}{2} = \frac{1}{2}(1 - \cos t),$$

$$\cos^2 \frac{t}{2} = \frac{1}{2}(1 + \cos t).$$

A trigonometric equation that is valid for all real numbers for which both sides of the equation are defined is called an **identity**.

The graph of $F(t) = a \sin(bt + c)$ or $G(t) = a \cos(bt + c)$, $b > 0$, can be obtained by shifting the graph of $f(t) = a \sin bt$ or $g(t) = a \cos bt$ horizontally a distance of $|c|/b$ units. If $c < 0$, the shift is the right, whereas if $c > 0$, the shift is to the left. The **amplitude** and **period** of both graphs are $|a|$ and $2\pi/b$, respectively. The number $|c|/b$ is called the **phase shift**.

When solving a **trigonometric equation**, care should be exercised to insure that *all* solutions are found. Usually there will be infinitely many solutions due to the periodic nature of the trigonometric functions.

By suitably restricting the domains of the trigonometric functions, the resulting functions are one-to-one. We can then define the **inverse trigonometric functions** as follows:

$$y = \arcsin x \quad \text{if and only if} \quad x = \sin y, \qquad -\frac{\pi}{2} \le y \le \frac{\pi}{2};$$

$$y = \arccos x \quad \text{if and only if} \quad x = \cos y, \qquad 0 \le y \le \pi;$$

$$y = \arctan x \quad \text{if and only if} \quad x = \tan y, \qquad -\frac{\pi}{2} < y < \frac{\pi}{2}.$$

Chapter 4 Review Exercises

In Problems 1–10, answer true or false.

1. A negative angle can be coterminal with a positive angle. _____

2. The product of the sine and cosine of an angle in the third quadrant is negative. _____

3. The reference angle of a fourth quadrant angle is a first quadrant angle. _____

4. The function $f(t) = \sin(t/2)$ is 4π-periodic. _____

5. There is an angle t such that $\sec t = \frac{1}{2}$. _____

6. If t is a third quadrant angle then $\cos(t/2)$ is positive. _____

7. The amplitude of $3 \sec 2t$ is 3. _____

8. If the sum of angles t_1 and t_2 is $\pi/2$, then $\tan t_1 = \cot t_2$. _____

9. If $\sin t = \frac{1}{2}$, then $\sin(t + 3\pi) = \frac{1}{2}$. _____

10. The arccosine of $-\sqrt{2}/2$ is $-\pi/4$. _____

11. Convert $-240°$ to radians.   12. Convert $98°12'$ to radians.

13. Convert $7\pi/6$ radians to degrees.   14. Convert -4π radians to degrees.

15. Find the exact value of $\tan 495°$.   16. Find the exact value of $\cos 495°$.

17. Find the exact value of $\sin(-7\pi/6)$.   18. Find the exact value of $\sec(-8\pi/3)$.

19. If t is a second quadrant angle and $\tan t = -4$, find the values of $\sin t, \cos t, \cot t, \sec t,$ and $\csc t$.

20. If $\cos t = \frac{4}{5}$ and t is in the fourth quadrant, find the values of $\sin t, \tan t, \cot t, \sec t,$ and $\csc t$.

21. Find the exact value of $\sin(5\pi/8)$.

22. Find the exact value of $\cot(-7\pi/12)$.

23. Show that $\cos(t + 3\pi/2) = \sin t$.

24. Show that $\tan u + \tan v = \dfrac{\sin(u + v)}{\cos u \cos v}$.

25. Show that $\tan t + \cot t = \sec t \csc t$ is an identity.

26. Show that $\dfrac{1 + \cos t}{\sin^3 t} = \dfrac{\sec t}{\tan t - \sin t}$ is an identity.

27. Show that $\sin(u + v) = \sin u + \sin v$ is not an identity.

28. Show that $\tan^2 t + \cot^2 t = 1$ is not an identity.

29. Graph $f(t) = \frac{1}{4}\cos \pi t$. Find the y intercept and the first three positive t intercepts.

30. Graph $f(t) = 4 \cos(-2t - \pi/3)$. State the amplitude, period, and phase shift. Find the y intercept and the first three positive t intercepts.

31. Find all solutions of $2 \cos^2 x - \cos x - 3 = 0$ without the aid of a calculator.

32. Find all solutions of $2 \sec^2 x - \sqrt{2} \sec x - 2 = 0$ without the aid of a calculator.

33. Use a calculator to find all solutions of $10 \sin^2 x + 11 \sin x = 6$ in $[0, 2\pi)$.

34. Use a calculator to find all solutions of $\tan^2 x = 3(1 - \sec x)$ in $[0, 2\pi)$.

35. Evaluate $\tan(\sec^{-1}(4/3))$ without the aid of a calculator.

36. Evaluate $\arccos(-\frac{1}{2})$ without the aid of a calculator.

37. Use a calculator to compute $\sin^{-1}(\tan 2.75)$.

38. Use a calculator to compute $\cos(\tan^{-1} 1.743)$.

39. Use a calculator to compute $\sec^{-1}(-3.1)$.

40. Express $\csc(\sin^{-1} x)$ in terms of x without any trigonometric functions.

5 Trigonometric Applications

5.1 Right-Triangle Trigonometry

Opposite and Adjacent Sides

The word *trigonometry** refers to the measurement of triangles. In fact, the trigonometric functions can be defined in terms of ratios of the lengths of the sides of a right triangle. As shown in Figure 5.1, if OAB is any right triangle, then the side AB is said to be the **side opposite** the angle θ. The side OA is called the **side adjacent** to the angle θ. The *hypotenuse OB* is opposite the right angle.

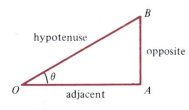

Figure 5.1

We now superimpose a Cartesian coordinate system on the triangle such that the origin is placed at the vertex O, and we let the coordinates of the vertex B be given by (a, b). See Figure 5.2(a). When a unit circle is drawn with center at the origin, we see from Figure 5.2(b) that there are two similar right triangles containing the same angle θ. Since corresponding sides of similar triangles are proportional, it follows that

$$(\sin \theta)/1 = b/r \quad \text{and} \quad (\cos \theta)/1 = a/r,$$

where $r = \sqrt{a^2 + b^2}$. That is,

$$\sin \theta = \frac{b}{r} = \frac{\text{side opposite}}{\text{hypotenuse}},$$

* From the Greek *trigonon* meaning triangle and *metria* meaning measurement.

285

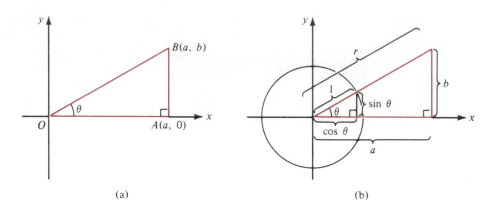

(a)   (b)

Figure 5.2

and
$$\cos \theta = \frac{a}{r} = \frac{\text{side adjacent}}{\text{hypotenuse}}.$$

Also we note that

$$\tan \theta = \frac{\sin \theta}{\cos \theta} = \frac{b/r}{a/r}$$

$$= \frac{b}{a} = \frac{\text{side opposite}}{\text{side adjacent}}.$$

Thus we can write each trigonometric function as a ratio of the lengths of sides in a right triangle.

We summarize the results.

$$\sin \theta = \frac{b}{r} = \frac{\text{side opposite}}{\text{hypotenuse}}$$

$$\cos \theta = \frac{a}{r} = \frac{\text{side adjacent}}{\text{hypotenuse}}$$

$$\tan \theta = \frac{b}{a} = \frac{\text{side opposite}}{\text{side adjacent}}$$

$$\cot \theta = \frac{a}{b} = \frac{\text{side adjacent}}{\text{side opposite}}$$

$$\sec \theta = \frac{r}{a} = \frac{\text{hypotenuse}}{\text{side adjacent}}$$

$$\csc \theta = \frac{r}{b} = \frac{\text{hypotenuse}}{\text{side opposite}}$$

(1)

Use of a Calculator

In the following examples the computations have been done with a scientific calculator. However, if Table IV is used to obtain the values of the trigonometric functions, the results may differ somewhat from those shown. This is due to the fact that the calculator is carrying eight or nine significant digits, while the table provides only four significant digits. To take full advantage of the calculator's greater accuracy, the computed values of the trigonometric functions must be retained in the calculator for subsequent use. If, instead, a displayed value is written down and then later a rounded version of that value is keyed into the calculator, the accuracy of the final result is likely to be diminished.

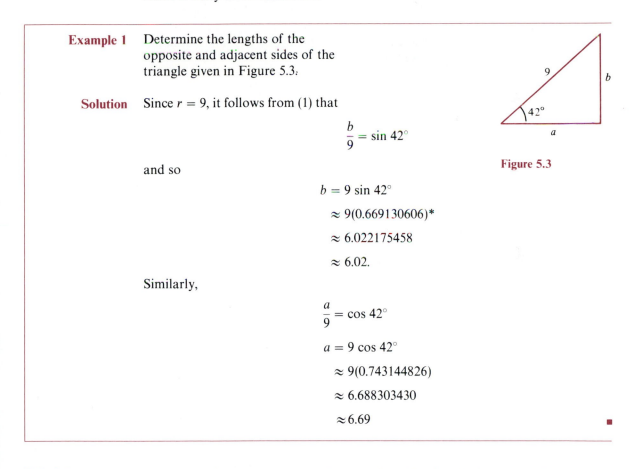

Example 1 Determine the lengths of the opposite and adjacent sides of the triangle given in Figure 5.3.

Figure 5.3

Solution Since $r = 9$, it follows from (1) that

$$\frac{b}{9} = \sin 42°$$

and so

$$b = 9 \sin 42°$$

$$\approx 9(0.669130606)*$$

$$\approx 6.022175458$$

$$\approx 6.02.$$

Similarly,

$$\frac{a}{9} = \cos 42°$$

$$a = 9 \cos 42°$$

$$\approx 9(0.743144826)$$

$$\approx 6.688303430$$

$$\approx 6.69$$

Calculator Comment

In the preceding example all the digits displayed by the calculator were shown. In the following examples all digits will be retained in the calculator for subsequent use, but only four decimal places will be shown in the text.

* At this point we are showing the value of the trigonometric function as a check for the reader. When working with a calculator this step would normally not be written down. We will continue to provide this check in subsequent examples.

Example 2   Consider the right triangle given in
Figure 5.4. Determine the remaining
sides.

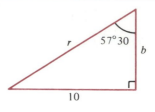

Figure 5.4

Solution   If r and b denote the lengths of the hypotenuse and side adjacent to the given angle, respectively, we can write

$$\frac{10}{r} = \sin 57°30',$$

or

$$r = \frac{10}{\sin 57.5°}$$

$$\approx \frac{10}{0.8434}$$

$$\approx 11.86.$$

We compute the value of b from

$$\frac{10}{b} = \tan 57°30'$$

or

$$b = \frac{10}{\tan 57.5°}$$

$$\approx \frac{10}{1.5697}$$

$$\approx 6.37.$$ ■

A word of caution is in order here. The symbol b in the preceding example denoted the side adjacent to the given angle, whereas in (1) the symbol b referred to the side opposite the angle θ. The reader should not merely memorize symbols; it is better to work with the concepts of opposite side, adjacent side, and hypotenuse.

Example 3 Two sides of a right triangle are given in Figure 5.5. Determine the length of the hypotenuse and the two acute angles α and β.

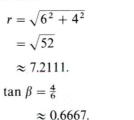

Solution From the Pythagorean theorem we have

$$r = \sqrt{6^2 + 4^2}$$

Figure 5.5

$$= \sqrt{52}$$

$$\approx 7.2111.$$

Now

$$\tan \beta = \tfrac{4}{6}$$

$$\approx 0.6667.$$

Using the arctangent function on a calculator, we find that $\beta \approx 33.69°$. Since the sum of the interior angles of any triangle is 180°, it follows that

$$\alpha + \beta + 90° = 180°$$

$$\alpha + \beta = 90°$$

$$\alpha = 90° - \beta$$

$$\approx 90° - 33.69°$$

$$= 56.31°.$$

■

Exercise 5.1

Calculator Problems

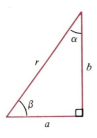

Figure 5.6

In Problems 1–27, find the indicated unknowns. Each problem refers to the triangle in Figure 5.6.

1. $a = 4, \beta = 27°; b, r$

2. $r = 10, \beta = 49°; a, b$

3. $b = 8, \beta = 34°20'; a, r$

4. $r = 25, \alpha = 50°; a, b$

5. $a = 6, \alpha = 61°10'; b, r$

6. $a = 5, b = 2; \alpha, \beta, r$

7. $b = 1.5, r = 3; \alpha, \beta, a$

8. $b = 4, \alpha = 58°; a, r$

9. $a = 4, b = 10; \alpha, \beta, r$

10. $b = 3, r = 6; \alpha, \beta, a$

11. $a = 9, r = 12; \alpha, \beta, b$

12. $a = 11, \alpha = 33.5°; b, r$

13. $b = 20, \alpha = 23°; a, r$

14. $r = 15, \beta = 31°40'; a, b$

15. $r = 12, \alpha = 22.7°; a, b$

16. $a = 3.7, \beta = 14°35'; \alpha, b, r$

17. $b = 4.3, \beta = 53.8°; \alpha, a, r$

18. $a = 5.4, r = 7.9; \alpha, \beta, b$

19. $r = 5.1, \alpha = 23°50'; \beta, a, b$

20. $a = 6, b = 3.7; \alpha, \beta, r$

21. $b = 17$, $\alpha = 15°5'$; β, a, r

22. $r = 12.2$, $\beta = 7°$; α, a, b

23. $a = 15.7$, $\alpha = 47°25'$; β, b, r

24. $b = 26$, $r = 37.5$; α, β, a

25. $a = 30$, $r = 50$; α, β, b

26. $b = 7.6$, $\beta = 41°$; α, a, r

27. $a = 4.8$, $\beta = 3.5°$; α, b, r

28. Show that the area of the triangle in Figure 5.6 is $\frac{1}{4}r^2 \sin 2\beta$.

5.2 Applications

Right-triangle trigonometry can be used to solve many practical problems, particularly those involving lengths, heights, and distances.

Example 1   A kite is caught in the top branches of a tree. If the 90-foot kite string makes an angle of 22° with the ground, find the approximate height of the tree by finding the distance from the kite to the ground.

Solution   Let h denote the height of the kite. From Figure 5.7 we see that

$$\frac{h}{90} = \sin 22°,$$

or

$$h = 90 \sin 22°$$

$$\approx 90(0.3746)$$

$$\approx 33.71 \text{ feet.}$$

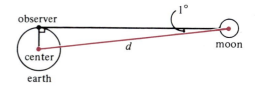

Figure 5.7

Example 2   The distance between the earth and the moon varies as the moon revolves around the earth. At a particular time an observer measures the angle of 1° shown in Figure 5.8.* Calculate the distance between the center of the earth and the center of the moon at this instant. Assume that the radius of the earth is 3963 miles.

Figure 5.8

---

* This angle is known as the **geocentric parallax**. The determination of this angle actually depends on two observations.

Solution Let d represent the distance between the center of the earth and the center of the moon. From the definition of the sine we have

$$\sin 1° = \frac{3963}{d},$$

or

$$d = \frac{3963}{\sin 1°}.$$

$$\approx \frac{3963}{0.0175}$$

$$\approx 227{,}075 \text{ miles.}$$ ∎

Example 3 A carpenter cuts the end of a 4-inch-wide board on a 25° bevel from the vertical, starting at a point $1\frac{1}{2}$ inches from the end of the board. Find the dimensions of the diagonal cut and the remaining side. See Figure 5.9.

Solution Let l, x, and y be the (unknown) dimensions as labeled in the figure. It then follows from the definition of the tangent of an angle that

$$\frac{x}{4} = \tan 25°$$

$$x = 4 \tan 25°$$

$$\approx 4(0.4663)$$

$$\approx 1.87 \text{ inches.}$$

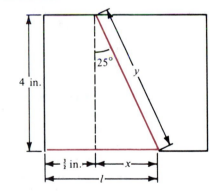

Also

$$\frac{y}{4} = \sec 25°$$

$$y = 4 \sec 25°$$

$$= \frac{4}{\cos 25°}$$

Figure 5.9

$$\approx \frac{4}{0.9063}$$

$$\approx 4.41 \text{ inches.}$$

Since $x \approx 1.87$, we see that

$$l = \tfrac{3}{2} + x \approx 3.37 \text{ inches.}$$ ∎

**Angles of
Elevation and
Depression**

Angles of **elevation** and of **depression** are measured between the line of sight and the horizontal.

Example 4   A surveyor uses an instrument, such as a theodolite, to measure the angle of elevation between the top of a mountain and ground level. At one point the angle of elevation is measured to be 41°. A half kilometer from the initial point and farther from the base of the mountain, the angle of elevation is measured to be 37°. How high is the mountain?

Solution   Let h represent the height. Since there are two right triangles sharing the common side h, it follows from Figure 5.10 that

$$\frac{h}{z} = \tan 41° \quad \text{and} \quad \frac{h}{z + \frac{1}{2}} = \tan 37°,$$

or

$$h = z \tan 41° \quad \text{and} \quad h = (z + \tfrac{1}{2})\tan 37°$$

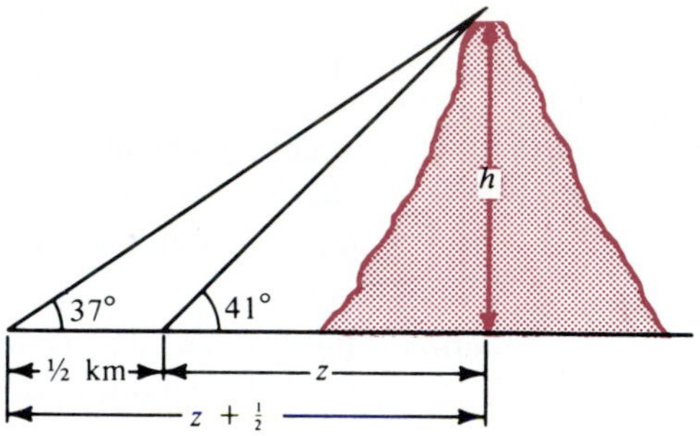

Equating the last two results gives an equation from which we can determine z:

$$(z + \tfrac{1}{2}) \tan 37° = z \tan 41°$$

$$z = \frac{-\frac{1}{2} \tan 37°}{\tan 37° - \tan 41°}$$

$$\approx 3.2556.$$

Figure 5.10

Therefore

$$h = z \tan 41°$$

$$\approx (3.2556)(0.8693)$$

$$\approx 2.83 \text{ kilometers}$$

$$= 2.83(1000) \text{ meters}$$

$$= 2830 \text{ meters.}$$

Example 5   Most airplanes approach San Francisco International Airport (SFO) on a straight 3° glide path starting at a point 5.5 miles from the field. In an experimental computerized technique, called the *two-segment approach*, a plane approaches the field on a 6° glide path starting at 5.5 miles out and then switches to a 3° glide path 1.5 miles from the point of touchdown. The purpose of this new approach is, of course, noise reduction. What is the height in feet of a plane P_1 using the experimental glide path when it switches to the 3° glide path? Compare the height of this plane with a plane P_2 using the standard 3° approach, when both planes are 5.5 miles from the field.

Solution For purposes of illustration, the angles and distances shown in Figure 5.11 are slightly exaggerated.

Let x be the height of P_1 at a point 1.5 miles out. From the figure we see that

$$\frac{x}{1.5} = \tan 3°,$$

or

$$x = 1.5 \tan 3°$$

$$\approx 1.5(0.0524)$$

$$\approx 0.0786 \text{ mile}$$

$$= 0.0786(5280) \text{ feet}$$

$$\approx 415 \text{ feet}.$$

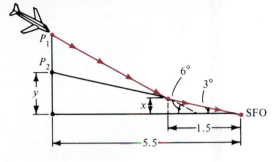

Figure 5.11

If y is the height of plane P_2, then

$$\frac{y}{5.5} = \tan 3°,$$

or

$$y = 5.5 \tan 3°$$

$$\approx 5.5(0.0524)$$

$$\approx 0.2882 \text{ mile}$$

$$= 0.2882(5280) \text{ feet}$$

$$\approx 1522 \text{ feet}.$$

Now, as Figure 5.12 shows, the height of P_1 at 5.5 miles out is given by

$$z = w + x,$$

where

$$\frac{w}{4} = \tan 6°$$

or

$$w = 4 \tan 6° \text{ mile}$$

$$\approx 4(0.1051)(5280) \text{ feet}$$

$$\approx 2220 \text{ feet}.$$

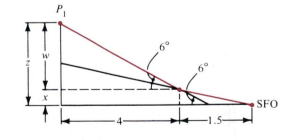

Figure 5.12

Therefore the approximate height of plane P_1 5.5 miles out is

$$z \approx 2220 + 415 = 2635 \text{ feet}.$$

∎

Exercise 5.2

1. A building casts a shadow 20 meters long. If the angle from the tip of the shadow to a point on top of the building is 69°, how high is the building?

2. Two trees are on opposite sides of a river as shown in Figure 5.13. A base line of 100 feet is measured from tree T_1, and from that position the angle β to T_2 is measured to be 29°42′. If the base line is perpendicular to the line segment between T_1 and T_2, find the distance between the two trees.

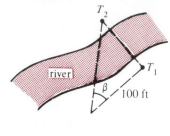

Figure 5.13

3. A 50-foot tower is located on the edge of a river. The angle of elevation between the opposite bank and the top of the tower is 37°. How wide is the river?

4. A surveyor uses a Geodimeter to measure the straight-line distance from a point G on the ground to a point on top of a mountain. Use the information given in Figure 5.14 to find the height of the mountain.

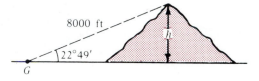

Figure 5.14

5. An observer on the roof of building A measures a 27° angle of depression between the horizontal and the base of building B. The angle of elevation from the same point to the roof of the second building is 41°25′. What is the height of building B if the height of building A is 150 feet?

6. Find the height of a mountain using the information given in Figure 5.15.

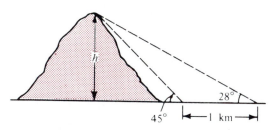

Figure 5.15

7. The top of a 20-foot ladder is leaning against the edge of the roof of a house. If the angle of inclination of the ladder from the horizontal is 51°, what is the approximate height of the house and how far is the bottom of the ladder from the base of the house?

8. An airplane at an altitude of 25,000 feet approaches a radar station located on a 2000-foot-high hill. At an instant in time, the angle between the radar dish pointed at the plane and the horizontal is 57°. What is the straight-line distance in miles between the airplane and the radar station at that particular time?

9. A 5-mile straight segment of a road climbs a 4000-foot hill. Determine the angle that the road makes with the horizontal.

10. A box has dimensions as shown in Figure 5.16. Find the length of the diagonal between the corners P and Q. What is the angle θ formed between the diagonal and the bottom edge of the box?

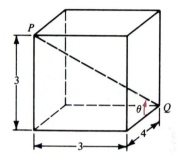

Figure 5.16

11. Observers in two towns A and B on either side of a 12,000-foot mountain measure the angles of elevation shown in Figure 5.17 between the ground and the top of the mountain. Assuming that the towns lie in the same vertical plane, find the horizontal distance between them.

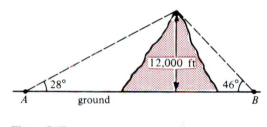

Figure 5.17

12. A draw bridge measures 7.5 meters from shore to shore, and when completely open it makes an angle of 43° with the horizontal. As shown in Figure 5.18, when the bridge is closed, the angle of depression from the shore to a point on the surface of the water below the opposite end is 27°. When the bridge is fully open, what is the distance d between the highest point of the bridge and the water below?

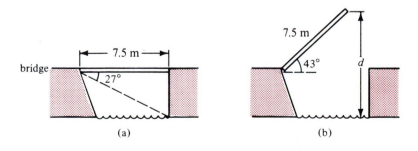

(a)　　　　(b)

Figure 5.18

13. A man standing 50 feet away from a 20-foot house looks up at a TV antenna located on the edge of the roof (Figure 5.19). If the angle between his line of sight to the edge of the roof and his line of sight to the top of the antenna is 12°, how tall is the antenna?

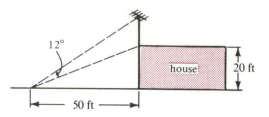

Figure 5.19

14. A 20-foot flagpole is located at the edge of a sheer 50-foot cliff at the bank of a river of unknown width x (Figure 5.20). An observer on the opposite side of the river measures an angle of 9° between her line of sight to the top of the flagpole and her line of sight to the top of the cliff. Find the value(s) of x.

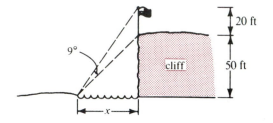

Figure 5.20

15. A vertical flagpole 45 feet tall casts a shadow 20 feet long on level ground. What is the angle of elevation of the sun?

16. A regular pentagon is inscribed in a circle of radius 5 inches. Find the length of a side of the pentagon.

17. Two cars leave an intersection at the same time and travel on straight roads at right angles to each other. If one car is travelling 40 miles per hour and the other 50 miles per hour, how far apart are the cars after 45 minutes?

18. An airplane flying at a speed of 400 miles per hour is climbing at an angle of 6° from the horizontal. When it passes directly over a car travelling 60 miles per hour, it is 2 miles above the car. Assuming that the airplane and the car remain in the same vertical plane, find the angle of elevation from the car to the airplane after 30 minutes.

19. Find the distance from the car to the airplane in Problem 18.

20. Derive a formula that expresses the area of a triangle in terms of the lengths of two sides and the included angle.

Review Problems

21. Find the hypotenuse of a right triangle with an acute angle of 20° and adjacent side 15.

22. Find the acute angles in the right triangle with sides 3, 4, and 5.

23. Find the acute angles in the right triangle with sides 5, 12, and 13.

5.3 The Law of Sines

We saw in Section 5.1 how to solve right triangles. In this section and the next we shall consider techniques for solving general triangles.

Law of Sines

Consider the triangle shown in Figure 5.21 with angles α, β, and γ, and opposite sides a, b, and c, respectively.

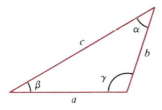

Figure 5.21

If we know the length of one side and two other parts of the triangle, we can then find the remaining three parts. This can be accomplished by using either the **law of sines**,

$$\frac{\sin \alpha}{a} = \frac{\sin \beta}{b} = \frac{\sin \gamma}{c},$$ (1)

or the law of cosines, which is developed in Section 5.4.

Although the law of sines is valid for any triangle, we shall derive it only for acute* triangles and leave the case of obtuse triangles as an exercise. See Problem 37.

As shown in Figure 5.22, let h be the altitude from angle α to side a.

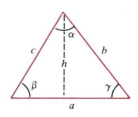

Figure 5.22

* An acute triangle is one in which all three angles are less than $90°$.

It follows that

$$\frac{h}{c} = \sin \beta,$$

or

$$h = c \sin \beta. \tag{2}$$

Similarly,

$$\frac{h}{b} = \sin \gamma,$$

or

$$h = b \sin \gamma. \tag{3}$$

Equating the expressions in (2) and (3) gives

$$c \sin \beta = b \sin \gamma,$$

so that

$$\frac{\sin \beta}{b} = \frac{\sin \gamma}{c}. \tag{4}$$

In an analogous manner, we can prove that

$$\frac{\sin \alpha}{a} = \frac{\sin \beta}{b}. \tag{5}$$

See Problem 36. Combining (4) and (5), we obtain

$$\frac{\sin \alpha}{a} = \frac{\sin \beta}{b} = \frac{\sin \gamma}{c}.$$

Example 1 Determine the remaining parts of the triangle shown in Figure 5.23.

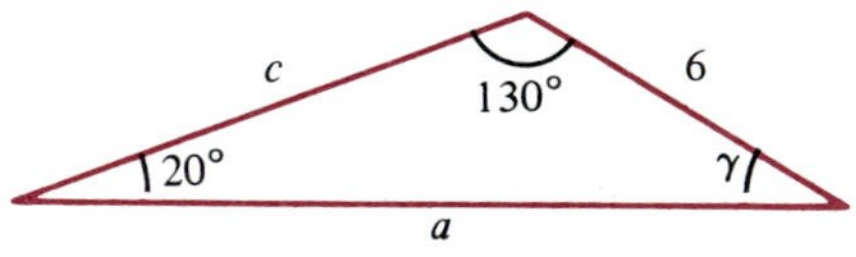

Figure 5.23

Solution Let $\beta = 20°$, $\alpha = 130°$, and $b = 6$. It follows immediately that

$$\gamma = 180° - 20° - 130°$$

$$= 30°$$

From the law of sines we see

$$\frac{\sin 130°}{a} = \frac{\sin 20°}{6}.$$

Solving for a then gives

$$a = 6\left(\frac{\sin 130°}{\sin 20°}\right) \approx 6\left(\frac{0.7660}{0.3420}\right) \approx 13.44.$$

To solve for c, we use

$$\frac{\sin 20°}{6} = \frac{\sin 30°}{c},$$

so that

$$c = 6\left(\frac{\sin 30°}{\sin 20°}\right)$$

$$\approx 6\left(\frac{0.5000}{0.3420}\right)$$

$$\approx 8.77. \qquad \blacksquare$$

Use of a Calculator

In the following example we encounter the expression $\sin \gamma \approx 0.7386$. If we use a calculator set in degree mode to solve this equation for γ we obtain

$$\gamma \approx \sin^{-1} 0.7386$$

$$\approx 47.61°.$$

Recall from Section 4.9 that the arcsine of a positive number less than 1 will always be a first quadrant angle. However, since γ is an angle in a triangle, we must consider the possibility that it is between 90° and 180°. In general, if x is a positive number less than 1 and $\sin \theta = x$, there will be a first quadrant angle θ satisfying this equation given by

$$\theta = \sin^{-1} x.$$

The second quadrant angle also satisfying the equation $\sin \theta = x$ is given by

$$\theta = 180° - \sin^{-1} x.$$

Whether the second quadrant angle also provides a solution of the triangle depends, as shown in the following example, on the sizes of the other angles.

Example 2 Determine the remaining parts of the triangle given in Figure 5.24.

Solution Let $\beta = 80°$, $b = 4$, and $c = 3$. From the law of sines it follows that

$$\frac{\sin \alpha}{a} = \frac{\sin 80°}{4} = \frac{\sin \gamma}{3}, \qquad (6)$$

and so

$$\frac{\sin 80°}{4} = \frac{\sin \gamma}{3}$$

$$\sin \gamma \approx \tfrac{3}{4}(0.9848)$$

$$\approx 0.7386.$$

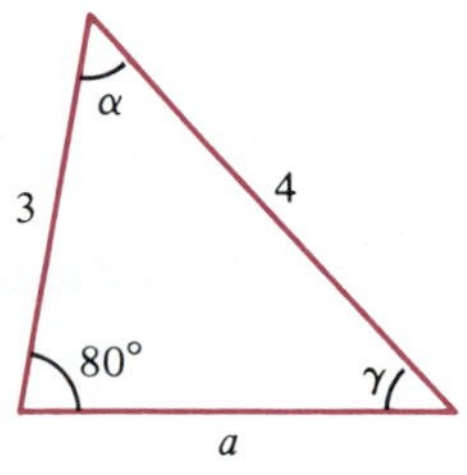

Figure 5.24

Based on the discussion preceding this example we conclude that γ can be either $47.61°$ or $180° - 47.61° = 132.39°$. Since the sum of the angles in a triangle is $180°$ and the given triangle contains an angle of $80°$, we can reject the value $\gamma \approx 132.39°$.

Now the angle α is determined from

$$\alpha = 180° - \beta - \gamma$$

$$\approx 180° - 80° - 47.61°$$

$$\approx 52.39°$$

Substituting this value in (6) implies that

$$\frac{\sin 52.39°}{a} = \frac{\sin 80°}{4},$$

or

$$a = 4\left(\frac{\sin 52.39°}{\sin 80°}\right)$$

$$\approx 4\left(\frac{0.7922}{0.9848}\right)$$

$$\approx 3.22.$$

∎

Solving Triangles: Four Cases

In general we may use the law of sines to solve triangles for which we know (a) two angles and any side, or (b) two sides and an angle opposite one of these sides. Triangles for which we know either (c) three sides, or (d) two sides and the included angle cannot be solved directly using the law of sines. In the next section we shall consider a method for solving these two cases.

In the two preceding examples each triangle had a unique solution. However, this may not always be true for triangles in which we are given two sides and an angle opposite one of these sides. As the examples below illustrate, this information may result in two solutions or no solution.

Example 3 Find the remaining parts of a triangle with $\beta = 50°$, $b = 5$, and $c = 6$.

Solution From the law of sines we have

$$\frac{\sin 50°}{5} = \frac{\sin \gamma}{6},$$

or

$$\sin \gamma = 6\left(\frac{\sin 50°}{5}\right) \approx 6\left(\frac{0.7660}{5}\right) \approx 0.9193.$$

Using a calculator, we find

$$\gamma \approx 66.82°$$

or

$$\gamma \approx 180° - 66.82° \approx 113.18°.$$

If $\gamma \approx 66.82°$, then

$$\alpha \approx 180° - 66.82° - 50° = 63.18°.$$

This triangle is shown in Figure 5.25. To find a, we use

$$\frac{\sin 63.18°}{a} = \frac{\sin 50°}{5},$$

which gives

$$a \approx 5\left(\frac{\sin 63.18°}{\sin 50°}\right)$$

$$\approx 5\left(\frac{0.8925}{0.7660}\right)$$

$$\approx 5.83.$$

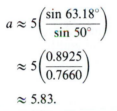

Figure 5.25

Now if $\gamma = 113.18°$, then

$$\alpha = 180° - 113.18° - 50° = 16.82°.$$

This triangle is shown in Figure 5.26. To find a, we use the law of sines:

$$\frac{\sin 16.82°}{a} = \frac{\sin 50°}{5}$$

$$a \approx 5\left(\frac{\sin 16.82°}{\sin 50°}\right)$$

$$\approx 5\left(\frac{0.2894}{0.7660}\right)$$

$$\approx 1.89.$$

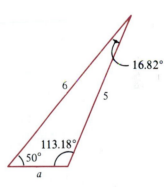

Figure 5.26

Example 4 Find the remaining parts of the triangle with $\gamma = 40°$, $b = 9$, and $c = 5$.

Solution From the law of sines we have

$$\frac{\sin \beta}{9} = \frac{\sin 40°}{5},$$

or

$$\sin \beta = 9 \left(\frac{\sin 40°}{5} \right) \approx 9 \left(\frac{0.6428}{5} \right) \approx 1.1570.$$

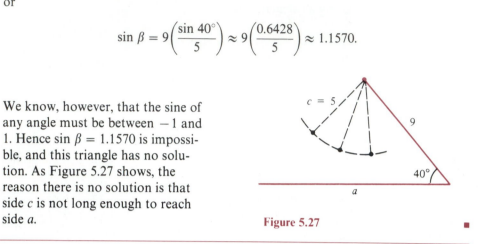

We know, however, that the sine of any angle must be between -1 and 1. Hence $\sin \beta = 1.1570$ is impossible, and this triangle has no solution. As Figure 5.27 shows, the reason there is no solution is that side c is not long enough to reach side a.

Figure 5.27

Of the four cases in solving triangles described on page 300, only (b) may yield more than one solution, and only (b) and (c) may yield no solution. For further discussion of this, see Problems 38–43.

Example 5   A building is situated on the side of a hill that slopes down at an angle of 15°. The sun is uphill from the building at an angle of elevation of 42°. Find the building's height if it casts a shadow 36 feet long.

Solution   Denote the height of the building on the downward slope by h and construct the right triangle QPS as shown in Figure 5.28. Now

$$\alpha + 15° = 42°,$$

so we have

$$\alpha = 42° - 15° = 27°.$$

Since QPS is a right triangle,

$$\gamma = 90° - 42° = 48°.$$

From the law of sines it follows that

$$\frac{\sin \alpha}{h} = \frac{\sin \gamma}{36},$$

so

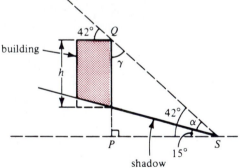

Figure 5.28

$$h = 36 \left(\frac{\sin \alpha}{\sin \gamma} \right) = 36 \left(\frac{\sin 27°}{\sin 48°} \right) \approx 36 \left(\frac{0.4540}{0.7431} \right) \approx 21.99 \text{ feet.}$$

Exercise 5.3

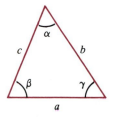

Figure 5.29

In Problems 1–26, solve the indicated triangle. The relative positions of α, β, γ, a, b, and c are shown in Figure 5.29.

1. $\alpha = 80°, \beta = 20°, b = 7$

2. $\alpha = 60°, \beta = 15°, c = 30$

3. $\beta = 37°, \gamma = 51°, a = 5$

4. $\alpha = 30°, \gamma = 75°, a = 6$

5. $\beta = 72°, b = 12, c = 6$

6. $\alpha = 120°, a = 9, c = 4$

7. $\gamma = 62°, b = 7, c = 4$

8. $\beta = 110°, \gamma = 25°, a = 14$

9. $\gamma = 15°, a = 8, c = 5$

10. $\alpha = 55°, a = 20, c = 18$

11. $\gamma = 150°, b = 7, c = 5$

12. $\alpha = 140°, \gamma = 20°, c = 12$

13. $\beta = 13°20', \gamma = 102°, b = 9$

14. $\alpha = 135°, a = 4, b = 5$

15. $\alpha = 20°, a = 8, c = 27$

16. $\beta = 47°10', b = 20, c = 25$

17. $\beta = 30°, a = 10, b = 7$

18. $\alpha = 75°, \gamma = 45°, b = 8$

19. $\gamma = 80°, b = 4, c = 8$

20. $\alpha = 43°, \beta = 62°, c = 7$

21. $\beta = 100°, a = 9, b = 20$

22. $\alpha = 35°, a = 9, b = 12$

23. $\beta = 115°, b = 11, c = 15$

24. $\alpha = 50°, a = 10, b = 15$

25. $\gamma = 95°, a = 20, c = 35$

26. $\gamma = 27.3°, b = 3, c = 2$

27. A 10-foot rope that is available to measure the length between two points A and B at opposite ends of a kidney-shaped swimming pool is not long enough. A third point C is found such that the distance from A to C is 10 feet. It is determined that angle ACB is 115° and angle ABC is 35°. Find the distance from A to B.

28. Two points A and B lie on opposite sides of a river. Another point, C, is located on the same side of the river as B at a distance of 230 feet from B. If angle ABC is 105° and angle ACB is 20°, find the distance across the river from A to B.

29. A telephone pole makes an angle of 82° with the level ground. The angle of elevation of the sun is 76°. Find the length of the telephone pole if its shadow is 3.5 meters. (Assume that the tilt of the pole is away from the sun and in the same plane as the pole and the sun.)

30. Suppose that a surveyor wants to find the straight-line distance between two points A and B at the same elevation on opposite sides of a mountain. The angle of elevation from A to the top of the mountain is 55°10' and the distance is 560 meters. (There are instruments that will measure this distance provided they can be carried to the top of the mountain.) If the angle of elevation from B to the top of the mountain is 48°, find the distance between A and B, assuming that they are in the same vertical plane as the top of the mountain.

31. The distance from the tee to the green on a golf hole is 370 yards. A golfer slices her drive and paces its distance off at 210 yards. From the point where the ball lies, she measures an angle of 160° from the tee to the green. Find the angle of her slice.

32. What is the distance from the ball to the green in Problem 31?

33. A man 5 feet 9 inches tall stands on a sidewalk which slopes down at a constant angle. A vertical street lamp directly behind him causes his shadow to be 15 feet long. The angle of depression from the top of the man to the tip of his shadow is 31°. Find the angle α, as shown in Figure 5.30, that the sidewalk makes with the horizontal.

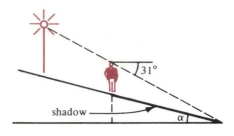

Figure 5.30

34. If the man in Problem 31 is 20 feet down the sidewalk from the lamppost, find the height of the light above the sidewalk.

35. Angles of elevation to an airplane are measured from the top and the base of a building 20 meters tall. The angle from the top of the building is 38°, and the angle from the base of the building is 40°. Find the altitude of the airplane.

36. Derive equation (5) using Figure 5.31.

37. Derive the law of sines for a triangle with an obtuse angle as shown in Figure 5.32.

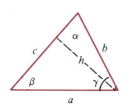

Figure 5.31

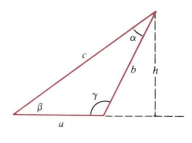

Figure 5.32

Problems 38–43 refer to the four situations described in (a), (b), (c), and (d) on page 300.

38. Explain why a unique triangle is determined if the sum of the two angles in (a) is less than 180°.

39. State a necessary and sufficient condition under which (c) will determine a triangle.

40. Explain why (d) will always determine a unique triangle provided that the included angle is less than 180°.

41. Let $\beta = 45°$ and $c = 5$, as shown in Figure 5.33. Find all values of b for which

  a. The triangle is a right triangle.
  b. The triangle has no solution.
  c. The triangle has two distinct solutions.
  d. The triangle has a unique solution.

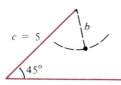

Figure 5.33

42. Let $0° < \beta < 90°$ and suppose that c is known. See Figure 5.34. Find all values of b for which
   a. The triangle is a right triangle.
   b. The triangle has no solution.
   c. The triangle has two distinct solutions.
   d. The triangle has a unique solution.

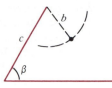

Figure 5.34

43. Let $90° < \beta < 180°$ and suppose that c is known. See Figure 5.35. Find all values of b for which
   a. The triangle has no solution.
   b. The triangle has a unique solution.

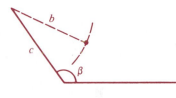

Figure 5.35

Review Problems

44. Find the two legs of a right triangle with hypotenuse 7 and one acute angle of $22°$.

45. Find the area of an isosceles triangle having equal angles of $25°$ with included side 12.

46. An equilateral triangle with perimeter 12 cm. is inscribed in a circle. Find the circumference of the circle.

47. Derive a formula for the area of the parallelogram shown in Figure 5.36.

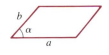

Figure 5.36

5.4 The Law of Cosines

In a right triangle, such as shown in Figure 5.37, the lengths of the sides are related by the Pythagorean theorem,

$$c^2 = a^2 + b^2. \tag{1}$$

Actually equation (1) is a special case of a more general formula that relates the lengths of the sides of *any* triangle.

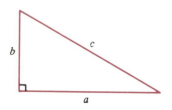

Figure 5.37

A Generalization of the Pythagorean Theorem

Suppose that the triangle in Figure 5.38(a) represents an arbitrary triangle, not necessarily a right triangle. If we introduce a Cartesian coordinate system with origin and x axis as shown in Figure 5.38(b), then the coordinates of the vertices are as shown. Now, by the distance formula, the length of the side opposite the angle γ is

$$c = \sqrt{(b \cos \gamma - a)^2 + (b \sin \gamma)^2},$$

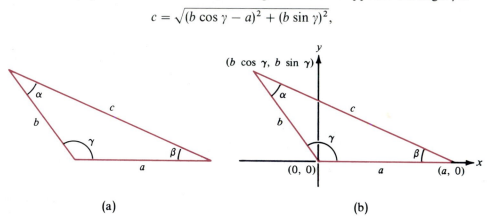

(a)   (b)

Figure 5.38

Thus,

$$c^2 = b^2\cos^2\gamma - 2ab \cos \gamma + a^2 + b^2\sin^2\gamma$$
$$= a^2 + b^2\underbrace{(\cos^2\gamma + \sin^2\gamma)}_{1} - 2ab \cos \gamma$$

$$= a^2 + b^2 - 2ab \cos \gamma. \tag{2}$$

Notice that equation (2) reduces to (1) when the angle γ is $90°$.

Since there is nothing special about placing the origin at the vertex of the angle γ, we can repeat the argument above two more times. For example, had we chosen the origin at the vertex of α and placed the x axis along the side of length b (see Figure 5.39), it would then follow, in exactly the same manner, that

$$a^2 = b^2 + c^2 - 2bc \cos \alpha. \tag{3}$$

By a similar argument we can express b in terms of a, c, and $\cos \beta$.

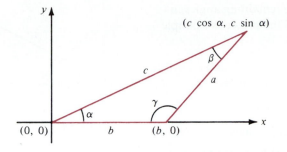

Figure 5.39

In general, for any triangle labeled as in Figure 5.38(a), we have

$$
\begin{aligned}
a^2 &= b^2 + c^2 - 2bc \cos \alpha \\
b^2 &= a^2 + c^2 - 2ac \cos \beta \\
c^2 &= a^2 + b^2 - 2ab \cos \gamma.
\end{aligned}
\tag{4}
$$

The equations above are known as the **law of cosines**.

We note that the results in (4) can be expressed as follows:

The square of the length of any side of a triangle equals the sum of the squares of the lengths of the other sides minus twice the product of these lengths and the cosine of the included angle.

Example 1 Determine the remaining side of the triangle given in Figure 5.40.

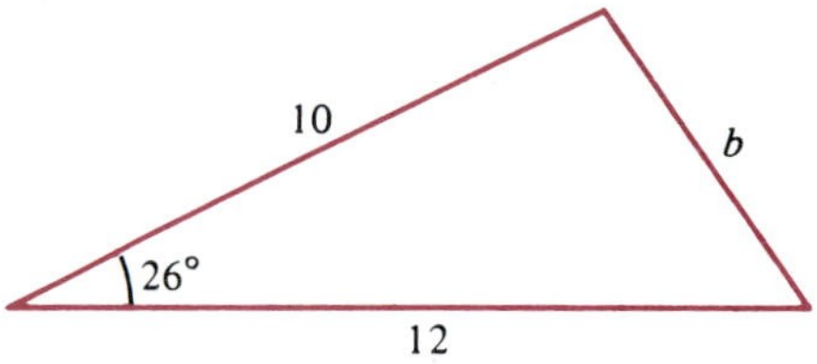

Solution If we call the unknown side b, then from the second equation of (4) we can write

$$b^2 = (10)^2 + (12)^2 - 2(10)(12)\cos 26°$$

$$\approx 100 + 144 - 240(0.8988)$$

$$\approx 244 - 215.7106$$

$$= 28.2894$$

and therefore

$$b \approx \sqrt{28.2894}$$

$$\approx 5.32.$$

Figure 5.40

Example 2   Determine the remaining angles in
the triangle of the previous example
(see figure).

Solution   We let $a = 12$, $b = 5.32$, and $c = 10$
in the third formula of (4).

$$(10)^2 = (12)^2 + (5.32)^2$$
$$- 2(12)(5.32)\cos \gamma$$

$$100 \approx 144 + 28.2894 - 127.65 \cos \gamma$$

$$127.65 \cos \gamma \approx 72.2894$$

$$\cos \gamma \approx \frac{72.2894}{127.65} \approx 0.5663.$$

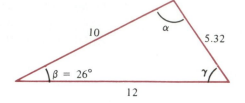

Figure 5.41

Then, using a calculator, we find that

$$\gamma \approx \cos^{-1}0.5663 \approx 55.51°.$$

Finally from $\alpha + \beta + \gamma = 180°$ it follows that

$$\alpha \approx 180° - 55.51° - 26°$$

$$= 180° - 81.51° = 98.49°.$$

Alternatively, since we know two sides and an angle opposite one of these sides, we
could have used the law of sines to find γ.   ∎

**Use of a
Calculator**

By using a calculator in the preceding example to evaluate $\alpha = \cos^{-1}0.5663$, we did
not need to consider as a special case the possibility that α could be a second quadrant
angle for the following reason. By definition, the arccosine of a number is always a
first or a second quadrant angle, and we obtain a second quadrant angle *only* when
we calculate the arccosine of a number between -1 and 0. In this case the number the
calculator displays is automatically the correct solution to the triangle. There is no
need to consider two possibilities as we did in the previous section when we were
working with the arcsine function.

Example 3   Determine the angles of the triangle
shown in Figure 5.42.

Solution   If we let $a = 7$, $b = 6$, and $c = 9$,
then from the second equation in (4)
we have

$$6^2 = 7^2 + 9^2 - 2(7)(9)\cos \beta.$$

Combining terms and rearranging
gives

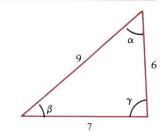

Figure 5.42

$$126 \cos \beta = 94$$

or

$$\cos \beta = \frac{94}{126}$$

$$\approx 0.7460.$$

Hence,

$$\beta \approx \cos^{-1} 0.7460$$

$$\approx 41.75°.$$

To find the angle γ we use the third equation in (4),

$$9^2 = 7^2 + 6^2 - 2(7)(6)\cos \gamma,$$

which implies that

$$\cos \gamma = \frac{4}{84}$$

$$\approx 0.0476.$$

Therefore

$$\gamma \approx \cos^{-1} 0.0476$$

$$\approx 87.27°$$

and

$$\alpha \approx 180° - 41.75° - 87.27°$$

$$= 50.98°.$$

Although we could have found γ from the law of sines, by using the law of cosines we were able to avoid an analysis of whether γ is a first- or second-quadrant angle. ■

Exercise 5.4

Calculator Problems

In Problems 1–15, solve the indicated triangle. The relative positions of α, β, γ, a, b, and c are shown in Figure 5.43.

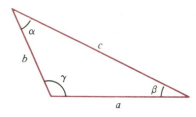

Figure 5.43

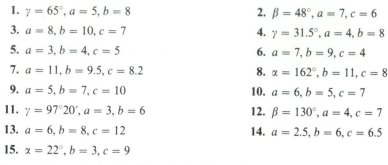

1. $\gamma = 65°$, $a = 5$, $b = 8$
2. $\beta = 48°$, $a = 7$, $c = 6$
3. $a = 8$, $b = 10$, $c = 7$
4. $\gamma = 31.5°$, $a = 4$, $b = 8$
5. $a = 3$, $b = 4$, $c = 5$
6. $a = 7$, $b = 9$, $c = 4$
7. $a = 11$, $b = 9.5$, $c = 8.2$
8. $\alpha = 162°$, $b = 11$, $c = 8$
9. $a = 5$, $b = 7$, $c = 10$
10. $a = 6$, $b = 5$, $c = 7$
11. $\gamma = 97°20'$, $a = 3$, $b = 6$
12. $\beta = 130°$, $a = 4$, $c = 7$
13. $a = 6$, $b = 8$, $c = 12$
14. $a = 2.5$, $b = 6$, $c = 6.5$
15. $\alpha = 22°$, $b = 3$, $c = 9$

16. A ship sails due west from a harbor for 22 nautical miles. It then sails 28° south of west for another 15 nautical miles. How far is the ship from the harbor?

17. A house measures 45 feet from front to back. The roof measures 32 feet from the front of the house to the peak of the roof and 18 feet from the peak to the back of the house. Find the angles of elevation of the front and back parts of the roof.

18. Two roads intersect at an angle of 75°. One car leaves the intersection at 1:00 P.M. and travels down the road at 40 miles per hour. Twenty minutes later a second car leaves the same intersection and travels down the other road at 35 miles per hour. What is the straight line distance between the two cars at 2:00 P.M.?

19. A slanted roof makes an angle of 35° with the horizontal and measures 28 feet from base to peak. A television antenna 16 feet high is to be attached to the peak of the roof and secured by a wire from the top of the antenna to the nearest point at the base of the roof. Find the length of wire required.

20. Two radar stations are situated 5000 meters from each other. An airplane passes directly over the line between the two stations. The distances from the stations to the plane are 2300 and 4000 meters. Find the altitude of the airplane.

21. A woman five feet tall is facing uphill on a sidewalk which slopes upward at an angle of 15° from the horizontal. The sun is behind the woman and causes her shadow to be 2 feet long. Find the angle of elevation of the sun measured from the horizontal.

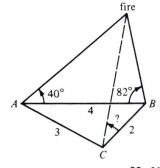

22. Two lookout towers are situated on mountain tops A and B, 4 miles from each other. A helicopter fire-fighting team is located in a valley at point C, 3 miles from A and 2 miles from B. Using the line between A and B as a reference, a fire is spotted at an angle of 40° from tower A and 82° from tower B. See Figure 5.44. At what angle, measured from CB, should the helicopter fly in order to head directly for the fire?

Figure 5.44

23. Use the law of cosines to derive **Heron's formula,***

$$A = \sqrt{s(s - a)(s - b)(s - c)},$$

for the area of a triangle with sides a, b, and c, where $s = \frac{1}{2}(a + b + c)$.

* This formula is named after the Greek mathematician Heron, but actually should be credited to Archimedes.

In Problems 24–26, use Heron's formula to find the area of the given triangle.

24. $a = 5, b = 8, c = 4$ **25.** $a = 12, b = 5, c = 13$ **26.** $\gamma = 25°, a = 7, b = 10$

27. Find the area of the quadrilateral shown in Figure 5.45.

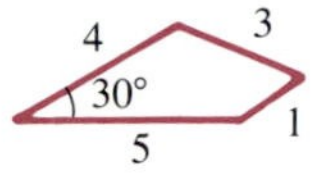

Figure 5.45

Review Problems

28. Find the acute angles in a right triangle with hypotenuse 7 and one side 4.

29. Find the angles in a triangle with sides 4, 5, and 8.

30. Let α and β be angles of a triangle with opposite sides of lengths a and b, respectively. Use the law of sines to show that

$$\frac{a - b}{a + b} = \frac{\sin \alpha - \sin \beta}{\sin \alpha + \sin \beta}.$$

31. A circle of radius 3 is inscribed in a regular octagon. Find the perimeter of the octagon.

32. From the top of a building 150 feet tall, the angle of depression to the top of a flagpole is 75.5°. From the base of the same building the angle of elevation to the top of the flagpole is 37°. How tall is the flagpole?

Chapter Review

In terms of the sides of a right triangle, the **sine, cosine,** and **tangent** of an angle θ are defined to be

$$\sin \theta = \frac{\text{side opposite the angle}}{\text{hypotenuse}},$$

$$\cos \theta = \frac{\text{side adjacent the angle}}{\text{hypotenuse}},$$

and $\tan \theta = \dfrac{\text{side opposite the angle}}{\text{side adjacent the angle}}.$

The other three trigonometric functions—cot θ, sec θ, and csc θ—are then defined in terms of the sine, cosine, and tangent.

When some of the sides and angles of a right triangle are given, the remaining sides and angles can be determined by using the six basic trigonometric functions. To solve an arbitrary triangle, we use either the **law of sines**

$$\frac{\sin \alpha}{a} = \frac{\sin \beta}{b} = \frac{\sin \gamma}{c},$$

or the **law of cosines**

$$a^2 = b^2 + c^2 - 2bc \cos \alpha$$

$$b^2 = a^2 + c^2 - 2ac \cos \beta$$

$$c^2 = a^2 + b^2 - 2ab \cos \gamma,$$

where a, b, c, α, β, and γ are as shown in Figure 5.43.

The law of sines can be used to find the remaining parts of a triangle when we know two angles and any side, or two sides and the angle opposite one of these sides. If we know three sides, or two sides and the angle between them, the law of cosines should be used.

Chapter 5 Review Exercises

In Problems 1–10, answer true or false.

1. Given two sides of a right triangle, the remaining side and angles can be found. _____

2. If A and B are two points not on the same horizontal line, with A below B, then the angle of elevation from A to B plus the angle of depression from B to A is 90°. _____

3. If three angles of a triangle are known, then its sides can be found by using the law of cosines. _____

4. The law of sines can be used to solve a triangle for which two angles and one side are given. _____

5. If a, b, and c are the lengths of the sides of a triangle, then $a > b + c$. _____

6. If a triangle has a known angle greater than 90° and the two adjacent sides are known, then the law of sines can be used to solve the triangle. _____

7. If two sides and the included angle are given, a unique triangle is always determined. _____

8. The Pythagorean Theorem is a special case of the law of cosines. _____

9. If two sides and an angle opposite one of these sides is given, a unique triangle is always determined. _____

10. If a is the side opposite angle α and b is the side opposite angle β in a triangle,

$$\alpha = \sin^{-1}[(a \sin \beta)/b]$$

for any choice of a, b, α, and β. _____

In Problems 11–16, find the indicated unknowns. Refer to the right triangle given in Figure 5.6.

11. $a = 30, b = 40; \alpha, \beta, r$ **12.** $a = 25, \alpha = 27°30'; b, r$ **13.** $b = 5, \alpha = 34°; a, r$

14. $b = 7, \beta = 45°; a, r$ **15.** $r = 10, \alpha = 41°40'; a, b$ **16.** $a = 3, r = 7; \alpha, \beta, b$

In Problems 17–22, determine, if possible, the remaining parts of a triangle. Refer to the triangle given in Figure 5.29.

17. $\alpha = 30°, \beta = 70°, b = 10$ **18.** $\gamma = 145°, a = 25, c = 20$ **19.** $\beta = 45°, b = 7, c = 8$

20. $\alpha = 51°, b = 20, c = 10$ **21.** $\gamma = 25°, a = 8, b = 5$ **22.** $a = 4, b = 6, c = 3$

23. A length of wire 14 inches long is bent into the shape of a triangle, two sides of which are 3 inches and 5 inches. Find the angles of the triangle.

24. Find the height and area of the trapezoid shown in Figure 5.46.

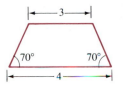

Figure 5.46

25. A 6-foot-tall man is standing near a 20-foot-tall streetlamp. Find the angle of elevation from the man's feet to the to the top of the streetlamp if his shadow is 10 feet long.

26. A man, 100 meters from the base of an overhanging cliff, measures a 28° angle of elevation from that point to the top of the cliff. If the cliff makes an angle of 65° with the horizontal ground, determine its approximate height h (see Figure 5.47).

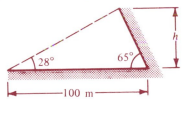

Figure 5.47

27. A rocket is launched from ground level at an angle of elevation of 43°. If the rocket hits a drone target plane flying at 20,000 feet, find the horizontal distance between the launching site and the point directly below the plane. What is the straight-line distance between the rocket launching site and the target plane?

28. A ship sails 15° north of east from a harbor for a distance of 10 nautical miles. At this point it changes its course to a direction 25° west of north and travels for 20 nautical miles. What is the straight-line distance from the harbor to this final point?

29. Determine the angles in the triangle with vertices at (0, 0), (5, 0), and (4, 3).

30. An entry in a soap-box derby rolls down a hill. Using the information given in Figure 5.48, find the total distance $d_1 + d_2$ that the soap box travels.

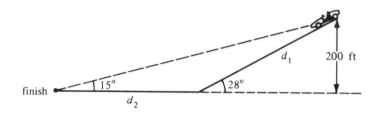

Figure 5.48

31. The angle between two sides of a parallelogram is 40°. If the lengths of the sides are 5 and 10 centimeters, find the lengths of the two diagonals.

32. Show that the area of a regular pentagon with sides having length s is

$$A = \tfrac{5}{4}s^2 \cot 36°.$$

33. Express the area of a triangle in terms of the lengths of sides a and c, and the angle β. (Refer to Figure 5.29.)

34. Show that the area of a triangle is also given by

$$A = a^2 \left(\frac{\sin \beta \sin \gamma}{2 \sin \alpha} \right).$$

35. Find a formula in terms of the radius r and the angle θ, measured in radians, for the area of the colored segment of the circle shown in Figure 5.49. [*Hint:* The area of a sector of a circle of radius r subtended by an angle θ, measured in radians, is $\tfrac{1}{2}r^2\theta$.]

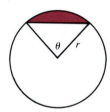

Figure 5.49

6 Conic Sections

6.1 Translation of Axes

In Section 1.2 we saw that the equation of the circle with center (2, 3) and radius 4 is

$$(x - 2)^2 + (y - 3)^2 = 4^2.$$

If we let

$$x' = x - 2 \quad \text{and} \quad y' = y - 3,$$

the equation becomes

$$x'^2 + y'^2 = 4^2.$$

This is the equation of a circle with radius 4 centered at the origin of the $x'y'$ coordinate system. In effect, we have superimposed a new coordinate system on the xy coordinate system, as shown in Figure 6.1.

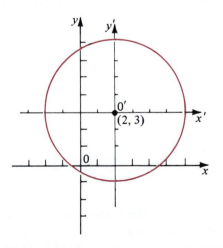

Figure 6.1

Translation

This idea is useful in the derivation of the equations of certain curves in the plane, as will become apparent in the following sections. We now formalize it. Suppose that an $x'y'$ coordinate system is superimposed on the xy coordinate system with the $x'y'$ origin at the point (h, k) in the xy system. Suppose further that the x' axis is parallel to, and has the same scale and same positive direction as, the x axis. We make the same assumptions regarding the relation between the y' axis and y axis. If these conditions hold, the $x'y'$ system is said to be a **translation** of the xy system. This concept is illustrated in Figure 6.2.

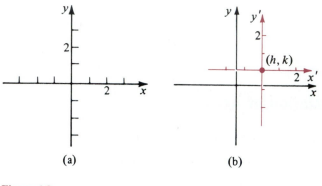

(a)   (b)

Figure 6.2

Example 1

Let the origin of the $x'y'$ system be at the point $(-2, 1)$ in the xy system. If P is a point in the plane with xy coordinates $(3, 4)$, find the $x'y'$ coordinates of P.

Solution

In Figure 6.3 we draw the two coordinate systems and plot the point P. To find x' coordinate of P, note that P is

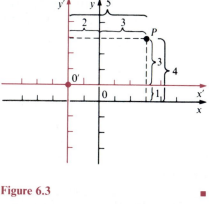

$$3 - (-2) = 5$$

units from the y' axis. Similarly, the y' coordinate of P is

$$4 - 1 = 3.$$

Thus, the $x'y'$ coordinates of P are $(5, 3)$.

Figure 6.3

∎

Translation Equations

We can generalize the techniques of the example above to obtain equations for converting from one coordinate system to another. Let the origin of the $x'y'$ system be at the point (h, k) in the xy system, and consider a point P whose xy coordinates are (x, y) and whose $x'y'$ coordinates are (x', y'). From the relations shown in Figure 6.4

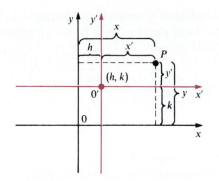

Figure 6.4

we obtain the **translation equations**

$$x = x' + h$$
$$y = y' + k. \tag{1}$$

Solving the equations in (1) for x' and y', we have

$$x' = x - h$$
$$y' = y - k. \tag{2}$$

Equations (1) and (2) are used to transform the coordinates of a point (or set of points) relative to one system to coordinates of the same point (or set of points) relative to a translated system.

Example 2   If we let $x = h$ and $y = k$ in (2), we obtain $x' = 0$ and $y' = 0$. That is, the point (h, k) in the xy system is the point $(0, 0)$ or the origin in the $x'y'$ system.   ∎

Example 3   Let the origin of the $x'y'$ system be at the point $(3, -1)$ in the xy system. If the $x'y'$ coordinates of a point P are $(6, -9)$, find the xy coordinates of this point.

Solution   Since we are given the $x'y'$ coordinates of a point, we use equation (1) to compute the xy coordinates.

$$x = x' + h = 6 + 3 = 9$$
$$y = y' + k = -9 + (-1) = -10.$$

The xy coordinates of P are $(9, -10)$.   ∎

The equations in (2) can be used to convert the equation of a curve in the $x'y'$ system to an equation of the same curve relative to the xy system.

Example 4   Let the origin of the $x'y'$ system $0'$ be located at $(1, 2)$ in the xy system. Find an equation of the circle centered at $0'$ with radius 3 in the $x'y'$ system and in the xy system.

Solution   The circle has its center at the origin of the $x'y'$ system. Thus, in this system its equation is $x'^2 + y'^2 = 9$. To obtain the equation of this same curve relative to the xy system, we use the translation equations (2) with $h = 1$ and $k = 2$. Thus we replace x' by $x - 1$ and y' by $y - 2$, and we obtain $(x - 1)^2 + (y - 2)^2 = 9$.   ∎

Example 5   Find an xy equation of the circle with xy center $(4, 2)$ and radius 3.

Solution   Let $h = 4$ and $k = 2$. Form the $x'y'$ system with origin at the xy point $(4, 2)$. The equation of the circle in the $x'y'$ system is

$$x'^2 + y'^2 = 9.$$

Using (2), we transform this equation to the xy system.

$$(x - 4)^2 + (y - 2)^2 = 9.$$

Of course, we could have obtained this result directly using the techniques in Section 1.2. See Figure 6.5.

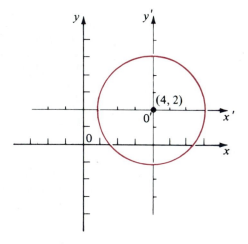

Figure 6.5   ∎

Example 6   Find an $x'y'$ equation of

$$2x + 3y + 4 = 0,$$

where the origin of the $x'y'$ system is at the xy point $(1, -2)$.

Solution   Let $h = 1$ and $k = -2$. From (1) we have

$$x = x' + 1$$

$$y = y' - 2.$$

Substituting these values into the equation of the line, we have

$$2(x' + 1) + 3(y' - 2) + 4 = 0$$

$$2x' + 2 + 3y' - 6 + 4 = 0$$

$$2x' + 3y' = 0.$$

See Figure 6.6.

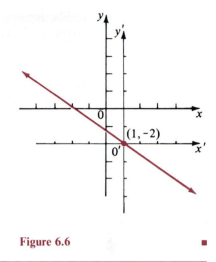

Figure 6.6

■

As an application of translation of axes, we derive a formula for the distance from a given point to a given line. By this we mean the *shortest* distance from a point (h, k) to a line $ax + by + c = 0$.

We first derive the formula for the case in which the point (h, k) is the origin. Recall that the shortest distance is measured along a line perpendicular to the given line l, as shown in Figure 6.7. Solving $ax + by + c = 0$ for y, we obtain the slope-intercept form of this equation.

$$y = -\frac{a}{b}x - \frac{c}{b}. \tag{3}$$

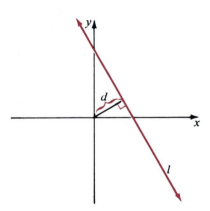

Figure 6.7

Since the slope of this line is $-a/b$, the slope of any perpendicular line is b/a (the negative reciprocal of $-a/b$). Because the particular perpendicular line we want passes through the origin, it has y intercept 0, and hence its equation is

$$y = \frac{b}{a}x. \tag{4}$$

In order to find the point of intersection of the two lines, substitute the value of y in equation (4) into equation (3). We then have

$$\frac{b}{a}x = -\frac{a}{b}x - \frac{c}{b}. \tag{5}$$

Solving (5) for x, we find

$$\frac{a}{b}x + \frac{b}{a}x = -\frac{c}{b}$$

$$\left(\frac{a}{b} + \frac{b}{a}\right)x = -\frac{c}{b}$$

$$\frac{a^2 + b^2}{ab}x = -\frac{c}{b}$$

$$x = \frac{-ac}{a^2 + b^2}. \tag{6}$$

Substituting (6) into (4) we obtain the y coordinate of the point of intersection.

$$y = \frac{b}{a}\left(\frac{-ac}{a^2 + b^2}\right) = \frac{-bc}{a^2 + b^2}.$$

Thus, the two lines intersect at the point

$$\left(\frac{-ac}{a^2 + b^2}, \frac{-bc}{a^2 + b^2}\right).$$

We use the distance formula to find the distance between the origin and this point. It follows that

$$d = \sqrt{\left(\frac{-ac}{a^2 + b^2} - 0\right)^2 + \left(\frac{-bc}{a^2 + b^2} - 0\right)^2}$$

$$= \sqrt{\frac{a^2c^2}{(a^2 + b^2)^2} + \frac{b^2c^2}{(a^2 + b^2)^2}}$$

$$= \sqrt{\frac{(a^2 + b^2)c^2}{(a^2 + b^2)^2}},$$

and hence

$$d = \frac{|c|}{\sqrt{a^2 + b^2}}. \tag{7}$$

Equation (7) is the formula for the distance from the origin to the line $ax + by + c = 0$.

It should be noted that we implicitly required $a \neq 0$ and $b \neq 0$ in our derivation. (Where?) If $a = 0$, the given line is horizontal; if $b = 0$, it is vertical. In these cases the fact that equation (7) holds must be derived separately. (See Problems 39 and 40.)

Example 7   Find the distance from the origin to the line

$$3x - 4y - 5 = 0.$$

Solution   Let $a = 3$, $b = -4$, $c = -5$, and use equation (7).

$$d = \frac{|-5|}{\sqrt{3^2 + (-4)^2}}$$

$$= \frac{5}{\sqrt{9 + 16}}$$

$$= \frac{5}{\sqrt{25}} = 1.$$

The line is shown in Figure 6.8.

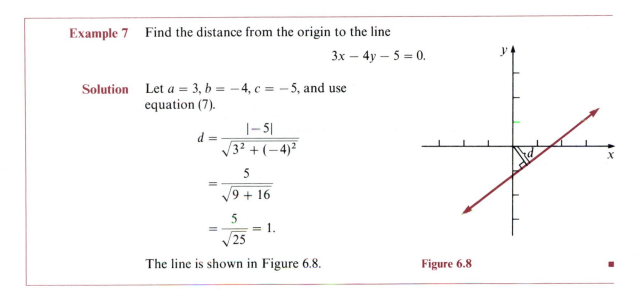

Figure 6.8

Distance from a Point to a Line

We now use translation of axes to derive a formula for the distance from the point $P(h, k)$ to the line $ax + by + c = 0$, which we denote by l in Figure 6.9. Let the $x'y'$ system be centered at $P(h, k)$ in the xy system. We use the equations in (1) to write the equation of the line l terms of $x'y'$ coordinates.

$$a(x' + h) + b(y' + k) + c = 0$$

$$ax' + ah + by' + bk + c = 0$$

$$ax' + by' + ah + bk + c = 0.$$

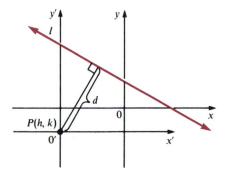

Figure 6.9

Now from equation (7), the distance from the origin $0'$ in the $x'y'$ system to the line l is

$$d = \frac{|ah + bk + c|}{\sqrt{a^2 + b^2}}.$$

Since the origin $0'$ in the $x'y'$ system is the point $P(h, k)$ in the xy system, we have:

> The distance from the point $P(h, k)$ to the line $ax + by + c = 0$ is
> $$d = \frac{|ah + bk + c|}{\sqrt{a^2 + b^2}}. \qquad (8)$$

Example 8   Find the distance from the point $(-2, 5)$ to the line

$$3x + 2y - 6 = 0.$$

Solution   Let $a = 3, b = 2, c = -6, h = -2,$
$k = 5,$ and use formula (8).

$$d = \frac{|3(-2) + 2(5) - 6|}{\sqrt{3^2 + 2^2}}$$

$$= \frac{|-6 + 10 - 6|}{\sqrt{9 + 4}}$$

$$= \frac{|-2|}{\sqrt{13}} = \frac{2}{\sqrt{13}}.$$

The line is shown in Figure 6.10.

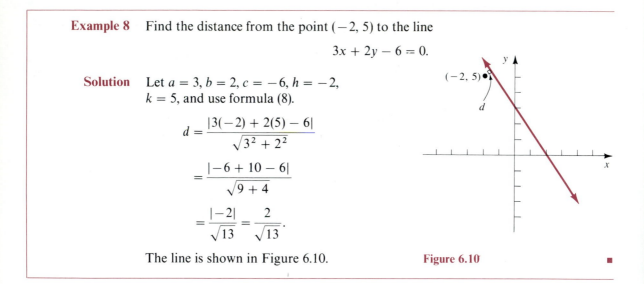

Figure 6.10

Exercise 6.1

Let the origin of the $x'y'$ system be at the xy point $(3, 2)$. In Problems 1–6, find the $x'y'$ coordinates of the given xy point.

1. $(5, 6)$ **2.** $(-2, 4)$ **3.** $(-3, -1)$ **4.** $(5, -2)$

5. $(3, -1)$ **6.** $(7, 2)$

Let the origin of the $x'y'$ system be at the xy point $(4, -3)$. In Problems 7–12, find the xy coordinates of the given $x'y'$ point.

7. $(2, 5)$ **8.** $(-3, 7)$ **9.** $(-8, -8)$ **10.** $(4, -3)$

11. $(0, 0)$ **12.** $(9, 0)$

In Problems 13–20, express the given xy equation in terms of the $x'y'$ system, where the origin of the $x'y'$ system is at the indicated xy point.

13. $(x - 2)^2 + (y + 5)^2 = 9; (2, -5)$

14. $(x + 1)^2 + (y + 2)^2 = 4; (-1, -2)$

15. $(x - 4)^2 + (y + 3)^2 = 5; (4, -3)$

16. $x^2 + y^2 - 2x - 8y + 7 = 0; (1, 4)$

17. $x^2 + y^2 + 6x - 2y + 2 = 0; (-3, 1)$

18. $2x - 5y + 7 = 0; (-1, 1)$

19. $3x + 2y - 6 = 0; (0, 3)$

20. $7x - 4y + 15 = 0; (-1, 2)$

In Problems 21–28, express the given $x'y'$ equation in terms of the xy system, where the origin of the $x'y'$ system is at the indicated xy point.

21. $x'^2 + y'^2 = 25; (-3, -1)$

22. $x'^2 + y'^2 = 7; (-2, 5)$

23. $x'^2 + y'^2 = 1; (4, -3)$

24. $x'^2 + y'^2 = r^2; (h, k)$

25. $2x' - 3y' = 0; (5, 0)$

26. $x' + 7y' = 0; (1, 1)$

27. $3x' + 5y' = 0; (5, -3)$

28. $ax' + by' + c = 0; (h, k)$

29. Let the origin of the $x'y'$ system be at the xy point $(2, 6)$. Show that the $x'y'$ equation of the straight line $3x - y + 2 = 0$ is $3x' - y' + 2 = 0$.

30. Let the origin of the $x'y'$ system be at the xy point (h, k). Show that the $x'y'$ equation of the line $y = mx + b$ will be $y' = mx' + b$ if $m = k/h$. Thus, a line has the same equation relative to both coordinate systems if the slope of the line is $m = k/h$.

In Problems 31–38, find the distance from the given point to the given line.

31. $(0, 0); 2x - 3y - 1 = 0$

32. $(0, 0); 4x + 5y = 8$

33. $(0, 0); 7x + 12y + 3 = 0$

34. $(2, 3); x - 4y + 7 = 0$

35. $(-1, -1); -3x + y + 2 = 0$

36. $(0, -2); x + y + 1 = 0$

37. $(3, -4); 2x + y = 2$

38. $(5, 0); -4x + 2y = 3$

39. Find the distance from the origin to the line $x = a$.

40. Find the distance from the origin to the line $y = b$.

41. Find an equation of the circle with center $(-1, 3)$ and tangent to $2x - y + 1 = 0$.

42. Find an equation of the circle with center $(2, 0)$ and tangent to $3x + 5y - 2 = 0$.

43. Find an equation of the circle with center (h, k) and tangent to $ax + by + c = 0$.

44. In this problem we outline an alternative derivation of formula (7).
   a. Verify that if (x_1, y_1) is any point on the line $ax + by + c = 0$, the coordinates of this point may be expressed as
$$\left(x_1, -\frac{a}{b}x_1 - \frac{c}{b}\right)$$
   b. Show that the distance between the point in part a and the origin is
$$d = \frac{1}{|b|}\sqrt{(a^2 + b^2)x_1^2 + 2acx_1 + c^2}.$$
   c. By factoring and completing the square, show that the equation in part b can be written as
$$d = \frac{\sqrt{a^2 + b^2}}{|b|}\sqrt{\left(x_1 + \frac{ac}{a^2 + b^2}\right)^2 + \left(\frac{bc}{a^2 + b^2}\right)^2}.$$

d. Explain why d will have minimum value when

$$x_1 = -\frac{ac}{a^2 + b^2}.$$

e. Use part d to show that the minimum value of d is

$$\frac{|c|}{\sqrt{a^2 + b^2}}.$$

The expression in part e is the distance from the origin to the line $ax + by + c = 0$.

6.2 The Parabola

In Section 2.6 we saw that the graph of any quadratic function is called a parabola. In this section we give a geometric definition of the parabola and discuss some of its elementary properties and applications.

> *Definition 6.1*   *A **parabola** is the set of all points in the plane which are equidistant from a fixed point F, called the **focus**, and a fixed line, called the **directrix**.*

Axis and Vertex

A parabola is shown in Figure 6.11. The line through the focus perpendicular to the directrix is called the **axis**. The point of intersection of the parabola and the axis is called the **vertex**, denoted by V in Figure 6.11.

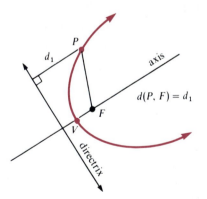

Figure 6.11

Equation of a Parabola

We shall first derive an equation of the parabola with focus at $F(0, p)$ and directrix $y = -p$, where $p > 0$. From Figure 6.12, it is clear that the axis of the parabola is

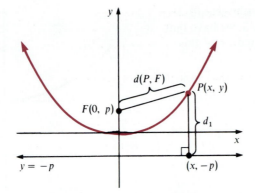

Figure 6.12

along the y axis. The origin is necessarily the vertex, since it lies on the axis p units from both the focus and the directrix.

If we let $P(x, y)$ be a point on the parabola, then the distance from P to the directrix is

$$d_1 = y - (-p) = y + p.$$

Using the distance formula, we find the distance from P to the focus:

$$d(F, P) = \sqrt{(x - 0)^2 + (y - p)^2}.$$

From Definition 6.1, then

$$\sqrt{(x - 0)^2 + (y - p)^2} = y + p.$$

Squaring both sides and simplifying, we have

$$x^2 + (y - p)^2 = (y + p)^2$$

$$x^2 + y^2 - 2py + p^2 = y^2 + 2py + p^2$$

$$x^2 = 4py. \tag{1}$$

Example 1 In Section 2.1 the curve $y = x^2$ was graphed. Since this has the form of equation (1) with $p = \frac{1}{4}$, we see now that it is a parabola with vertex at the origin and focus $(0, \frac{1}{4})$. The graph is shown again in Figure 6.13.

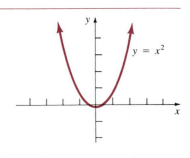

Figure 6.13 ∎

Example 2   For the parabola with focus $(0, 3)$ and directrix $y = -3$, we have that $p = 3$. Thus, the equation of this parabola is

$$x^2 = 4(3)y$$

or $$x^2 = 12y.$$

The graph is shown in Figure 6.14.

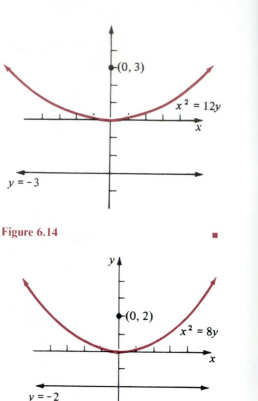

Figure 6.14

Example 3   For the parabola $x^2 = 8y$, we have $4p = 8$, so $p = 2$. Thus, the focus is at $(0, 2)$ and the directrix is $y = -2$. See Figure 6.15.

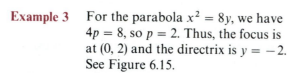

Figure 6.15

Equation (1) does not depend on the assumption $p > 0$. (See Problem 55.) However, the direction in which the parabola opens does depend on the sign of p, since the parabola opens around the focus. Specifically, if $p > 0$, the parabola opens up, and if $p < 0$, the parabola opens down. This is illustrated in Figure 6.16.

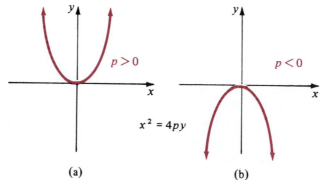

(a)   (b)

Figure 6.16

Example 4 Find an equation of the parabola with directrix $y = 2$ and focus $(0, -2)$. Graph the parabola.

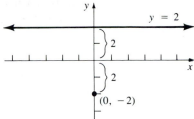

Figure 6.17

Solution In Figure 6.17 we have graphed the directrix and the focus. We see that the form of the equation is

$$x^2 = 4py.$$

Since $p = -2$, the equation must be

$$x^2 = 4(-2)y \quad \text{or} \quad x^2 = -8y.$$

To graph the parabola, we first plot the vertex at $(0, 0)$ and then locate another pair of points on the parabola. If $y = -2$, then

$$x^2 = -8(-2) = 16$$

or

$$x = \pm 4.$$

Thus, the points $(4, -2)$ and $(-4, -2)$ lie on the parabola. See Figure 6.18.

Figure 6.18

In general, for the parabola $x^2 = 4py$, the choice of $y = p$ gives $x = \pm 2p$, and thus the points $(\pm 2p, p)$ lie on the parabola. (See Problem 60.)

Example 5 Find the focus and equation of the parabola with vertex at the origin and directrix $y = -3$. Graph the parabola.

Solution The directrix is 3 units from the vertex, so $p = 3$. Since the directrix lies below the vertex, the focus is 3 units above the vertex at $(0, 3)$, as shown in Figure 6.19. From (1) the equation is

$$x^2 = 4(3)y \quad \text{or} \quad x^2 = 12y.$$

To sketch the graph, plot the vertex and the two additional points $(\pm 6, 3)$ on the parabola.

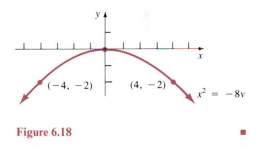

Figure 6.19

If the focus of a parabola lies on the x axis at $F(p, 0)$ and the directrix is $x = -p$, then the x axis is the axis of the parabola and the vertex is at $(0, 0)$. As shown in Figure 6.20, if $p > 0$, the parabola opens to the right, whereas if $p < 0$, it opens to the left. The equation in these cases can be shown to be

$$y^2 = 4px. \tag{2}$$

(See Problem 56.)

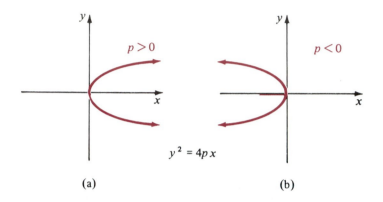

Figure 6.20

Equations of a Parabola   An equation of a parabola with focus $(0, p)$ and directrix $y = -p$ is

$$x^2 = 4py. \tag{3}$$

An equation of a parabola with focus $(p, 0)$ and directrix $x = -p$ is

$$y^2 = 4px. \tag{4}$$

The vertex of each of these parabolas is at the origin.

Example 6   Find the focus, vertex, directrix, and axis of the parabola $y^2 = -6x$. Graph the parabola and indicate the focus and directrix.

Solution   The equation has the form given in (4). Thus, the vertex is at the origin, the axis is the x axis, and

$$4p = -6 \quad \text{or} \quad p = -\tfrac{3}{2}.$$

Since $p < 0$, the focus is to the left of the vertex at $(-\tfrac{3}{2}, 0)$ and the directrix is $x = \tfrac{3}{2}$. To graph the parabola, let $x = -\tfrac{3}{2}$. Then

$$y^2 = -6(-\tfrac{3}{2}) = 9,$$

and

$$y = \pm 3.$$

Thus, the points $(-\frac{3}{2}, \pm 3)$ lie on the parabola which is sketched in Figure 6.21.

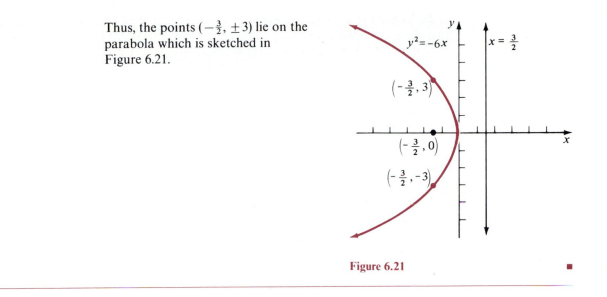

Figure 6.21

■

Parabola with Vertex at (h, k)

By using translation of axes we can readily find the equation of a parabola with vertex (h, k) and axis parallel to one of the coordinate axes. First we assume, as shown in Figure 6.22, that the axis is vertical and the focus is at $F(h, k + p)$. If we take $V(h, k)$ to be the origin of the $x'y'$ coordinate system, then the equation of the parabola relative to this system is

$$x'^2 = 4py'.$$

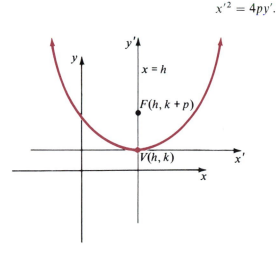

Figure 6.22

Using the translation equations (2) in Section 6.1, we obtain

$$\boxed{(x - h)^2 = 4p(y - k),}$$ (5)

which is the equation of the parabola relative to the xy coordinate system. The directrix $y' = -p$ becomes

$$y - k = -p \quad \text{or} \quad y = k - p$$

in the xy system.

Similarly, the equation of the parabola with vertex (h, k), axis parallel to the x axis, and focus at $(h + p, k)$ is

$$\boxed{(y - k)^2 = 4p(x - h).} \tag{6}$$

In this case the directrix is $x = h - p$. (See Problem 57.)

In Figure 6.23 we illustrate the four types of parabolas that result from taking p positive or negative and the axis either horizontal or vertical.

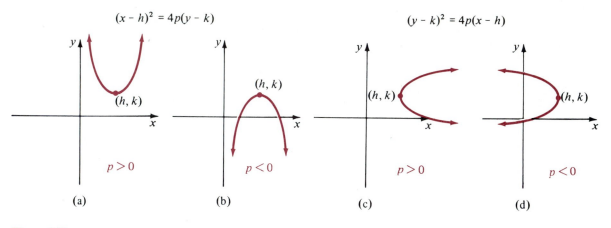

Figure 6.23

Example 7   Find an equation of the parabola with vertex $(-3, -1)$ and directrix $y = 2$.

Solution   If we graph the vertex and directrix as in Figure 6.24, it is easy to see that $p = -3$, and the form of the equation must be

$$(x - h)^2 = 4p(y - k).$$

Substituting $h = -3$, $k = -1$, and $p = -3$ then gives

$$[x - (-3)]^2 = 4(-3)[y - (-1)]$$

or

$$(x + 3)^2 = -12(y + 1).$$

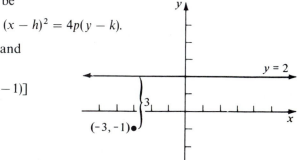

Figure 6.24

Example 8 Find the vertex, focus, axis, and directrix of the parabola $y^2 - 4y - 8x - 28 = 0$. Graph.

Solution First, complete the square in y:

$$(y^2 - 4y \quad) = 8x + 28$$

$$(y^2 - 4y + 4) = 8x + 28 + 4$$

$$(y - 2)^2 = 8x + 32$$

$$(y - 2)^2 = 8(x + 4). \tag{7}$$

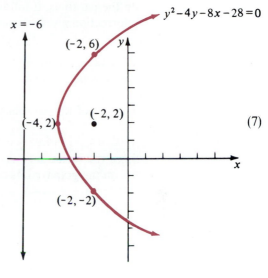

Thus, we see that the vertex is $(-4, 2)$. Also, $4p = 8$, so $p = 2$. Using Figure 6.23c as a guide, we can locate the focus at $(-2, 2)$, the directrix $x = -6$, and the axis $y = 2$. See Figure 6.25. Using $x = -2$ in equation (7), we find that $y = 6$ and $y = -2$. Thus, the points $(-2, -2)$ and $(-2, 6)$ lie on the graph of the parabola.

Figure 6.25

Example 9 Find the vertex and focus of the parabola $2x^2 + 4x - y + 5 = 0$. Graph.

Solution In order to complete the square the coefficient of x^2 must be 1, so write the equation as

$$2(x^2 + 2x \quad) = y - 5.$$

Now complete the square.

$$2(x^2 + 2x + 1) = y - 5 + 2. \tag{8}$$

Note that we added 2 to the right-hand side of (8), since we added $2(1) = 2$ to the left-hand side. Simplifying (8), we obtain

$$2(x + 1)^2 = y - 3. \tag{9}$$

In order to put equation (9) in the same form as (5), we divide both sides of (9) by 2:

$$(x + 1)^2 = \tfrac{1}{2}(y - 3).$$

Now we see that

$$4p = \tfrac{1}{2}, \quad \text{or} \quad p = \tfrac{1}{8}.$$

The vertex is $(-1, 3)$ and the focus is $(-1, \tfrac{25}{8})$. If $x = 0$, we see that $y = 5$, so the point $(0, 5)$ is on the parabola.

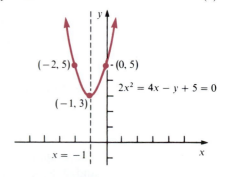

Figure 6.26

Using the fact that a parabola is symmetric with respect to its axis, we can locate another point on the parabola. Since the axis of this parabola is $x = -1$, and $(0, 5)$ lies on the parabola, it follows by symmetry that $(-2, 5)$ is also on the parabola. This information gives an accurate graph of the parabola. See Figure 6.26. ∎

Example 10 Find an equation of the parabola with vertex $(-2, -3)$, axis parallel to the y axis, and passing through the point $(0, -8)$.

Solution The axis of the parabola is vertical, so the form of the equation is

$$(x - h)^2 = 4p(y - k).$$

Since the vertex is $(-2, -3)$, we have $h = -2$ and $k = -3$. Now the point $(0, -8)$ lies on the graph, so its coordinates must satisfy

$$(x + 2)^2 = 4p(y + 3)$$

Thus, we solve

$$(0 + 2)^2 = 4p(-8 + 3)$$

for p:

$$4 = 4p(-5)$$

$$-\tfrac{1}{5} = p.$$

Therefore, the equation of the parabola is

$$(x + 2)^2 = -\tfrac{4}{5}(y + 3).$$ ∎

Example 11 Find the x and y intercepts and graph the parabola

$$x^2 - 6x - 8y - 15 = 0.$$

Solution We find the x intercepts by setting $y = 0$.

$$x^2 - 6x - 15 = 0$$

$$x = \frac{6 \pm \sqrt{36 + 60}}{2}$$

$$= \frac{6 \pm \sqrt{96}}{2}$$

$$= \frac{6 \pm 4\sqrt{6}}{2}$$

$$= 3 \pm 2\sqrt{6}.$$

The x intercepts are $(3 - 2\sqrt{6}, 0)$ and $(3 + 2\sqrt{6}, 0)$. The y intercept is found by setting $x = 0$.

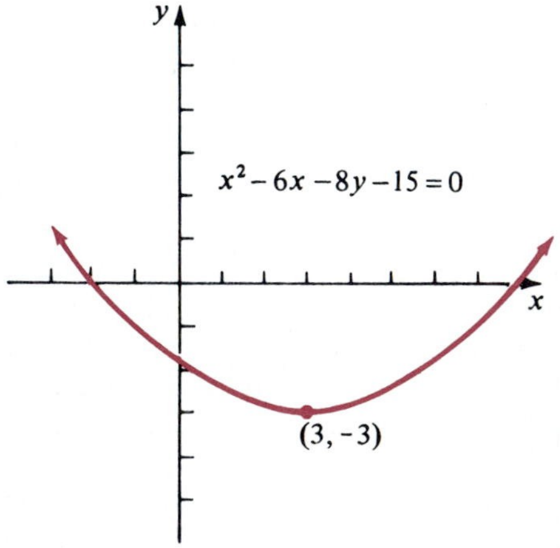

$x^2 - 6x - 8y - 15 = 0$

$(3, -3)$

Figure 6.27

$$-8y - 15 = 0$$

$$y = -\tfrac{15}{8}.$$

The vertex is found by completing the square in x.

$$(x^2 - 6x \quad) = 8y + 15$$

$$(x^2 - 6x + 9) = 8y + 15 + 9$$

$$(x - 3)^2 = 8(y + 3).$$

Thus, the vertex is at $(3, -3)$. To graph the parabola we use $3 + 2\sqrt{6} \approx 7.9$, $3 - 2\sqrt{6} \approx -1.9$, and $-\tfrac{15}{8} \approx -1.9$. See Figure 6.27. ■

Applications of the Parabola

The parabola has many interesting properties that make it suitable for certain applications. The design of mirrors for telescopes and certain lighting systems is based on an important reflection property of parabolas. As illustrated in Figure 6.28, a ray of light from a point source located at the focus of a parabola will be reflected along a line parallel to the axis.

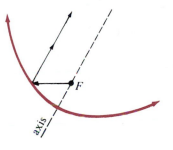

Figure 6.28

Thus, the shape of the reflecting surface in most searchlights, automobile headlights, and flashlights is obtained by rotating a parabola about its axis. The light source is placed at the focus. Then, theoretically, the result of this design is a beam of light parallel to the axis. See Figure 6.29(a). Of course, in reality some dispersion will occur, since there is no point source of light.

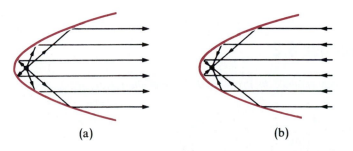

(a)   (b)

Figure 6.29

Conversely, if an incoming ray is parallel to the axis of a parabola, it will be reflected along a line passing through the focus. Reflecting telescopes and radar antennae utilize this property by placing the eyepiece of the telescope and the receiving equipment for the antenna at the focus of a parabolic reflector. See Figure 6.29(b). The proof of this reflection property of parabolas is outlined in Problem 64.

Parabolas are also important in the design of suspension bridges. The main cables of suspension bridges are usually parabolic in shape, since it can be shown that if the weight of a bridge is distributed uniformly along its length, then a cable in the shape of a parabola will bear the load evenly.

In addition, the path of a projectile will be a parabola if the motion is considered to be in a plane and air resistance is neglected.

Example 12   A reflecting telescope has a parabolic mirror that is 20 feet across the top and 5 feet deep at the center. Where should the eyepiece be located?

Solution   The eyepiece should be located at the focus because by the reflection property of parabolic mirrors discussed above, all of the rays will be concentrated at the focus. If we superimpose a coordinate system on the parabola as shown in Figure 6.30, the parabola will have the standard form $x^2 = 4py$ and pass through (10, 5). Substituting $x = 10$ and $y = 5$ into the standard form we have

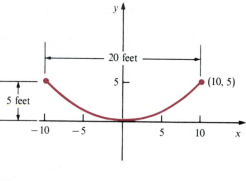

Figure 6.30

$$(10)^2 = 4p(5)$$

or

$$5 = p.$$

Thus, the focus is at (0, 5) and the eyepiece should be located at the center 5 feet above the vertex.   ∎

Exercise 6.2

For each parabola in Problems 1–24, find the vertex, focus, directrix, and axis. Graph the parabola.

1. $y^2 = 4x$
2. $y^2 = \frac{7}{2}x$
3. $y^2 = -\frac{4}{3}x$
4. $y^2 = -10x$
5. $x^2 = -16y$
6. $x^2 = \frac{1}{10}y$
7. $x^2 = 28y$
8. $x^2 = -64y$
9. $(y - 1)^2 = 16x$
10. $(y + 3)^2 = -8(x + 2)$
11. $(x + 5)^2 = -4(y + 1)$

12. $(x - 2)^2 + y = 0$ **13.** $y^2 + 12y - 4x + 16 = 0$ **14.** $x^2 + 6x + y + 11 = 0$

15. $x^2 + 5x - \frac{1}{4}y + 6 = 0$ **16.** $x^2 - 2x - 4y + 17 = 0$ **17.** $y^2 - 8y + 2x + 10 = 0$

18. $y^2 - 4y - 4x + 3 = 0$ **19.** $4x^2 = 2y$ **20.** $3(y - 1)^2 = 9x$

21. $-2x^2 + 12x - 8y - 18 = 0$ **22.** $4y^2 + 16y - 6x - 2 = 0$

23. $6y^2 - 12y - 24x - 42 = 0$ **24.** $3x^2 + 30x - 8y + 75 = 0$

In Problems 25–44, find an equation of the parabola satisfying the given conditions.

25. Focus at $(0, 7)$, directrix $y = -7$ **26.** Focus at $(0, -5)$, directrix $y = 5$

27. Focus at $(-4, 0)$, directrix $x - 4 = 0$ **28.** Focus at $(\frac{3}{2}, 0)$, directrix $x + \frac{3}{2} = 0$

29. Focus at $(\frac{5}{2}, 0)$, vertex at $(0, 0)$ **30.** Focus at $(0, -10)$, vertex at $(0, 0)$

31. Focus at $(2, 3)$, directrix $y = -3$ **32.** Focus at $(1, -7)$, directrix $x = -5$

33. Focus at $(-1, 4)$, directrix $x = 5$ **34.** Focus at $(-2, 0)$ directrix $y = \frac{3}{2}$

35. Focus at $(1, 5)$, vertex at $(1, -3)$ **36.** Focus at $(-2, 3)$, vertex at $(-2, 5)$

37. Focus at $(8, -3)$, vertex at $(0, -3)$ **38.** Focus at $(1, 2)$, vertex at $(7, 2)$

39. Vertex at $(0, 0)$, directrix $y = -\frac{7}{4}$ **40.** Vertex at $(0, 0)$, directrix $x = 6$

41. Vertex at $(5, 1)$, directrix $y = 7$ **42.** Vertex at $(-1, 4)$, directrix $x = 0$

43. Vertex at $(0, 0)$, axis along the y axis, through $(-2, 8)$

44. Vertex at $(0, 0)$, axis along the x axis, through $(1, \frac{1}{4})$

In Problems 45 and 46, find equations of the parabolas satisfying the given conditions. Graph the parabolas.

45. Vertex at $(2, -3)$, passing through $(3, 9)$, with axis parallel to a coordinate axis. (*Note:* There will be two parabolas satisfying these conditions.)

46. Vertex at $(7, -1)$, passing through $(19, 2)$, with axis parallel to a coordinate axis.

47. A large spotlight is designed so that a cross section through its axis is a parabola, and the light source is at the focus. Find the position of the light source if the spotlight is 4 feet across at the opening and 2 feet deep.

48. An automobile headlight is designed so that a cross section through its axis is a parabola, and the light source is at the focus. Find the location of the light source if the headlamp is 6 inches across and 2 inches deep.

49. Suppose that a light ray emanating from the focus of the parabola $y^2 = 4x$ strikes the parabola at $(1, -2)$. What is the equation of the reflected ray?

50. Suppose that two towers of a suspension bridge are 350 feet apart and the vertex of the parabolic cable is tangent to the road midway between the towers. If the cable is 1 foot above the road at a point 20 feet from the vertex, find the height of the towers above the road.

51. Two 75-foot towers of a suspension bridge with a parabolic cable are 250 feet apart. The vertex of the parabola is tangent to the road midway between the towers. Find the height of the cable above the roadway at a point 50 feet from one of the towers.

52. Assume that the water gushing from the end of a horizontal pipe follows a parabolic arc with vertex at the end of the pipe. The pipe is 20 meters above the ground. At a point

2 meters below the end of the pipe, the horizontal distance from the water to a vertical line through the end of the pipe is 4 meters. Where does the water strike the ground?

53. A dart thrower releases a dart 5 feet above the ground. The dart is thrown horizontally and follows a parabolic path. It hits the ground $10\sqrt{10}$ feet from the dart thrower. At a distance of 10 feet from the dart thrower, how high should a bullseye be placed in order that the dart hit it?

54. The vertical position of a projectile is given by $y = -16t^2$ and the horizontal position by $x = 40t$ for $t \geq 0$. By eliminating t between the two equations, show that the path of the projectile is a parabolic arc. Graph the path of the projectile.

55. Derive equation (1) for $p < 0$.

56. Derive equation (2).

57. Derive equation (6) and show that the equation of the directrix of the parabola $(y - k)^2 = 4p(x - h)$ is $x = h - p$.

If the directrix of a parabola is neither vertical nor horizontal, then none of the standard forms in this section apply. In Problems 58 and 59, use the definition of a parabola and the formula for the distance from a point to a line to find the equation of the given parabola.

58. Focus at (2, 5) and directrix $3x - 4y - 16 = 0$

59. Focus at $(-1, 2)$ and directrix $2x + 5y - 5 = 0$

60. The **focal width** of a parabola is the length of the line segment through the focus, perpendicular to the axis, with endpoints on the parabola. See Figure 6.31.
     This line segment is called the **focal chord** or **latus rectum**.
   a. Find the focal width of the parabola $x^2 = 8y$.
   b. Show that in general the focal width of the parabola $x^2 = 4py$ and $y^2 = 4px$ is $4|p|$.
   c. For the parabolas $x^2 = 4py$ and $y^2 = 4px$, show that the endpoints of the focal chord are $(\pm 2p, p)$ and $(p, \pm 2p)$, respectively.

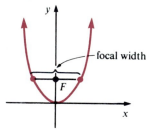

focal width

Figure 6.31

61. Prove that on a parabola the point closest to the focus is the vertex.

62. The orbit of a comet is a parabola with the sun at the focus. When the comet is 50-million kilometers from the sun, the line from the comet to the sun is perpendicular to the axis of the parabola. Use the result of Problem 60(b) to write an equation of the comet's path.

63. For the comet in Problem 62 use the result of Problem 61 to determine the shortest distance between the sun and the comet.

64. In this problem we outline a proof of the reflection property of parabolas.
     By a law of physics a light ray is reflected from a surface at an angle equal to the angle at which it strikes the surface; that is, the angle of incidence equals the angle of reflection (see Figure 6.32).

angle of incidence

angle of reflection

Figure 6.32

Now suppose that we have the parabola $y^2 = 4px$ with focus $(p, 0)$. As shown in Figure 6.33, let $P(x_0, y_0)$ denote a point on the parabola and let l be the horizontal line through P. Now draw the tangent line t to the curve $y^2 = 4px$ at the point $P(x_0, y_0)$.

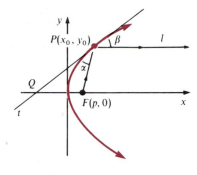

In order to verify the reflection property of the parabola we must show the angle β formed by the tangent line t and the horizontal line l equals the angle of incidence α formed by the tangent line t and the ray of light FP. It will then follow that β is the angle of reflection and so the reflected ray is parallel to the axis of the parabola.

Figure 6.33

We need the equation of the tangent line t. Calculus provides a simple method for finding the slope of a tangent line. With some effort the slope of t can be obtained without the use of calculus as outlined below in parts a–c.

Denote the slope of t by m; then the equation of t is $y - y_0 = m(x - x_0)$ or, solving for x,

$$x = \frac{1}{m}(y - y_0) + x_0.$$

a. Substitute this value of x into the equation of the parabola and write the result as a quadratic expression in y.

   The values of y in part a represent the ordinates of the points of intersection of t and the parabola. Since a tangent line will intersect the graph of a parabola in exactly one point, the discriminant of the quadratic must be 0.

b. Set the discriminant equal to 0 and solve for m.

c. Use the fact that (x_0, y_0) is on the graph of $y^2 = 4px$ to express m strictly in terms of y_0 and p. The result is $2p/y_0$.

d. Using the slope of the tangent line t found in part c and the point $P(x_0, y_0)$, write the equation of the tangent line t.

e. From this equation find the x coordinate of the x intercept Q.

f. Using the fact that the coordinates of $P(x_0, y_0)$ must satisfy $y^2 = 4px$, show that the x coordinate of Q can be written strictly in terms of y_0 and p. The result is $-y_0^2/4p$.

g. Verify $d(Q, P) = d(F, P)$.

h. Deduce that the angle PQF equals the angle α.

i. Conclude $\alpha = \beta$.

In Problems 65–68, find the x and y intercepts and graph the given parabola.

65. $(y + 4)^2 = 4(x + 1)$

66. $x^2 + 8y - 8 = 0$

67. $x^2 + 4x - 48y - 44 = 0$

68. $y^2 - 6y - 4x + 17 = 0$

69. By completing the square, rewrite $y = ax^2 + bx + c$ in the standard form $(x - h)^2 = 4p(y - k)$ and identify the vertex, directrix, focus, and axis.

Review Problems

70. Express $x^2 + y^2 - 2x + 6y + 2 = 0$ in terms of the $x'y'$ coordinate system where the origin of the $x'y'$ system is at the xy point $(1, -3)$.

71. If $(x - 5)^2 + (y + 2)^2 = 4$ has the $x'y'$ equation $x'^2 + y'^2 = 4$, where in the xy system is the origin of the $x'y'$ system?

72. Find the distance from $(-1, 5)$ to $3x - 5y + 1 = 0$.

73. Find an equation of the circle centered at the origin which is tangent to the line $4x - 7y + 1 = 0$.

6.3 The Ellipse

In this section we define the ellipse and discuss some of its properties and applications.

> **Definition 6.2**   An **ellipse** is the set of all points P in the plane such that the sum of the distances between P and two fixed points F_1 and F_2 is constant. The two fixed points F_1 and F_2 are called **foci** (plural of **focus**). The midpoint of the line segment joining the foci is called the **center**.

An ellipse is shown in Figure 6.34.

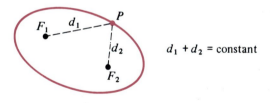

$d_1 + d_2$ = constant

Figure 6.34

Equation of an Ellipse

We now derive an equation of the ellipse with foci $F_1(-c, 0)$ and $F_2(c, 0)$ on the x axis. The center of this ellipse is at the origin. See Figure 6.35. For convenience we denote the constant distance sum by $2a$, as this choice gives a simple form for the final equation. For any point $P(x, y)$ on the ellipse we have, from Definition 6.2,

$$d(F_1, P) + d(F_2, P) = 2a.$$

Using the distance formula, we have

$$\sqrt{(x + c)^2 + (y - 0)^2} + \sqrt{(x - c)^2 + (y - 0)^2} = 2a$$

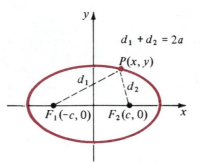

Figure 6.35

or
$$\sqrt{(x + c)^2 + y^2} = 2a - \sqrt{(x - c)^2 + y^2}.$$

We square both sides and simplify:

$$(x + c)^2 + y^2 = 4a^2 - 4a\sqrt{(x - c)^2 + y^2} + (x - c)^2 + y^2$$

$$4a\sqrt{(x - c)^2 + y^2} = 4a^2 + (x - c)^2 + y^2 - (x + c)^2 - y^2$$

$$4a\sqrt{(x - c)^2 + y^2} = 4a^2 + x^2 - 2cx + c^2 - x^2 - 2cx - c^2$$

$$a\sqrt{(x - c)^2 + y^2} = a^2 - cx.$$

Squaring again, we obtain

$$a^2[(x - c)^2 + y^2] = a^4 - 2a^2cx + c^2x^2,$$

or
$$(a^2 - c^2)x^2 + a^2y^2 = a^2(a^2 - c^2).$$

Dividing both sides by $a^2(a^2 - c^2)$ gives

$$\frac{x^2}{a^2} + \frac{y^2}{a^2 - c^2} = 1. \tag{1}$$

Since the sum of two sides of a triangle is greater than the third side, we see from triangle F_1PF_2 in Figure 6.35 that $2a > 2c$ or $a > c$. Since a and c are both positive, $a^2 - c^2 > 0$. Thus, we may set $b = \sqrt{a^2 - c^2}$ and write equation (1) as

$$\frac{x^2}{a^2} + \frac{y^2}{b^2} = 1. \tag{2}$$

Major and Minor Axes; Vertices

The line segment through the foci with endpoints on the ellipse is called the **major axis**. The line segment with endpoints on the ellipse which is perpendicular to the major axis at the center is called the **minor axis**. The endpoints of the axes are called **vertices**. For this ellipse the major and the minor axis lie on the x and y axis, respectively. Therefore, to determine the coordinates of the vertices, we simply find the x and y intercepts of (2). Letting $y = 0$, for the x intercepts, we have $x^2/a^2 = 1$ or $x = \pm a$. Similarly, setting $x = 0$ we find that $y = \pm b$. Thus, as indicated in Figure 6.36(a), the

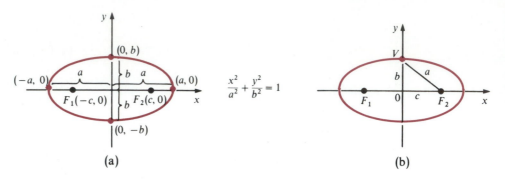

Figure 6.36

vertices of the ellipse are $(\pm a, 0)$ and $(0, \pm b)$. It is easy to see that the length of the major axis is $2a$ and the length of the minor axis is $2b$. Another important observation is that, since $a^2 = b^2 + c^2$, we have $a > b$, and thus the major axis is longer than the minor axis. The relationship $a^2 = b^2 + c^2$ is easily remembered if one observes that a is the hypotenuse of the right triangle F_2OV in Figure 6.36(b) and b and c are its legs.

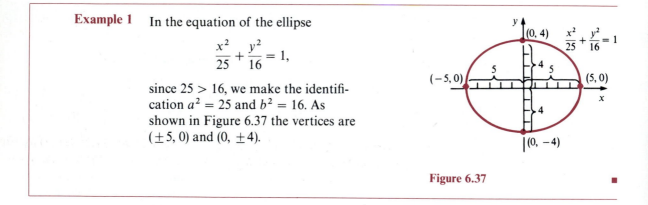

Example 1   In the equation of the ellipse

$$\frac{x^2}{25} + \frac{y^2}{16} = 1,$$

since $25 > 16$, we make the identification $a^2 = 25$ and $b^2 = 16$. As shown in Figure 6.37 the vertices are $(\pm 5, 0)$ and $(0, \pm 4)$.

Figure 6.37   ■

If the ellipse is positioned so that the foci are on the y axis at $F_1(0, -c)$ and $F_2(0, c)$ a similar derivation then yields the equation

$$\frac{x^2}{b^2} + \frac{y^2}{a^2} = 1. \qquad (3)$$

(See Problem 46.) Here again $a > c$, $b^2 = a^2 - c^2$, and $a > b$. The vertices are located at $(0, \pm a)$ and $(\pm b, 0)$ and the center is at the origin. As shown in Figure 6.38 the major axis is vertical and the minor axis is horizontal for an ellipse of this type.

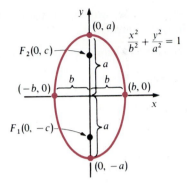

Figure 6.38

> **Equations of an Ellipse** An equation of the ellipse with foci $(-c, 0)$ and $(c, 0)$ and vertices $(\pm a, 0)$ and $(0, \pm b)$ is
>
> $$\frac{x^2}{a^2} + \frac{y^2}{b^2} = 1, \qquad \text{where } b^2 = a^2 - c^2. \tag{4}$$
>
> An equation of the ellipse with foci $(0, -c)$ and $(0, c)$ and vertices $(0, \pm a)$ and $(\pm b, 0)$ is
>
> $$\frac{x^2}{b^2} + \frac{y^2}{a^2} = 1, \qquad \text{where } b^2 = a^2 - c^2.$$

Example 2 Find the vertices and foci for the ellipse $9x^2 + 3y^2 = 27$. Graph the ellipse.

Solution Dividing by 27, we find that

$$\frac{x^2}{3} + \frac{y^2}{9} = 1$$

or

$$\frac{x^2}{(\sqrt{3})^2} + \frac{y^2}{(3)^2} = 1.$$

Since $3 > \sqrt{3}$, we have $a = 3$ and $b = \sqrt{3}$. The denominator of the y^2 term is then a^2, and the major axis is vertical. Thus, the vertices are $(0, \pm 3)$ and $(\pm\sqrt{3}, 0)$. Since

$$c^2 = a^2 - b^2 = 9 - 3 = 6,$$

we have $c = \sqrt{6}$. The foci are then on the y axis at $(0, \pm\sqrt{6})$. The ellipse

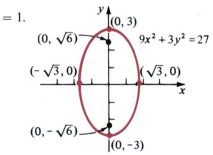

Figure 6.39

is sketched in Figure 6.39. Although the foci are shown, the reader should be aware that they do *not* lie on the graph of the ellipse. ∎

Example 3   Find an equation of the ellipse with vertices at $(0, \pm 5)$ and foci at $(0, \pm 2)$. Graph the ellipse.

Solution   In Figure 6.40 we have plotted the vertices and the foci. We see that $a = 5$ and $c = 2$. Since

$$b^2 = a^2 - c^2,$$

we have $b^2 = 25 - 4 = 21$,

and so $b = \sqrt{21}$. Since the foci lie on the y axis and the center is at the origin, the equation of the ellipse is

$$\frac{x^2}{b^2} + \frac{y^2}{a^2} = 1.$$

Substituting the values of a and b gives

$$\frac{x^2}{21} + \frac{y^2}{25} = 1.$$

The graph is shown in Figure 6.41

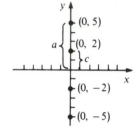

Figure 6.40

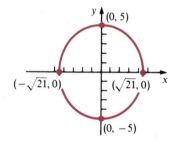

Figure 6.41   ∎

Example 4   Find an equation of the ellipse with horizontal major axis and vertices $(\pm 4, 0)$ which passes through the point $(-2\sqrt{3}, 1)$.

Solution   This ellipse has the form $x^2/a^2 + y^2/b^2 = 1$. Since the vertices $(\pm 4, 0)$ lie at the endpoints of the major axis, $2a = 8$, so that $a = 4$. Since the point $(-2\sqrt{3}, 1)$ is on the graph, the coordinates must satisfy the equation

$$\frac{x^2}{(4)^2} + \frac{y^2}{b^2} = 1.$$

Therefore,

$$\frac{(-2\sqrt{3})^2}{(4)^2} + \frac{(1)^2}{b^2} = 1$$

or

$$\frac{1}{b^2} = 1 - \frac{12}{16} = \frac{4}{16} = \frac{1}{4}.$$

Thus $b^2 = 4$ and the equation of the ellipse is

$$\frac{x^2}{16} + \frac{y^2}{4} = 1.$$

∎

Example 5   Find an equation of the ellipse with foci at $(0, \pm\sqrt{3})$ such that the length of the major axis is 12.

Solution   Because the foci are on the y axis, the equation of the ellipse must be

$$\frac{x^2}{b^2} + \frac{y^2}{a^2} = 1.$$

Since the length of the major axis is $2a = 12$, we have $a = \frac{12}{2}$, or 6. Now from $c = \sqrt{3}$ it follows that

$$b^2 = a^2 - c^2$$

$$= 36 - 3$$

$$= 33$$

and the equation of the ellipse is

$$\frac{x^2}{33} + \frac{y^2}{36} = 1.$$

∎

An Ellipse Centered at (h, k)

Up to now we have considered only ellipses centered at the origin. We can obtain the equation of an ellipse centered at (h, k) with major and minor axes parallel to the coordinate axes by using translation of axes. (See Problems 47 and 48.)

   If the center of the ellipse is (h, k) and the foci are located at $(h - c, k)$ and $(h + c, k)$, then from equation (2) the equation of this ellipse is found to be

$$\frac{(x - h)^2}{a^2} + \frac{(y - k)^2}{b^2} = 1. \tag{5}$$

In the ellipse (5) the major axis is horizontal and the minor axis is vertical. Similarly, the equation of the ellipse with center (h, k) and foci located at $(h, k - c)$ and $(h, k + c)$ is found to be

$$\frac{(x - h)^2}{b^2} + \frac{(y - k)^2}{a^2} = 1. \tag{6}$$

In this ellipse the major axis is vertical and the minor axis is horizontal. As before, $a > c$ and $b^2 = a^2 - c^2$, so that $a > b$. The vertices can be obtained by setting $x = h$ and $y = k$ in (5) and (6) and solving for y and x, respectively. They are labeled in Figure 6.42. Note also that the length of the major axis is $2a$ and the length of the minor axis is $2b$.

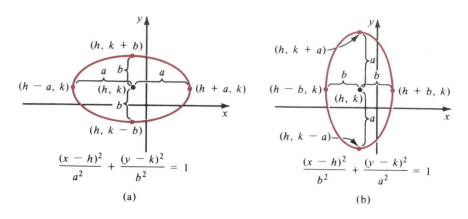

Figure 6.42

Example 6   Find the foci and vertices of the ellipse $4x^2 + 16y^2 - 8x - 96y + 84 = 0$. Graph.

Solution   To obtain the equation we must complete the square in x and y. But first we factor 4 from both x^2 and x and 16 from both y^2 and y. (Recall from Section 0.7 that the quadratic terms x^2 and y^2 should have the coefficient 1 in order to complete the square.)

$$4(x^2 - 2x \quad) + 16(y^2 - 6y \quad) = -84$$

$$4(x^2 - 2x + 1) + 16(y^2 - 6y + 9) = -84 + 4(1) + 16(9).$$

(Observe that we have been careful to add 4(1) and 16(9) to *both* sides of the equality.)

$$4(x - 1)^2 + 16(y - 3)^2 = 64.$$

Dividing by 64, we obtain

$$\frac{(x - 1)^2}{16} + \frac{(y - 3)^2}{4} = 1. \tag{7}$$

Thus, the center of the ellipse is $(1, 3)$. The major axis is horizontal and $a = 4$ and $b = 2$. In Figure 6.43 we have located the vertices at $(-3, 3), (5, 3), (1, 1)$, and $(1, 5)$. This was done by measuring $a = 4$ units to the left and right of the center and $b = 2$ units up and down from the center. (Alternatively, we could obtain the vertices by first setting $x = 1$ and then $y = 3$ in (7) and

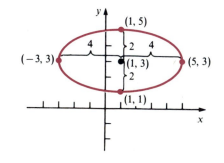

Figure 6.43

solving

$$\frac{(y-3)^2}{4} = 1 \quad \text{and} \quad \frac{(x-1)^2}{16} = 1,$$

respectively.) To find the foci, we observe that

$$c^2 = a^2 - b^2 = 16 - 4 = 12,$$

and so $c = \sqrt{12} = 2\sqrt{3}$. Thus the foci are $(1 - 2\sqrt{3}, 3)$ and $(1 + 2\sqrt{3}, 3)$. ∎

Example 7 Find an equation of the ellipse with center at $(-2, 1)$, vertical major axis of length 4, and minor axis of length 1.

Solution The length of the major axis is $2a = 4$; hence, $a = 2$. Similarly, the length of the minor axis is $2b = 1$, so that $b = \frac{1}{2}$. By sketching the center and the axes, we see from Figure 6.44 that the vertices are $(-\frac{5}{2}, 1), (-\frac{3}{2}, 1), (-2, -1)$, and $(-2, 3)$. Since the major axis is vertical, the equation is

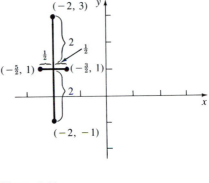

Figure 6.44

$$\frac{[x-(-2)]^2}{(\frac{1}{2})^2} + \frac{(y-1)^2}{(2)^2} = 1$$

or $\qquad \dfrac{(x+2)^2}{\frac{1}{4}} + \dfrac{(y-1)^2}{4} = 1.$ ∎

Applications of the Ellipse

There are numerous applications of the ellipse. The paths of planets around the sun are approximately elliptical, as are the orbits of our satellites about the earth. Sometimes gears are elliptical in shape, and elliptical arches are used in architecture. Ellipses have a reflection property analogous to the one discussed in Section 6.2 for the parabola. It can be shown that any ray of light emanating from one focus and reflecting off the ellipse will pass through the other focus. This principle holds for sound waves as well as light waves, and it is the basis for the design of a "whisper chamber." See Problem 45.

Example 8 The orbit of the planet Mercury is an ellipse with the sun at one focus. This ellipse has a major axis of length 72 million miles and a minor axis of length 70.4 million miles. What is the least distance (perihelion) between Mercury and the sun? What is the greatest distance (aphelion)?

Solution In Figure 6.45 we have sketched the orbit of Mercury (scale slightly exaggerated) and superimposed the xy coordinate system with origin at the center and x axis on the major axis of the ellipse. Therefore, the equation of the ellipse must be

$$\frac{x^2}{a^2} + \frac{y^2}{b^2} = 1,$$

where $2a = 72$ million miles and $2b = 70.4$ million miles. Since the least distance between the sun and Mercury is $a - c$ and the largest distance is $a + c$, we must find c. We have

$$c = \sqrt{a^2 - b^2}$$
$$= \sqrt{(36)^2 - (35.2)^2}$$
$$\approx 7.5 \text{ million miles.}$$

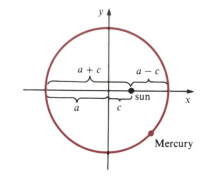

Figure 6.45

From $a = 36$ million miles we find that the greatest distance is

$$a + c = 36 + 7.5 = 43.5 \text{ million miles,}$$

and the least distance is

$$a - c = 36 - 7.5 = 28.5 \text{ million miles.}$$

Exercise 6.3

In Problems 1–20, find the center, foci, and vertices of the given ellipse. Graph.

1. $\dfrac{x^2}{25} + \dfrac{y^2}{9} = 1$ **2.** $\dfrac{x^2}{16} + \dfrac{y^2}{4} = 1$ **3.** $x^2 + \dfrac{y^2}{16} = 1$

4. $\dfrac{x^2}{4} + \dfrac{y^2}{10} = 1$ **5.** $9x^2 + 16y^2 = 144$ **6.** $4x^2 + 2y^2 = 8$

7. $9x^2 + 4y^2 = 36$ **8.** $x^2 + 4y^2 = 4$

9. $\dfrac{(x-1)^2}{49} + \dfrac{(y-3)^2}{36} = 1$ **10.** $\dfrac{(x+1)^2}{25} + \dfrac{(y-2)^2}{36} = 1$

11. $(x+5)^2 + \dfrac{(y+2)^2}{16} = 1$ **12.** $\dfrac{(x-3)^2}{64} + \dfrac{(y+4)^2}{81} = 1$

13. $4x^2 + (y + \frac{1}{2})^2 = 4$ **14.** $36(x+2)^2 + (y-4)^2 = 72$

15. $5(x-1)^2 + 3(y+2)^2 = 45$ **16.** $6(x-2)^2 + 8y^2 = 48$

17. $25x^2 + 9y^2 - 100x + 18y - 116 = 0$ **18.** $9x^2 + 5y^2 + 18x - 10y - 31 = 0$

19. $x^2 + 3y^2 + 18y + 18 = 0$ **20.** $4y^2 + 12x^2 - 4y - 24x + 1 = 0$

In Problems 21–39, find an equation of the ellipse satisfying the given conditions.

21. Vertices at $(\pm 5, 0)$, foci at $(\pm 3, 0)$

22. Vertices at $(\pm 9, 0)$, foci at $(\pm 2, 0)$

23. Vertices at $(0, \pm 3)$, foci at $(0, \pm 1)$

24. Vertices at $(0, \pm 7)$, foci at $(0, \pm 3)$

25. Vertices at $(0, \pm 3)$, $(\pm 1, 0)$

26. Vertices at $(0, \pm 2)$, $(\pm 4, 0)$

27. Vertices at $(-3, -3), (5, -3), (1, -1), (1, -5)$

28. Vertices at $(1, -6), (1, 2), (-2, -2), (4, -2)$

29. One focus at $(0, -2)$, center at origin, $b = 3$

30. One focus at $(1, 0)$, center at origin, $a = 3$

31. Foci at $(\pm \sqrt{2}, 0)$, length of minor axis 6

32. Foci at $(0, \pm \sqrt{5})$, length of major axis 16

33. Foci at $(0, \pm 3)$, passing through $(-1, 2\sqrt{2})$

34. Vertices at $(\pm 5, 0)$, passing through $(\sqrt{5}, 4)$

35. Vertices at $(\pm 4, 1)$, passing through $(2\sqrt{3}, 2)$

36. Center at $(1, -1)$, one focus at $(1, 1)$, $a = 5$

37. Center at $(1, 3)$, one focus at $(1, 0)$, one vertex at $(1, -1)$

38. Endpoints of minor axis at $(0, 5)$ and $(0, -1)$, one focus at $(6, 2)$

39. Endpoints of major axis at $(2, 4)$ and $(13, 4)$, one focus at $(4, 4)$

40. The planet Pluto has an elliptical orbit about the sun with the sun at one focus. If the length of the major axis is 7350 million miles and the length of the minor axis is 7117 million miles, find the smallest and the largest distances between Pluto and the sun.

41. A satellite orbits the earth in an elliptical path with center of earth at one focus. It has a minimum altitude of 200 miles and a maximum altitude of 1000 miles above the surface of the earth. If the radius of the earth is 4000 miles, what is the equation of its path?

42. An elliptical gear rotates about its center and is always kept in mesh with a circular gear that is free to move horizontally. See Figure 6.46. If the origin of the xy coordinate system is placed at the center of the ellipse, the equation of the ellipse in its present position is $3x^2 + 8y^2 = 24$. The diameter of the circular gear equals the length of the minor axis of the elliptical gear. If the units are centimeters, how far does the center of the circular gear move horizontally during the rotation from one vertex of the elliptical gear to the next?

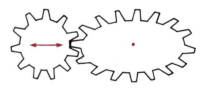

Figure 6.46

43. An archway is semielliptical with a vertical major axis. The base of the arch is 10 feet across and the highest part of the arch is 15 feet. Find the height of the arch above the point on the base of the arch 3 feet from the center.

44. From Definition 6.2 we can develop a technique for drawing ellipses. Place two tacks in a piece of cardboard or plywood. Tie the ends of a length of string to each tack. Using a pencil, stretch the string taut and draw a curve with the point of the pencil. This curve will be an ellipse with

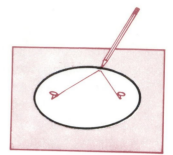

$$2c = \text{distance between the tacks}$$

and

Figure 6.47

$$2a = \text{length of the string.}$$

See Figure 6.47.

45. Suppose that a room is constructed on a flat elliptical base by rotating a semiellipse 180° about its major axis. Then, by the reflection property of the ellipse, anything whispered at one focus will be distinctly heard at the other focus. If the height of the room is 16 feet and the length is 40 feet, find the location of the whispering and listening posts.

46. Derive equation (3).

47. Derive equation (5) and verify Figure 6.42(a). [*Hint:* Let $x'y'$ be a translation of the xy coordinate system to a new origin at (h, k). Consider the ellipse $(x')^2/a^2 + (y')^2/b^2 = 1$. Find its equation in the xy system. Find the vertices and foci of this ellipse in the xy system.]

48. Derive equation (6) and verify Figure 6.42(b).

49. Find an equation of the ellipse with foci at $(0, 2)$ and $(8, 6)$ and fixed distance sum $2a = 12$. [*Hint:* Here the major axis is neither horizontal nor vertical; thus, none of the standard forms from this section apply. Use the definition of the ellipse.]

50. Find an equation of the ellipse with foci at $(-1, -3)$ and $(-5, 7)$ and fixed distance sum $2a = 20$.

51. The **focal width** of the ellipse is the length of a line segment, perpendicular to the major axis, through a focus with endpoints on the ellipse. See Figure 6.48.

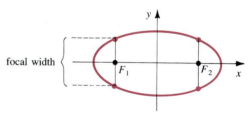

a. Find the focal width of the ellipse $x^2/9 + y^2/4 = 1$.

b. Show that, in general, the focal width of the ellipse

Figure 6.48

$$\frac{x^2}{a^2} + \frac{y^2}{b^2} = 1$$

is $2b^2/a$.

Review Problems

52. Find the vertex, focus and directrix for the parabola $x = -y^2/8$. Graph.

53. If the xy point $(2, -4)$ has $x'y'$ coordinates $(4, 6)$, at what xy point is the origin of the $x'y'$ system located?

54. In Problem 53 what are the $x'y'$ coordinates of the origin of the xy system?

55. Find the vertex, axis, focus, and directrix of the parabola $x^2 - 8x = 4y - 8$.

6.4 The Hyperbola

In this section we define the hyperbola and study some of its elementary properties and applications.

> **Definition 6.3** A **hyperbola** is the set of all points P in the plane such that the absolute value of the difference of the distances between P and two fixed points F_1 and F_2 is constant. The two fixed points F_1 and F_2 are called **foci**. The midpoint of the line segment joining the foci is called the **center**.

A hyperbola is sketched in Figure 6.49. Observe that the graph has two distinct parts called **branches**.

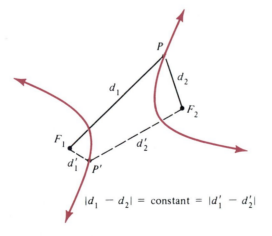

$$|d_1 - d_2| = \text{constant} = |d_1' - d_2'|$$

Figure 6.49

Equation of a Hyperbola

We now derive an equation of the hyperbola with foci $F_1(-c, 0)$ and $F_2(c, 0)$ on the x axis and center at the origin. Let $P(x, y)$ be a point on the right-hand branch of the graph and let $d_1 = d(P, F_1)$ and $d_2 = d(P, F_2)$, as shown in Figure 6.50. (The derivation for points on the other branch is similar. See Problem 51.) Clearly, $d_1 > d_2$, so $d_1 - d_2 > 0$, and thus

$$|d_1 - d_2| = d_1 - d_2.$$

If we now let the fixed constant distance be $2a$, then by Definition 6.3

$$d_1 - d_2 = 2a.$$

From the distance formula we have

$$\sqrt{(x+c)^2 + y^2} - \sqrt{(x-c)^2 + y^2} = 2a,$$

or

$$\sqrt{(x+c)^2 + y^2} = 2a + \sqrt{(x-c)^2 + y^2}.$$

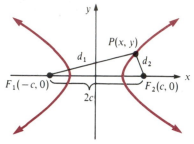

Figure 6.50

Squaring both sides and simplifying gives

$$cx - a^2 = a\sqrt{(x - c)^2 + y^2}.$$

Squaring and simplifying, we find

$$(c^2 - a^2)x^2 - a^2 y^2 = a^2(c^2 - a^2). \qquad (1)$$

From the triangle $F_1 P F_2$ in Figure 6.50 we see that

$$d_2 + 2c > d_1,$$

and so

$$2c > d_1 - d_2$$

$$2c > 2a$$

$$c > a.$$

Since $c > a > 0$, $c^2 > a^2$ and so $c^2 - a^2$ is positive. Thus we may set

$$b = \sqrt{c^2 - a^2}.$$

Hence equation (1) may be written as

$$b^2 x^2 - a^2 y^2 = a^2 b^2,$$

or

$$\frac{x^2}{a^2} - \frac{y^2}{b^2} = 1. \qquad (2)$$

Transverse Axis and Vertices

The line segment with endpoints on the hyperbola and lying on the line through the foci is called the **transverse axis**; its endpoints are called **vertices**. For the hyperbola described by (2) the transverse axis lies on the x axis. Therefore to determine the coordinates of the vertices, we simply find the x intercepts. Letting $y = 0$ gives $x^2/a^2 = 1$, or $x = \pm a$. Thus, as indicated in Figure 6.51, the vertices of the hyperbola are $(\pm a, 0)$, and the length of the transverse axis is $2a$.

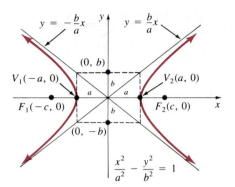

Figure 6.51

Conjugate Axis

The line segment through the center of the hyperbola, perpendicular to the transverse axis, and with endpoints $(0, -b)$ and $(0, b)$ is called the **conjugate axis**. The endpoints of the conjugate axis are not on the hyperbola, but they are useful in drawing its graph.

Asymptotes

Solving equation (2) for y in terms of x, we obtain

$$y = \pm \frac{bx}{a} \sqrt{1 - \frac{a^2}{x^2}}.$$

As $|x|$ increases, we note that a^2/x^2 gets closer to 0, and thus $\sqrt{1 - a^2/x^2}$ approaches 1. Therefore for large values of $|x|$, points on the graph of the hyperbola are close to points on the lines

$$y = \frac{b}{a}x \quad \text{and} \quad y = -\frac{b}{a}x.$$

These lines are called the **asymptotes** of the hyperbola. As shown in Figure 6.51, the asymptotes intersect at the origin, which is the center of the hyperbola. Note also that the asymptotes are the extended diagonals of the rectangle centered at the origin with width $2a$ and height $2b$. This rectangle is called the **auxiliary rectangle** for the hyperbola. Now since

$$y = \pm \frac{b}{a}x$$

is equivalent to

$$\frac{y^2}{b^2} = \frac{x^2}{a^2},$$

the asymptotes of the hyperbola

$$\frac{x^2}{a^2} - \frac{y^2}{b^2} = 1$$

are given by the single (and easy to remember) equation,

$$\frac{x^2}{a^2} - \frac{y^2}{b^2} = 0. \tag{3}$$

Alternatively, the equations of the asymptotes of the hyperbola (2) can always be found by recalling that they pass through the origin and opposite corners of the auxiliary rectangle.

Example 1 For the hyperbola

$$\frac{x^2}{25} - \frac{y^2}{9} = 1$$

we see that $a^2 = 25$ and $b^2 = 9$, and so $a = 5$ and $b = 3$. Thus from (3) we find that asymptotes are $y = \pm 3x/5$. Figure 6.52 shows the graph of the hyperbola along with the transverse and conjugate axes, the asymptotes, and the auxiliary rectangle.

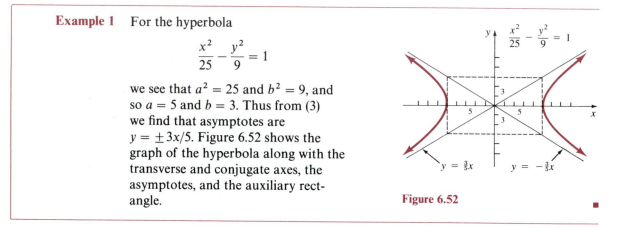

Figure 6.52

When the foci are on the y axis at $F_1(0, -c)$ and $F_2(0, c)$, it can be shown that the equation of the hyperbola is

$$\frac{y^2}{a^2} - \frac{x^2}{b^2} = 1. \tag{4}$$

(See Problem 45.) Here again $c > a$ and

$$b^2 = c^2 - a^2.$$

As indicated in Figure 6.53, the vertices are $(0, \pm a)$ and the transverse axis is on the

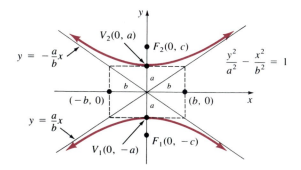

Figure 6.53

y axis. In this case the equations of the asymptotes are

$$y = \pm \frac{a}{b}x,$$

or, equivalently,

$$\frac{y^2}{a^2} - \frac{x^2}{b^2} = 0. \tag{5}$$

Equations of a Hyperbola An equation of the hyperbola with foci $(\pm c, 0)$ and vertices $(\pm a, 0)$ is

$$\frac{x^2}{a^2} - \frac{y^2}{b^2} = 1. \tag{6}$$

The asymptotes are given by

$$\frac{x^2}{a^2} - \frac{y^2}{b^2} = 0.$$

An equation of the hyperbola with foci $(0, \pm c)$ and vertices $(0, \pm a)$ is

$$\frac{y^2}{a^2} - \frac{x^2}{b^2} = 1. \tag{7}$$

The asymptotes of this hyperbola are given by

$$\frac{y^2}{a^2} - \frac{x^2}{b^2} = 0.$$

For both hyperbolas we have the relation

$$b^2 = c^2 - a^2.$$

Note that for the hyperbola (unlike the ellipse) there is no relation between the relative sizes of a and b; rather, a^2 is always the denominator of the positive term and the vertices have $\pm a$ as a coordinate.

Example 2 Find the vertices, foci, and asymptotes of the hyperbola

$$16y^2 - 9x^2 = 144.$$

Graph.

Solution Dividing by 144, we obtain

$$\frac{16y^2}{144} - \frac{9x^2}{144} = 1,$$

or

$$\frac{y^2}{3^2} - \frac{x^2}{4^2} = 1.$$

Since the y^2 term has the positive coefficient, we identify $a = 3$ and $b = 4$. From $b^2 = c^2 - a^2$ we see that

$$c = \sqrt{a^2 + b^2} = \sqrt{9 + 16} = 5.$$

Thus the vertices and foci are $(0, \pm 3)$ and $(0, \pm 5)$, respectively. To find the asymptotes, we solve

$$\frac{y^2}{3^2} - \frac{x^2}{4^2} = 0$$

for y:

$$\frac{y^2}{9} = \frac{x^2}{16}$$

$$y^2 = \tfrac{9}{16}x^2$$

$$y = \pm \tfrac{3}{4}x.$$

To graph the hyperbola, we first graph the asymptotes as follows. We draw a rectangle of height

$$2a = 2(3) = 6$$

and width

$$2b = 2(4) = 8$$

centered at the origin. The extensions of the diagonals of this rectangle are the asymptotes. Now we draw the branches of the hyperbola by starting at the vertices $(0, 3)$ and $(0, -3)$, sketching toward the asymptotes. The resulting graph is shown in Figure 6.54.

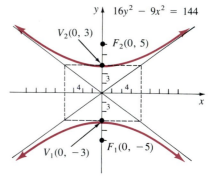

Figure 6.54

Example 3   Find an equation of the hyperbola with $b = 2$ and foci at $(\pm 3, 0)$.

Solution   Since the foci are on the x axis and the center is at the origin, the equation of the hyperbola has the form

$$\frac{x^2}{a^2} - \frac{y^2}{b^2} = 1.$$

We know that $b = 2$, and from the given foci it follows that $c = 3$. Thus, using

$$b^2 = c^2 - a^2,$$

we have

$$a^2 = c^2 - b^2 = 3^2 - 2^2 = 9 - 4 = 5.$$

Therefore, the equation of the hyperbola is

$$\frac{x^2}{5} - \frac{y^2}{4} = 1.$$

■

Example 4 Find an equation of the hyperbola with foci $(0, \pm 6)$ and vertices $(0, \pm 4)$.

Solution From the location of the foci we see that the equation of the hyperbola has the form

$$\frac{y^2}{a^2} - \frac{x^2}{b^2} = 1.$$

Now $a = 4$ and $c = 6$, so we have

$$b^2 = c^2 - a^2 = 36 - 16 = 20.$$

Hence, we find that

$$\frac{y^2}{16} - \frac{x^2}{20} = 1.$$ ∎

Example 5 Find an equation of the hyperbola with vertices $(0, \pm 4)$ and asymptotes $y = \pm \frac{1}{2} x$.

Solution Since the vertices are on the y axis, we see from equation (5) that the asymptotes have the form

$$\frac{y^2}{a^2} - \frac{x^2}{b^2} = 0,$$

or, solving for y,

$$y = \pm \frac{a}{b} x.$$

Thus, in this case, the fraction a/b is $\frac{1}{2}$. (It is *not* true that $a = 1$ and $b = 2$, but only that the ratio of a to b is $\frac{1}{2}$.) From the fact that the vertices are $(0, \pm 4)$, we see that $a = 4$. Therefore,

$$\frac{4}{b} = \frac{1}{2}$$

or $b = 8$, and the equation of the hyperbola is

$$\frac{y^2}{4^2} - \frac{x^2}{8^2} = 1.$$ ∎

Hyperbola Centered at (h, k)

To obtain an equation of a hyperbola with center at (h, k) and foci at $(h + c, k)$ and $(h - c, k)$ we use translation of axes as discussed in Section 6.1. The resulting equation is

$$\frac{(x - h)^2}{a^2} - \frac{(y - k)^2}{b^2} = 1. \tag{8}$$

As shown in Figure 6.55(a), this hyperbola has a horizontal transverse axis with vertices at $(h - a, k)$ and $(h + a, k)$. The asymptotes are the straight lines given by

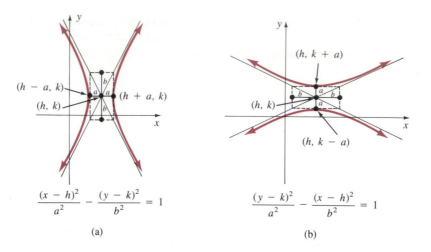

$$\frac{(x - h)^2}{a^2} - \frac{(y - k)^2}{b^2} = 1$$

(a)

$$\frac{(y - k)^2}{a^2} - \frac{(x - h)^2}{b^2} = 1$$

(b)

Figure 6.55

$$\frac{(x - h)^2}{a^2} - \frac{(y - k)^2}{b^2} = 0. \tag{9}$$

Similarly, an equation of the hyperbola with center (h, k) and foci at $(h, k - c)$ and $(h, k + c)$ is found to be

$$\boxed{\frac{(y - k)^2}{a^2} - \frac{(x - h)^2}{b^2} = 1.} \tag{10}$$

(See Problems 46 and 47.)

As shown in Figure 6.55(b), this hyperbola has a vertical transverse axis and vertices at $(h, k - a)$ and $(h, k + a)$. The asymptotes are given by

$$\frac{(y - k)^2}{a^2} - \frac{(x - h)^2}{b^2} = 0. \tag{11}$$

As before, a^2 is the denominator of the term with the positive coefficient, and we also retain the relationship

$$b^2 = c^2 - a^2.$$

Example 6   Find the center, vertices, foci, and asymptotes of the hyperbola

$$4x^2 - y^2 - 8x - 4y - 4 = 0.$$

Graph.

Solution   Before completing the square in x and y we remember to factor 4 from the x^2 and x terms and to factor -1 from the y^2 and y terms, so that the lead coefficients in each expression are 1. Then we have

$$4(x^2 - 2x \quad) - (y^2 + 4y \quad) = 4$$

$$4(x^2 - 2x + 1) - (y^2 + 4y + 4) = 4 + 4 - 4$$

$$4(x - 1)^2 - (y + 2)^2 = 4$$

$$\frac{(x - 1)^2}{1} - \frac{(y + 2)^2}{4} = 1.$$

We see now that the center is $(1, -2)$. Since the x^2 term has the positive coefficient, the transverse axis is horizontal, and we identify $a = 1$ and $b = 2$. As indicated in Figure 6.55(a), we can then locate the vertices by measuring 1 unit to the left and right of the center. Thus the vertices are $(2, -2)$ and $(0, -2)$. From

$$b^2 = c^2 - a^2$$

we have

$$c^2 = a^2 + b^2 = 1 + 4 = 5$$

and so $c = \sqrt{5}$. Hence the foci are $(1 - \sqrt{5}, -2)$ and $(1 + \sqrt{5}, -2)$. To find the asymptotes, we solve

$$\frac{(x - 1)^2}{1} - \frac{(y + 2)^2}{4} = 0$$

for y:

$$\frac{(y + 2)^2}{4} = (x - 1)^2$$

$$\frac{y + 2}{2} = \pm(x - 1)$$

$$y + 2 = \pm(2x - 2)$$

$$y = \pm(2x - 2) - 2.$$

Thus $y = 2x - 4$ and $y = -2x$. The graph is drawn by locating the center $(1, -2)$ and using the values $a = 1$ and $b = 2$ to draw the rectangle that determines the asymptotes. See Figure 6.56.

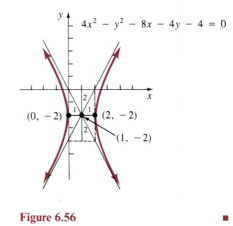

Figure 6.56

■

Example 7   Find an equation of the hyperbola with foci at $(-7, 1)$ and $(5, 1)$ and one vertex at $(3, 1)$. Graph.

Solution　We begin by locating the given foci and vertex as in Figure 6.57. Since the center of the hyperbola is the midpoint of the line segment joining the foci, it has coordinates $\left(\dfrac{-7+5}{2}, \dfrac{1+1}{2}\right)$ or $(-1, 1)$. From Figure 6.57 we have $c = 6$ and $a = 4$. Thus, from

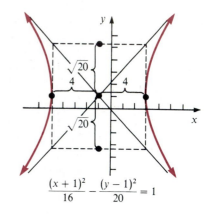

Figure 6.57

$$b^2 = c^2 - a^2,$$

we obtain

$$b^2 = 36 - 16$$

$$= 20.$$

The transverse axis is horizontal, so the equation of the hyperbola is

$$\frac{[x-(-1)]^2}{16} - \frac{(y-1)^2}{20} = 1$$

or

$$\frac{(x+1)^2}{16} - \frac{(y-1)^2}{20} = 1.$$

The graph is given in Figure 6.58.

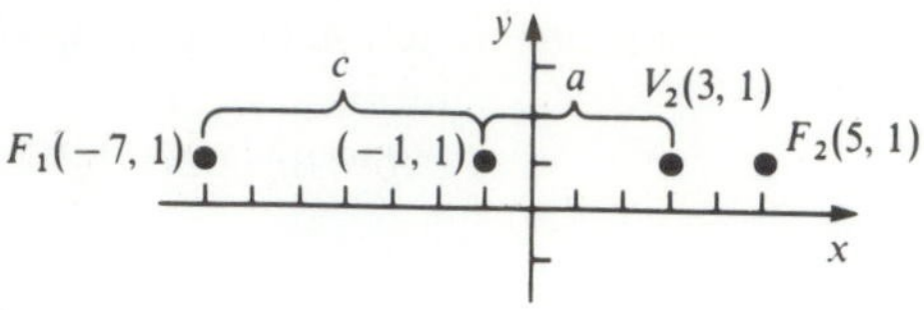

Figure 6.58

∎

Example 8　Find an equation of the hyperbola with center $(2, -3)$, passing through the point $(4, 1)$, and having one vertex at $(2, 0)$.

Solution　Since the distance from the center to one vertex is a, we have $a = 3$. From the given location of the center and vertex it follows that the transverse axis is vertical. Therefore from (10) we know that the equation is

$$\frac{(y+3)^2}{3^2} - \frac{(x-2)^2}{b^2} = 1, \tag{12}$$

where b^2 is yet to be determined. Since the point $(4, 1)$ is on the graph of the hyperbola, its coordinates must satisfy equation (12). Hence

$$\frac{(1+3)^2}{9} - \frac{(4-2)^2}{b^2} = 1$$

$$\frac{16}{9} - \frac{4}{b^2} = 1$$

$$\frac{7}{9} = \frac{4}{b^2}$$

and so $$b^2 = \tfrac{36}{7}.$$

We conclude that an equation of the hyperbola is

$$\frac{(y+3)^2}{9} - \frac{(x-2)^2}{\frac{36}{7}} = 1.$$

∎

Applications of the Hyperbola

The hyperbola has several important applications in situations involving sounding techniques. In particular, several navigational systems utilize the hyperbola as follows. Two fixed radio transmitters at a known distance from each other transmit synchronized signals. The difference in reception times by a navigator determines the difference $2a$ of the distances from the navigator to the two transmitters. This information locates the navigator somewhere on the hyperbola with foci at the transmitters and fixed difference in distances from the foci equal to $2a$. By using two pairs of transmitters the navigator's actual position can be determined as the intersection of two hyperbolas. See Figure 6.59.

In addition, the hyperbola has a reflection property analogous to that of the parabola and the ellipse. See Problem 44.

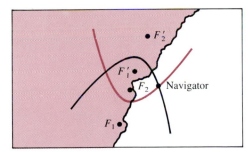

Figure 6.59

Example 9
The sound of a dynamite blast is heard at different times at two points A and B. From this it is determined that the blast occurred 1000 feet closer to A than B. If A and B are 2600 feet apart, show that the location of the blast lies on a particular branch of a hyperbola and find the equation of the branch.

Solution
In Figure 6.60 we have placed the points A and B on the x axis at $(1300, 0)$ and $(-1300, 0)$ respectively. If $P(x, y)$ denotes the location of the blast then

$$d(P, B) - d(P, A) = 1000.$$

From Definition 6.3 and the derivation following it, we see that this is the equation for the right branch of a hyperbola with fixed distance difference $2a = 1000$ and

$c = 1300$. Thus, the equation has the form

$$\frac{x^2}{a^2} - \frac{y^2}{b^2} = 1, \quad \text{where } x \geq 0$$

or, equivalently,

$$x = a\sqrt{1 + \frac{y^2}{b^2}}.$$

With $a = 500$ and $c = 1300$, we have

$$b^2 = c^2 - a^2$$
$$= (1300)^2 - (500)^2$$
$$= (1200)^2.$$

Thus, an equation of the right branch of the hyperbola is

$$x = 500\sqrt{1 + \frac{y^2}{(1200)^2}}$$

or

$$x = \frac{5}{12}\sqrt{(1200)^2 + y^2}.$$

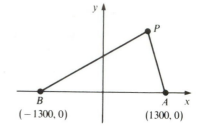

Figure 6.60

Exercise 6.4

In Problems 1–20, find the center, vertices, foci, and asymptotes of the given hyperbola. Graph.

1. $\dfrac{x^2}{16} - \dfrac{y^2}{25} = 1$

2. $\dfrac{x^2}{4} - \dfrac{y^2}{4} = 1$

3. $\dfrac{y^2}{64} - \dfrac{x^2}{9} = 1$

4. $\dfrac{y^2}{6} - 4x^2 = 1$

5. $4x^2 - 16y^2 = 64$

6. $5x^2 - 5y^2 = 25$

7. $2y^2 - 10x^2 = 40$

8. $9x^2 - 16y^2 + 144 = 0$

9. $\dfrac{(x-5)^2}{4} - \dfrac{(y+1)^2}{49} = 1$

10. $\dfrac{(x+2)^2}{10} - \dfrac{(y+4)^2}{25} = 1$

11. $\dfrac{(y-4)^2}{36} - x^2 = 1$

12. $\dfrac{(y-\frac{1}{4})^2}{4} - \dfrac{(x+3)^2}{9} = 1$

13. $25(x-3)^2 - 5(y-1)^2 = 125$

14. $10(x+1)^2 - 2(y-\frac{1}{2})^2 = 100$

15. $8(x+4)^2 - 5(y-7)^2 + 40 = 0$

16. $9(x-1)^2 - 81(y-2)^2 = 9$

17. $5x^2 - 6y^2 - 20x + 12y - 16 = 0$

18. $16x^2 - 25y^2 - 256x - 150y + 399 = 0$

19. $4x^2 - y^2 - 8x + 6y - 4 = 0$

20. $2y^2 - 9x^2 - 18x + 20y + 5 = 0$

In Problems 21–42, find an equation of the hyperbola satisfying the given conditions.

21. Foci at $(\pm 5, 0)$, $a = 3$

22. Foci at $(\pm 10, 0)$, $b = 2$

23. Foci at $(0, \pm 4)$, one vertex at $(0, -2)$

24. Foci at $(0, \pm 3)$, one vertex at $(0, -\frac{3}{2})$

25. Foci at $(\pm 4, 0)$, length of transverse axis 6

26. Foci at $(0, \pm 7)$, length of transverse axis 10

27. Center at $(0, 0)$, one vertex at $(0, \frac{5}{2})$, one focus at $(0, -3)$

28. Center at $(0, 0)$, one vertex at $(7, 0)$, one focus at $(9, 0)$

29. Center at $(0, 0)$, one vertex at $(-2, 0)$, one focus at $(-3, 0)$

30. Center at $(0, 0)$, one vertex at $(1, 0)$, one focus at $(5, 0)$

31. Vertices at $(0, \pm 8)$, asymptotes $y = \pm 2x$

32. Foci at $(0, \pm 3)$, asymptotes $y = \pm \frac{3}{2}x$

33. Vertices at $(\pm 2, 0)$, asymptotes $y = \pm \frac{4}{3}x$

34. Foci at $(\pm 5, 0)$, asymptotes $y = \pm \frac{3}{5}x$

35. Center at $(1, -3)$, one focus at $(1, -6)$, one vertex at $(1, -5)$

36. Center at $(2, 3)$, one focus at $(0, 3)$, one vertex at $(3, 3)$

37. Foci at $(-4, 2)$ and $(2, 2)$, one vertex at $(-3, 2)$

38. Vertices at $(2, 5)$ and $(2, -1)$, one focus at $(2, 7)$

39. Vertices at $(\pm 2, 0)$, passing through $(2\sqrt{3}, 4)$

40. Vertices at $(0, \pm 3)$, passing through $(16/3, 5)$

41. Center $(-1, 3)$, one vertex at $(-1, 4)$, passing through $(-5, 3 + \sqrt{5})$

42. Center $(3, -5)$, one vertex at $(3, -2)$, passing through $(1, -1)$

43. Three points are located at $A(-10, 16)$, $B(-2, 0)$, and $C(2, 0)$, where the units are kilometers. An artillery gun is known to lie on the line segment between A and C, and by sounding techniques it is determined that it is 2 kilometers closer to B than to C. Find the point where the gun is located.

44. It can be shown that a ray of light emanating from one focus of a hyperbola will be reflected back along the line from the opposite focus. See Figure 6.61. A light ray from the left focus of the hyperbola

$$\frac{x^2}{16} - \frac{y^2}{20} = 1$$

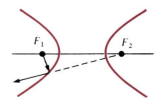

Figure 6.61

strikes the hyperbola at $(-6, -5)$. Find an equation of the reflected ray.

45. Derive equation (4).

46. Derive equation (8) and verify Figure 6.55(a). [*Hint:* Let $x'y'$ be a translation of the xy coordinate system to a new origin at (h, k). Consider the hyperbola

$$\frac{(x')^2}{a^2} - \frac{(y')^2}{b^2} = 1.$$

Find its equation in the xy system. Find the vertices, foci, and asymptotes of this hyperbola relative to the xy system.]

47. Derive equation (10) and verify Figure 6.55(b).

48. Find the equation of the hyperbola with foci at $(0, -2)$ and $(8, 4)$ and fixed distance difference $2a = 8$.

49. The **focal width** of a hyperbola is the length of a line segment, perpendicular to the line containing the transverse axis and through a focus, with endpoints on the hyperbola. See Figure 6.62.

    a. Find the focal width of the hyperbola

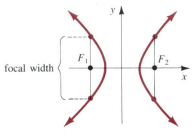

$$\frac{x^2}{4} - \frac{y^2}{9} = 1.$$

    b. Show that, in general, the focal width of the hyperbola

$$\frac{x^2}{a^2} - \frac{y^2}{b^2} = 1$$

Figure 6.62

    is $2b^2/a$.

50. Two hyperbolas are said to be **conjugate** if the transverse axis of each hyperbola is the conjugate axis of the other.

    a. Find an equation of the hyperbola that is conjugate to

$$\frac{x^2}{25} - \frac{y^2}{144} = 1.$$

    Graph both hyperbolas on the same set of axes.

    b. Find an equation of the hyperbola that is conjugate to

$$\frac{x^2}{a^2} - \frac{y^2}{b^2} = 1.$$

51. As shown in Figure 6.63, let $P(x, y)$ be a point on the left-hand branch of the hyperbola. Then

$$d_1 < d_2$$

so

$$d_1 - d_2 < 0$$

and

$$|d_1 - d_2| = d_2 - d_1.$$

Carry out the derivation of the equation of the hyperbola in this case.

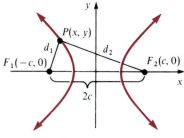

Figure 6.63

Review Problems

52. Find the distance from the point $(4, 1)$ to the line $3x + 5y - 15 = 0$.

53. Find the vertex, focus, directrix, and axis of the parabola $(y - 2)^2 + x = 0$. Graph.

54. Find an equation of the ellipse with foci at $(0, \pm\sqrt{2})$ and length of major axis 5.

55. Find the center, foci, and vertices of the ellipse $5(x - 1)^2 + 10y^2 = 50$. Graph.

56. Find an equation of the parabola with vertex at the origin, axis along the x axis, and passing through $(-4, 10)$.

6.5 Conic Sections

Conic Sections

The circle, parabola, ellipse, and hyperbola are called **conic sections** since these curves can be obtained by intersecting a right circular cone with a plane as shown in Figure 6.64. The intersection of the cone with a plane perpendicular to the axis of the cone yields a *circle*. If the plane is tilted slightly, the resulting curve is an *ellipse*. When a plane is parallel to a line on the cone, the curve of intersection is a *parabola*. Finally, if the plane intersects both halves of the cone, the curve is a *hyperbola*. The conic sections were first studied in this form by the Greek mathematician Appolonius in the third century B.C.

It can be shown that for appropriate choices of constants, A, C, D, E, and F, the graph of the equation

$$Ax^2 + Cy^2 + Dx + Ey + F = 0 \tag{1}$$

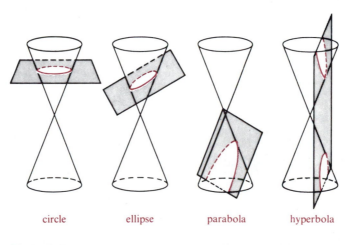

circle   ellipse   parabola   hyperbola

Figure 6.64

will be one of the conic sections described above. For other values of these constants, equation (1) gives the *degenerate conic sections* (two intersecting lines, two parallel lines, one line, a point) or no graph. (See Problems 18–20.) In fact it can be shown that the graph of the general second-degree equation $Ax^2 + Bxy + Cy^2 + Dx + Ey + F = 0$ will be a conic section. Equations containing an xy term will be discussed in the next section.

Example 1 Determine whether the graph of the equation

$$9x^2 + 16y^2 - 36x - 32y - 92 = 0$$

is a circle, parabola, ellipse, or hyperbola. Graph.

Solution We complete the square in x and y to determine which conic section this equation represents:

$$9(x^2 - 4x \quad) + 16(y^2 - 2y \quad) = 92$$

$$9(x^2 - 4x + 4) + 16(y^2 - 2y + 1) = 92 + 36 + 16$$

$$9(x - 2)^2 + 16(y - 1)^2 = 144.$$

Dividing by 144, we have

$$\frac{(x - 2)^2}{16} + \frac{(y - 1)^2}{9} = 1,$$

which we recognize as the equation of an ellipse with center $(2, 1)$, major axis horizontal, and $a^2 = 16$ and $b^2 = 9$. Thus, the vertices are $(-2, 1), (6, 1), (2, 4)$, and $(2, -2)$. Since $b^2 = a^2 - c^2$, we have $c = \sqrt{a^2 - b^2} = \sqrt{7}$. Hence, the foci are $(2 \pm \sqrt{7}, 1)$. See Figure 6.65 for the graph of the ellipse.

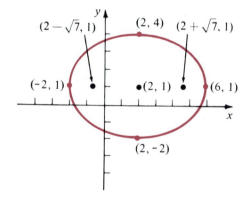

Figure 6.65

Example 2 Graph the conic section given by $4x^2 + 8x - y^2 - 8y - 12 = 0$.

Solution We complete the square in x and y:

$$4(x^2 + 2x \quad) - (y^2 + 8y \quad) = 12$$

$$4(x^2 + 2x + 1) - (y^2 + 8y + 16) = 12 + 4 - 16$$

$$4(x + 1)^2 - (y + 4)^2 = 0$$

or
$$(x + 1)^2 - \frac{(y + 4)^2}{4} = 0. \tag{2}$$

Since the terms on the left-hand side of (2) have opposite signs, one expects this to be a hyperbola. However it is impossible to place this equation into standard form since the right-hand side is zero. Thus, this is a degenerate conic section. Since (2) is equivalent to

$$(x + 1)^2 = \frac{(y + 4)^2}{4}$$

or

$$\pm(x + 1) = \frac{y + 4}{2},$$

we find that the graph consists of the two straight lines $y = 2x - 2$ and $y = -2x - 6$. These are sketched in Figure 6.66. Note that this degenerate form of the hyperbola is obtained as a conic section when the plane intersecting the cone contains the axis of the cone.

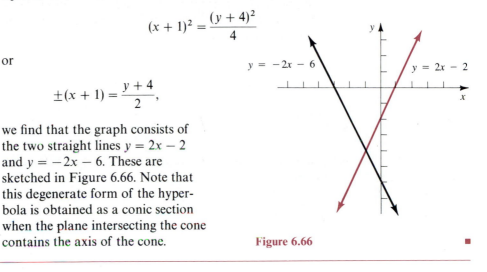

Figure 6.66

Applications

It is of interest to note that the conic sections had been studied for almost 2000 years before any significant applications were discovered. In the seventeenth century Galileo Galilei showed that in the absence of air resistance the path of a projectile follows a parabolic arc. At about the same time Johannes Kepler hypothesized that the orbits of the planets about the sun are ellipses with the sun at one focus. This was later verified by Newton using the methods of the newly discovered calculus. Kepler also experimented with the reflecting properties of parabolic mirrors; these investigations speeded the development of reflecting telescopes. In 1789 Sir William Herschel completed construction on a telescope, the largest to that time, which utilized a parabolic reflecting mirror 4 feet in diameter.

As we have seen in the preceding sections numerous applications of the conic sections are found in the diverse areas of architecture, astronomy, engineering, optics, navigation, and military science.

Exercise 6.5

In Problems 1–16, determine whether the graph of the given equation is a circle, parabola, ellipse, hyperbola, or a degenerate conic. Graph.

1. $y^2 - 6x + 2y = 0$

2. $4x^2 - y^2 + 16x - 2y + 19 = 0$

3. $2x^2 + 2y^2 - 8x + 4y + 5 = 0$

4. $3x^2 - 4y^2 + 12x - 8y - 16 = 0$

5. $4x^2 + 3y^2 - 8x + 6y + 7 = 0$

6. $x^2 + y^2 + 2x - 10y + 30 = 0$

7. $4x^2 + 6y^2 + 8x - 12y - 2 = 0$

8. $x^2 - y^2 - 10x + 6y + 16 = 0$

9. $3x^2 + y^2 - 6x = 0$

10. $2x^2 + 4x - 3y + 6 = 0$

11. $2x^2 + y^2 + 24x + 12y + 109 = 0$

12. $-x^2 - y^2 + 4x - 10y - 4 = 0$

13. $-x^2 - y^2 + 6x + 4y - 13 = 0$

14. $-x^2 + y^2 - 4x + 14y + 45 = 0$

15. $-2x^2 + 6y^2 + 8x - 12y = 0$

16. $-2x^2 - 4y^2 + 8x - 4y + 3 = 0$

17. If a plane perpendicular to the axis of a cone is positioned so that the two halves of the cone are on opposite sides of the plane, the resulting conic section is simply a point. What are the other possible conic sections besides the circle, parabola, ellipse, hyperbola, and point?

In Problems 18–20, refer to the equation

$$Ax^2 + Cy^2 + Dx + Ey + F = 0. \tag{3}$$

18. Show that if A and C have the same signs, the graph of (3) is an ellipse, a circle, a point, or does not exist.

19. Show that if A and C have opposite signs, the graph of (3) is a hyperbola or a pair of intersecting lines.

20. Show that if either $A = 0$ or $C = 0$, the graph of (3) is a parabola, two parallel lines, one line, or does not exist.

Review Problems

21. Find an equation of the parabola with vertex $(2, -3)$ and directrix $y = 4$.

22. Find an equation of the ellipse with center $(0, 0)$, focus $(2, 0)$, and vertex $(5, 0)$.

23. Find the vertices, foci, and asymptotes of the hyperbola $x^2 - y^2/9 = 1$.

24. Find the $x'y'$ coordinates of the xy point $(2, -10)$ if the origin of the $x'y'$ system is at the xy point $(2, 5)$.

25. Find an $x'y'$ equation of $3x - 4y + 5 = 0$ if the origin of the $x'y'$ system is at the xy point $(-2, 1)$.

6.6 Rotation of Axes

In Section 6.1 we derived the equations

$$x = x' + h$$
$$y = y' + k,$$

which express the coordinates of a point relative to the xy system in terms of the coordinates of the same point relative to the translated $x'y'$ system. Suppose now that

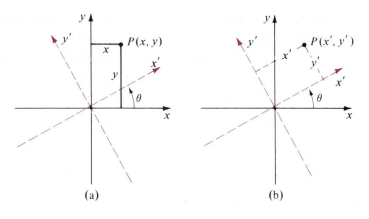

Figure 6.67

instead of translating the coordinate axes, we rotate them through an angle θ as indicated in Figure 6.67. A careful analysis of these figures yields the equations relating the xy coordinates of a point P to its $x'y'$ coordinates. (See Problem 33.)

$$\boxed{\begin{aligned} x &= x' \cos \theta - y' \sin \theta \\ y &= x' \sin \theta + y' \cos \theta. \end{aligned}}$$

(1)

Example 1 Suppose that the x' axis makes an angle of $30°$ with the x axis. Find the xy coordinates of the $x'y'$ point $P(3, -5)$.

Solution We use $\theta = 30°$, $x' = 3$, and $y' = -5$ in equations (1).

$$x = 3 \cos 30° - (-5)\sin 30°$$

$$= 3\left(\frac{\sqrt{3}}{2}\right) + 5\left(\frac{1}{2}\right)$$

$$= \frac{3\sqrt{3} + 5}{2} \approx 5.1$$

$$y = 3 \sin 30° + (-5)\cos 30°$$

$$= 3\left(\frac{1}{2}\right) - 5\left(\frac{\sqrt{3}}{2}\right)$$

$$= \frac{3 - 5\sqrt{3}}{2} \approx -2.8.$$

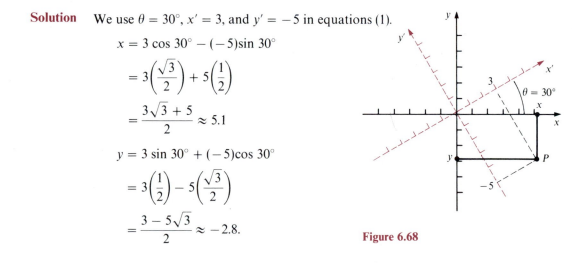

Figure 6.68

Thus the xy coordinates of P are approximately $(5.1, -2.8)$. See Figure 6.68. ∎

If we solve the equations in (1) for x' and y', we obtain another set of rotation equations:

$$x' = x \cos \theta + y \sin \theta$$
$$y' = -x \sin \theta + y \cos \theta.$$

(2)

These are useful in finding the equations of curves whose orientations are "skewed" with respect to the xy coordinate axes.

Example 2 Find an equation of the parabola with vertex at the origin and focus at xy point $(1, \sqrt{3})$.

Solution We first derive the equation of this parabola relative to the rotated coordinate system with x' axis through the focus F. Recall from Section 6.2 that the distance from the vertex to the focus of a parabola with equation of the form $y'^2 = 4px'$ is p. In this case we find p using the Pythagorean theorem.

$$p^2 = (\sqrt{3})^2 + 1^2$$
$$= 3 + 1 = 4,$$
$$p = 2.$$

Thus an $x'y'$ equation of the parabola is

$$y'^2 = 4(2)x'$$

or $\qquad y'^2 = 8x'.$ (3)

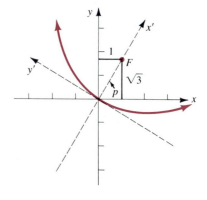

Figure 6.69

We see from Figure 6.69 that the angle θ measured from the x axis to the x' axis is

$$\theta = \arctan \frac{\sqrt{3}}{1} = \frac{\pi}{3} \quad \text{or} \quad \theta = 60°.$$

Now, to find an xy equation of the parabola we substitute equations (2) into equation (3).

$$(-x \sin 60° + y \cos 60°)^2 = 8(x \cos 60° + y \sin 60°)$$

$$\left(-\frac{\sqrt{3}}{2}x + \frac{1}{2}y \right)^2 = 8\left(\frac{1}{2}x + \frac{\sqrt{3}}{2}y \right)$$

$$\frac{3}{4}x^2 - \frac{\sqrt{3}}{2}xy + \frac{1}{4}y^2 = 4x + 4\sqrt{3}y$$

$$3x^2 - 2\sqrt{3}xy + y^2 - 16x - 16\sqrt{3}y = 0. \qquad \blacksquare$$

As in the preceding example, the appearance of an xy term in the equation of a conic section is an indication that the orientation of the curve is skewed with respect to the coordinate axes. If we know the angle of rotation θ for an $x'y'$ system relative to which the curve is *not* skewed, we may then use equations (1) to find an $x'y'$ equation of the curve. The result will be an equation without an $x'y'$ term that can be put in one of the standard forms for a conic section.

Example 3 Find a representation of the equation

$$xy = 1$$

which does not contain an xy term.

Solution The graph of $xy = 1$ shown in Figure 6.70 appears to be that of a hyperbola with transverse axis $y = x$. We thus try $\theta = 45°$. Substituting equations (1) into $xy = 1$, we have

$$(x' \cos 45° - y' \sin 45°)$$

$$\times (x' \sin 45° + y' \cos 45°) = 1$$

$$\left(\frac{\sqrt{2}}{2} x' - \frac{\sqrt{2}}{2} y'\right)$$

$$\times \left(\frac{\sqrt{2}}{2} x' + \frac{\sqrt{2}}{2} y'\right) = 1$$

$$\left(\frac{\sqrt{2}}{2}\right)^2 (x' - y')(x' + y') = 1$$

$$\frac{1}{2}(x'^2 - y'^2) = 1$$

$$\frac{x'^2}{2} - \frac{y'^2}{2} = 1$$

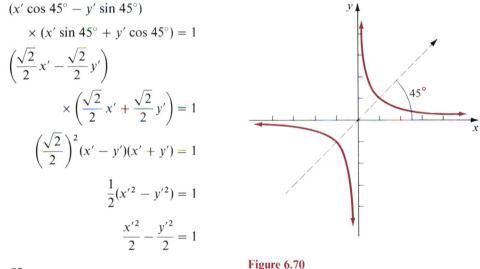

Figure 6.70

or

$$\frac{x'^2}{(\sqrt{2})^2} - \frac{y'^2}{(\sqrt{2})^2} = 1.$$

We see then that the curve is indeed a hyperbola with transverse axis $y = x$. ∎

To eliminate the xy term, it may not always be so apparent what angle of rotation θ to use. We will now discuss how to select the angle of rotation θ so that any equation of the form

$$Ax^2 + Bxy + Cy^2 + Dx + Ey + F = 0, \tag{4}$$

where $B \neq 0$, can be transformed into an equation in x' and y' with no $x'y'$ term. If we rotate the axes through an angle θ, then substituting for x and y from the rotation equations (1), we obtain

$$A(x' \cos \theta - y' \sin \theta)^2 + B(x' \cos \theta - y' \sin \theta)(x' \sin \theta + y' \cos \theta)$$

$$+ \, C(x' \sin \theta + y' \cos \theta)^2 + D(x' \cos \theta - y' \sin \theta) + E(x' \sin \theta + y' \cos \theta) + F = 0.$$

This equation can be written in the form

$$A'(x')^2 + B'x'y' + C'(y')^2 + D'x' + E'y' + F' = 0,$$

where
$$B' = 2(C - A)\sin \theta \cos \theta + B(\cos^2 \theta - \sin^2 \theta).$$

Therefore, in order to eliminate the $x'y'$ term we must select θ so that $B' = 0$, that is,

$$2(C - A)\sin \theta \cos \theta + B(\cos^2 \theta - \sin^2 \theta) = 0.$$

By the double-angle formulas this is equivalent to

$$(C - A)\sin 2\theta + B \cos 2\theta = 0$$

or
$$\cot 2\theta = \frac{A - C}{B}. \tag{5}$$

Thus the choice of the rotation angle (5) will eliminate the xy term from the equation (4). From this result and the discussion of Section 6.5 it follows that any equation of the form (4) is either a conic section or a degenerate conic.

Example 4   After a suitable rotation of axes, identify and sketch the graph of

$$5x^2 + 3xy + y^2 = 22.$$

Solution   If we identify $A = 5$, $B = 3$, $C = 1$ in (4), then the desired rotation angle satisfies

$$\cot 2\theta = \frac{5 - 1}{3} = \tfrac{4}{3}.$$

Since $\cot 2\theta$ is positive, we may choose 2θ such that $0 < 2\theta < 90°$. Thus using half-angle formulas, we find

$$\sin \theta = \sqrt{\frac{1 - \cos 2\theta}{2}} = \sqrt{\frac{1 - \frac{4}{5}}{2}} = \frac{1}{\sqrt{10}},$$

$$\cos \theta = \sqrt{\frac{1 + \cos 2\theta}{2}} = \sqrt{\frac{1 + \frac{4}{5}}{2}} = \frac{3}{\sqrt{10}}.$$

Equations (1) are then

$$x = \frac{3}{\sqrt{10}}x' - \frac{1}{\sqrt{10}}y'$$

$$y = \frac{1}{\sqrt{10}}x' + \frac{3}{\sqrt{10}}y'.$$

Substituting these into the given equation, we have

$$5\left(\frac{3}{\sqrt{10}}x' - \frac{1}{\sqrt{10}}y'\right)^2 + 3\left(\frac{3}{\sqrt{10}}x' - \frac{1}{\sqrt{10}}y'\right)\left(\frac{1}{\sqrt{10}}x' + \frac{3}{\sqrt{10}}y'\right)$$
$$+ \left(\frac{1}{\sqrt{10}}x' + \frac{3}{\sqrt{10}}y'\right)^2 = 22$$

$$5\left(\frac{9}{10}x'^2 - \frac{6}{10}x'y' + \frac{1}{10}y'^2\right) + 3\left(\frac{3}{10}x'^2 + \frac{8}{10}x'y' - \frac{3}{10}y'^2\right)$$
$$+ \left(\frac{1}{10}x'^2 + \frac{6}{10}x'y' + \frac{9}{10}y'^2\right) = 22$$

$$45x'^2 - 30x'y' + 5y'^2 + 9x'^2 + 24x'y' - 9y'^2 + x'^2 + 6x'y' + 9y'^2 = 220.$$

The last equation simplifies to

$$\frac{x'^2}{4} + \frac{y'^2}{44} = 1.$$

We recognize this to be the equation of an ellipse, and use the new axes to sketch the graph as shown in Figure 6.71.

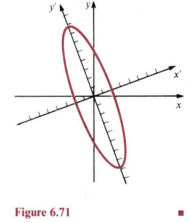

Figure 6.71

Exercise 6.6

In Problems 1–6, find the xy coordinates of the given x'y' point. Use the specified angle of rotation θ.

1. $(2, -3)$, $\theta = 30°$ **2.** $(-5, 1)$, $\theta = 45°$ **3.** $(0, 4)$, $\theta = 90°$ **4.** $(3, 0)$, $\theta = 60°$

5. $(4, 2)$, $\theta = 15°$ **6.** $(1, 1)$, $\theta = -30°$

In Problems 7–12, find the x'y' coordinates of the given xy point. Use the specified angle of rotation θ.

7. $(2, -2)$, $\theta = 45°$ **8.** $(1, 1)$, $\theta = 60°$

9. $(5, 0)$, $\theta = 45°$ **10.** $(\sqrt{3}, 1)$, $\theta = 30°$

11. $(4, 8)$, $\theta = -60°$ **12.** $(6, -3)$, $\theta = 75°$

In Problems 13–18, use rotation of axes to find an equation of the given curve relative to the xy coordinate system.

13. A parabola with vertex $(0, 0)$ and focus $(2\sqrt{2}, 2\sqrt{2})$

14. An ellipse with center $(0, 0)$, focus $(\sqrt{3}, 1)$, and vertex $\left(\dfrac{3\sqrt{3}}{2}, \dfrac{3}{2}\right)$

15. A hyperbola with center $(0, 0)$, focus $(-\sqrt{2}, \sqrt{2})$, and vertex $(-1, 1)$

16. A parabola with vertex $(0, 0)$ and directrix $y = \sqrt{3}x - 2$

17. An ellipse with center $(0, 0)$, focus $(-1, \sqrt{3})$, and vertex $\left(-\dfrac{5}{2}, \dfrac{5\sqrt{3}}{2}\right)$

18. A hyperbola with center $(0, 0)$, focus $(1, \sqrt{3})$, and vertex $\left(\dfrac{1}{3}, \dfrac{\sqrt{3}}{3}\right)$

In Problems 19–24, use the given angle θ to represent the equation relative to the x'y' system without an x'y' term.

19. $2x^2 + 4xy - y^2 = 48$, $\theta = \arctan \dfrac{1}{2}$

20. $x^2 + xy + y^2 = 24$, $\theta = 45°$

21. $x^2 + 2xy + y^2 + 16\sqrt{2}y = 0$, $\theta = \dfrac{\pi}{4}$

22. $3xy + 4y^2 + 18 = 0$, $\theta = \arctan 3$

23. $3x^2 + 4xy - 16 = 0$, $\theta = \arctan \dfrac{1}{2}$

24. $2x^2 + 4xy + 5y^2 = 36$, $\theta = \arctan 2$

In Problems 25–28, determine the rotation angle θ such that an x'y' equation contains no x'y' term. Find an x'y' equation.

25. $6x^2 + 3xy + 2y^2 = 234$

26. $4x^2 + 4xy + y^2 + 2\sqrt{5}x - 4\sqrt{5}y = 0$

27. $2x^2 - 4xy + 2y^2 + 8\sqrt{2}x + 8\sqrt{2}y = 0$

28. $-7x^2 + 5xy + 5y^2 = 165$

In Problems 29–32, for the given xy equation perform a suitable rotation of axes so that the resulting x'y' equation has no x'y' term. Sketch the graph.

29. $x^2 + xy + y^2 = 4$

30. $x^2 - 2xy - y^2 - 8x - 8y = 0$

31. $17x^2 - 12xy + 8y^2 - 80 = 0$

32. $41x^2 - 24xy + 34y^2 - 25 = 0$

33. a. Using Figure 6.72 show that $x' = d \cos \phi$, $y' = d \sin \phi$ and $x = d \cos(\theta + \phi)$, $y = d \sin(\theta + \phi)$;
b. Using the results given in part (a) derive the rotation equations (1) and (2).

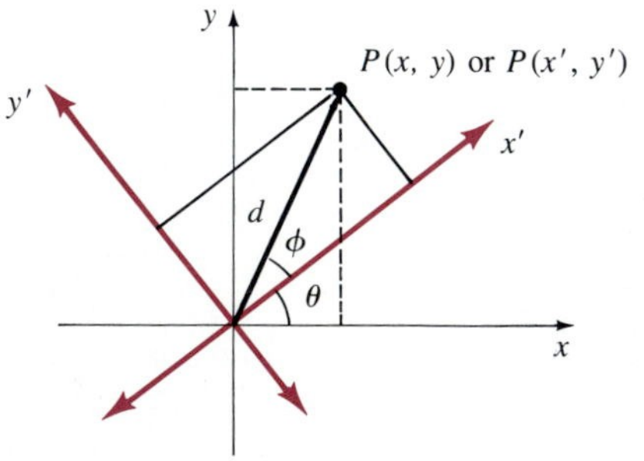

Figure 6.72

34. Prove that, except for degenerate cases, the graph of $Ax^2 + Bxy + Cy^2 + Dx + Ey + F = 0$ is
  a. a parabola if $B^2 - 4AC = 0$;
  b. an ellipse if $B^2 - 4AC < 0$;
  c. a hyperbola if $B^2 - 4AC > 0$.

In Problems 35 and 36, use the result of Problem 34 to determine the type of graph without graphing.

35. $4x^2 + 4xy + y^2 - 6 = 0$ **36.** $17x^2 - 12xy + 8y^2 - 68x + 24y - 12 = 0$

Review Problems

37. Find an equation of the hyperbola with foci $(5, -2)$ and $(5, 4)$ and one vertex at $(5, 3)$.

38. A small decorative archway in a garden is a semi-ellipse with a span of 9 feet and height at the center 3 feet. What is the height of the arch 1.5 feet from the center?

39. Find an equation of the parabola with directrix $x = 15$ and vertex $(3, 5)$.

40. Find the vertices, foci, and asymptotes of the hyperbola $y(y - 4) = x(x - 4) + 1$.

41. Find an $x'y'$ equation of the line $13x - 7y + 6 = 0$ if the origin of the $x'y'$ system is at the xy point $(2, 3)$.

Chapter Review

If the origin of a **translated coordinate system** is at the xy point (h, k), then the **translation equations**

$$x = x' + h$$
$$y = y' + k \tag{1}$$

are used to express the coordinates of a point given in the xy system relative to the $x'y'$ system. Solving the equations in (1) for x' and y', we obtain

$$x' = x - h$$
$$y' = y - k,$$

which are then used to convert the equations of a curve in the $x'y'$ system to an equation of the same curve in the xy system.

An equation of the **parabola** with **vertex** (h, k), **directrix** $y = k - p$, **focus** $(h, k + p)$, and **axis** $x = h$ is

$$(x - h)^2 = 4p(y - k). \tag{2}$$

An equation of the parabola with vertex (h, k), directrix $x = h - p$, focus $(h + p, k)$, and axis $y = k$ is

$$(y - k)^2 = 4p(x - h). \tag{3}$$

The equation $Ax^2 + Dx + Ey + F = 0$ can be put in the form of equation (2) by completing the square. Similarly, $Cy^2 + Dx + Ey + F = 0$ can be put in the form of equation (3).

An equation of the **ellipse** with **center** (h, k) and a horizontal **major axis** is

$$\frac{(x - h)^2}{a^2} + \frac{(y - k)^2}{b^2} = 1. \tag{4}$$

The **vertices** are $(h \pm a, k)$ and $(h, k \pm b)$ and the **foci** are located at $(h \pm c, k)$, where a, b, and c are related by

$$a^2 = b^2 + c^2$$

If the center is (h, k) and the major axis is vertical, an equation of the ellipse is

$$\frac{(x - h)^2}{b^2} + \frac{(y - k)^2}{a^2} = 1. \tag{5}$$

In this case the vertices are $(h \pm b, k)$ and $(h, k \pm a)$ and the foci are $(h, k \pm c)$. As before, $a^2 = b^2 + c^2$. In equations (4) and (5) a^2 is always the larger denominator.

An equation of the **hyperbola** with **center** (h, k) and a horizontal **transverse axis** is

$$\frac{(x - h)^2}{a^2} - \frac{(y - k)^2}{b^2} = 1. \tag{6}$$

The **vertices** are $(h \pm a, k)$ and the **foci** are located at $(h \pm c, k)$, where a, b, and c are related by

$$a^2 + b^2 = c^2.$$

If the center is (h, k) and the transverse axis is vertical, an equation of the hyperbola is

$$\frac{(y - k)^2}{a^2} - \frac{(x - h)^2}{b^2} = 1. \tag{7}$$

In this case the vertices are $(h, k \pm a)$ and the foci are $(h, k \pm c)$. For both equations (6) and (7), a^2 is the denominator of the positive term and $a^2 + b^2 = c^2$. To obtain the **asymptotes** of the hyperbola in (6) or (7), simply set the left-hand side of the equation equal to 0 and solve for y.

The circle, parabola, ellipse, and hyperbola are **conic sections**. They are among the graphs of the second-degree equation

$$Ax^2 + Cy^2 + Dx + Ey + F = 0.$$

The conic sections have numerous applications in architecture, astronomy, engineering, and physics.

The xy term can be eliminated from an equation of the form

$$Ax^2 + Bxy + Cy^2 + Dx + Ey + F = 0,$$

by rotating the axes through an angle θ for which

$$\cot 2\theta = \frac{A - C}{B}.$$

Chapter 6 Review Exercises

In Problems 1–10, answer true or false.

1. The center of the $x'y'$ system must be at $(1, -2)$ if the circle $(x - 1)^2 + (y + 2)^2 = 4$ is to have the equation $x'^2 + y'^2 = 4$. _____

2. The parabola $y^2 = 10x$ has a horizontal axis. _____

3. The foci of the ellipse $4x^2 + y^2 = 1$ are $(0, \pm 2)$. _____

4. For a hyperbola the length of the transverse axis is always greater then that of the conjugate axis. _____

5. For the parabola $x^2 = 8y$ the distance from the focus to the vertex is 2. _____

6. The length of the major axis of the ellipse $\dfrac{x^2}{9} + \dfrac{y^2}{16} = 1$ is 8. _____

7. The asymptotes of a hyperbola always pass through the origin. _____

8. The directrix of a parabola must be either horizontal or vertical. _____

9. If $(4, 0)$, $(6, 3)$, $(4, 6)$, and $(2, 3)$ are vertices of an ellipse, then its major axis is horizontal. _____

10. Rotating the axes $45°$ will eliminate the xy term from $x + xy + y = 1$. _____

11. Find the $x'y'$ coordinates of the xy point $(10, -4)$ if the origin of the $x'y'$ system is at the xy point $(\sqrt{2}, 4)$.

12. Express the equation $x^2 + y^2 - 10x + 2y + 6 = 0$ in terms of the $x'y'$ system, where the origin of the $x'y'$ system is at the xy point $(5, -1)$.

13. Express the equation $x' - 3y' = 0$ in terms of the xy system, where the origin of the $x'y'$ system is at the xy point $(4, 2)$.

14. Find the vertex, focus, directrix, and axis of the parabola $x^2 - 2x + 4y + 1 = 0$. Graph.

In Problems 15 and 16, find an equation of the parabola that satisfies the given conditions.

15. Focus at $(3, -1)$, vertex at $(0, -1)$

16. Vertex at $(1, 2)$, passing through $(4, 5)$, with a vertical axis

17. Find the center, foci, and vertices of the ellipse $4x^2 + y^2 + 8x - 6y + 9 = 0$. Graph.

In Problems 18 and 19, find an equation of the ellipse satisfying the given conditions.

18. Vertices at $(0, \pm 4)$, foci at $(\pm 5, 0)$

19. Vertices at $(\pm 2, -2)$, passing through $\left(1, \dfrac{\sqrt{3}-4}{2}\right)$

20. Find the center, vertices, foci, and asymptotes of the hyperbola $9x^2 - y^2 - 54x - 2y + 71 = 0$. Graph.

In Problems 21 and 22, find an equation of the hyperbola satisfying the given conditions.

21. Foci at $(\pm 2\sqrt{5}, 0)$, asymptotes $y = \pm 2x$

22. Vertices at $(-3, 2)$ and $(-3, 4)$, one focus at $(-3, 3 + \sqrt{2})$

In Problems 23–26, determine the conic section which has the given equation. Graph.

23. $x^2 - 6x - 2y + 8 = 0$

24. $x^2 + y^2 - 8x - 2y + 10 = 0$

25. $x^2 - y^2 - 8x - 2y + 11 = 0$

26. $4x^2 + 5y^2 - 20y = 0$

27. A satellite orbits the planet Neptune in an elliptical orbit with the center of the planet at one focus. If the length of the major axis of the orbit is 2×10^9 m and the length of the minor axis is 6×10^8 m, find the maximum distance between the satellite and the center of Neptune.

28. A parabolic mirror has a depth of 7 cm at the center and the distance across the top of the mirror is 20 cm. Find the distance from the vertex to the focus.

In Problems 29 and 30, perform a suitable rotation of axes so that the resulting $x'y'$ equation has no $x'y'$ term. Sketch the graph.

29. $xy = -3$

30. $8x^2 - 4xy + 5y^2 = 36$

31. For the equation $2x^2 - 4xy - y^2 + 20x - 2y + 17 = 0$ (a) rotate the axes to an $x'y'$ system so that the $x'y'$ term is missing and then (b) translate the axes so that the first degree terms are eliminated.

Appendix

A.1   Review of Basic Mathematics

Sets

A **set** is a collection of objects or **elements** each having a certain property. We can specify a set in two different ways: by stating a property that an element of the set must possess or by simply listing the elements. We usually designate a set by a capital letter such as A or B and an element of the set by a lowercase letter such as x.

Example 1   The set

is the same as

$$A = \{x \mid x = 2n, n \text{ an integer}\}$$

$$A = \{\ldots, -4, -2, 0, 2, 4, 6, \ldots\}.$$

■

To indicate that x is an element of the set A, we use the symbolism

$$x \in A \qquad \text{whereas} \qquad x \notin A$$

means that x is *not* an element of A.

Example 2   If A denotes the set given in the first example, then

$$14 \in A \qquad \text{but} \qquad 13 \notin A.$$

■

If every element of a set A is also an element of a set B, we say that A is a **subset** of B and write

$$A \subset B.$$

It follows, then, that every set is a subset of itself.

Example 3 If
$$I = \{\ldots, -3, -2, -1, 0, 1, 2, 3, \ldots\}$$
is the set of all integers, and
$$A = \{\ldots, -4, -2, 0, 2, 4, 6, \ldots\},$$
then $A \subset I$. Also, $A \subset A$ and $I \subset I$. ∎

The set containing no elements is said to be **empty** and is denoted by the symbol $\varnothing$. We note that the set $\{\varnothing\}$ is *not* empty, since it contains one element, the empty set. Also, the empty set is a subset of every set.

The **union** of two sets A and B is the set
$$A \cup B = \{x \mid x \in A \text{ or } x \in B\}.$$

The **intersection** of two sets A and B is the set
$$A \cap B = \{x \mid x \in A \text{ and } x \in B\}.$$

The accompanying figures show the union and the intersection of two sets by means of a **Venn diagram**. The intersection of two sets consists of the elements that are common to both sets. If there are no common elements, that is, $A \cap B = \varnothing$, the sets are said to be **disjoint**.

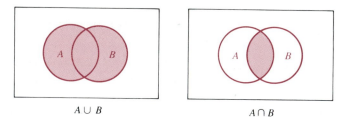

$A \cup B$ $A \cap B$

Example 4 If $A = \{0, 6, 7, 9, 10\}$, $B = \{-1, 5, 8, 11\}$, $C = \{-1, 3, 7, 10\}$,
then
$$A \cup B = \{-1, 0, 5, 6, 7, 8, 9, 10, 11\}$$
$$A \cap B = \varnothing$$
$$A \cap C = \{7, 10\}$$
$$B \cap C = \{-1\}.$$
∎

The set of all possible elements under consideration throughout a discussion is sometimes called the **universe**. For example, when working with sets of integers, fractions, or irrational numbers, it is natural to assume that the universe is the set R of real numbers.

The **complement** of a set A, denoted by A', is the set

$$A' = \{x \mid x \notin A, \text{ but } x \text{ is an element of the universe}\}.$$

Example 5 If the universe is R, and Q and H are the sets of rational and irrational numbers, respectively, then

$$Q' = H \qquad \text{and} \qquad H' = Q. \qquad\blacksquare$$

Properties of Real Numbers

We now list some **basic properties** of the real number system:

1. $a + b = b + a$ (commutative law of addition)
2. $a + (b + c) = (a + b) + c$ (associative law of addition)
3. $ab = ba$ (commutative law of multiplication)
4. $a(bc) = (ab)c$ (associative law of multiplication)
5. $a(b + c) = ab + ac$ (distributive law)
6. $a + 0 = a$ (additive identity)
7. $a \cdot 1 = a$ (multiplicative identity)
8. $a + (-a) = a - a = 0$ (additive inverse)
9. $a \cdot a^{-1} = a \cdot \left(\dfrac{1}{a}\right) = 1,\ a \neq 0$ (multiplicative inverse)

Fractions

The rules for combining fractions are given below. It is understood that any letter in the denominator represents a nonzero number.

1. $\dfrac{a}{b} + \dfrac{c}{b} = \dfrac{a + c}{b}$

2. $\dfrac{a}{b} + \dfrac{c}{d} = \dfrac{ad + bc}{bd}$ (common denominator)

3. $\dfrac{a}{b} \cdot \dfrac{c}{d} = \dfrac{ac}{bd}$

4. $\dfrac{a}{b} \div \dfrac{c}{d} = \dfrac{\dfrac{a}{b}}{\dfrac{c}{d}} = \dfrac{a}{b} \cdot \dfrac{d}{c} = \dfrac{ad}{bc}$

5. $\dfrac{ka}{b} \cdot \dfrac{c}{kd} = \dfrac{ac}{bd}$ (cancellation of $k \neq 0$)

Division of Polynomials

In the study of calculus it is sometimes necessary to divide one polynomial by another. The following example illustrates the method of long division.

Example 6   Perform the indicated division:

$$\frac{2x^5 - 3x^3 + 5x + 6}{x^2 + 4x}.$$

Solution

$$
\begin{array}{r}
2x^3 - 8x^2 + 29x - 116 \\
x^2 + 4x\,\overline{\smash{\big)}\,2x^5 + 0x^4 - 3x^3 + 0x^2 + 5x + 6} \\
\underline{2x^5 + 8x^4} \\
-8x^4 - 3x^3 \\
\underline{-8x^4 - 32x^3} \\
29x^3 + 0x^2 \\
\underline{29x^3 + 116x^2} \\
-116x^2 + 5x \\
\underline{-116x^2 - 464x} \\
469x + 6 \quad \text{(remainder)}
\end{array}
$$

When a polynomial $P(x)$ of degree n is divided by a linear term $x - a$, we always obtain a constant remainder r. That is,

$$\frac{P(x)}{x - a} = Q(x) + \frac{r}{x - a}$$

or

$$P(x) = (x - a)Q(x) + r,$$

where $Q(x)$ is a polynomial of degree $n - 1$. Thus, $x - a$ is a factor of $P(x)$ if and only if $r = 0$. This is the same as saying that $x = a$ is a root of the equation $P(x) = 0$ when $r = 0$.

In the case of the linear divisor $x - a$, we can use **synthetic division** as a shortcut to the actual division. This technique is illustrated in the following example.

Example 7   Perform the indicated division:

$$\frac{2x^3 + 7x^2 - 4x - 20}{x + 2}.$$

Solution

$$
\begin{array}{r|rrrr}
-2 & 2 & 7 & -4 & -20 \\
   &   & -4 & -6 & 20 \\
\hline
   & 2 & 3 & -10 & 0 \quad \text{remainder}
\end{array}
$$

coefficients of $Q(x)$

The numbers on the second line indicated by the arrowheads are obtained by multiplying the numbers on the third line by -2. The fact that $r = 0$ means that $x - (-2) = x + 2$ is a factor of $2x^3 + 7x^2 - 4x - 20$. We have

$$\frac{2x^3 + 7x^2 - 4x - 20}{x + 2} = 2x^2 + 3x - 10$$

and so

$$2x^3 + 7x^2 - 4x - 20 = (x + 2)(2x^2 + 3x - 10).$$

We now review two important concepts from geometry.

Pythagorean Theorem

The sides of a right triangle are related by the **Pythagorean theorem**,

$$c^2 = a^2 + b^2,$$

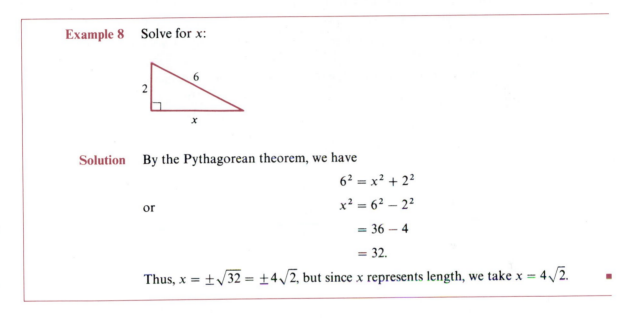

where c is the length of the hypotenuse and a and b are lengths of the other two sides. Conversely, if the sides of a triangle satisfy this equation, we then have a right triangle.

Example 8 Solve for x:

By the Pythagorean theorem, we have

$$6^2 = x^2 + 2^2$$

or

$$x^2 = 6^2 - 2^2$$

$$= 36 - 4$$

$$= 32.$$

Thus, $x = \pm\sqrt{32} = \pm4\sqrt{2}$, but since x represents length, we take $x = 4\sqrt{2}$. ∎

Similar Triangles

The corresponding sides of **similar triangles** are proportional. That is, if

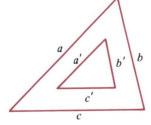

then $\dfrac{a}{a'} = \dfrac{b}{b'} = \dfrac{c}{c'}.$

Exercise A.1

Let the universe be $\{1, 2, 3, \ldots, 14, 15\}$ *and*

$$A = \{1, 4, 6, 8, 10, 15\}, \qquad B = \{3, 9, 11, 12, 14\}, \qquad C = \{1, 2, 5, 7, 8, 13, 14\}.$$

In Problems 1–16, find the indicated set.

1. $A \cup B$ **2.** $A \cup C$ **3.** $B \cup C$ **4.** $A \cap B$

5. $A \cap C$ **6.** $B \cap C$ **7.** $(A \cap B) \cup B$ **8.** $A \cup (B \cup C)$

9. A' **10.** B' **11.** C' **12.** $(A \cup C)'$

13. $A' \cup B'$ **14.** $(A \cap B)'$ **15.** $A \cup (B' \cap C')$ **16.** $(A' \cap C)'$

In Problems 17–22, list the elements of the given set.

17. $\{x \mid x = 2k - 3, k \text{ an integer}\}$ **18.** $\{x \mid x^2 - 9 = 0\}$

19. $\{(x, y) \mid xy = 3, x \text{ and } y \text{ integers}\}$ **20.** $\{(x, y) \mid y = 2x, x \text{ an integer}\}$

21. $\{(x, y) \mid x = 4k, y = -k, k \text{ an integer}\}$ **22.** $\{(x, y) \mid x + y = 0 \text{ and } x - y = 2\}$

In Problems 23 and 24, use long division to perform the indicated operation.

23. $\dfrac{4x^3 - 2x^2 + x - 1}{2x + 1}$ **24.** $\dfrac{x^4 - x^3 + x^2 - 2x + 3}{x^2 - x + 1}$

In Problems 25 and 26, use synthetic division to perform the indicated operation.

25. $\dfrac{5x^3 - 2x^2 + x - 6}{x - 3}$ **26.** $\dfrac{x^3 + 4x^2 + 8x + 8}{x + 2}$

In Problems 27–30, find the indicated unknown.

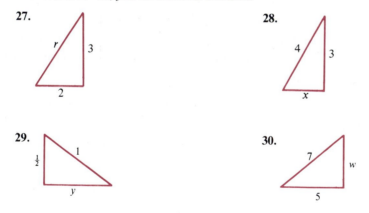

27.

28.

29.

30.

In Problems 31 and 32, determine x and y for the given similar triangles.

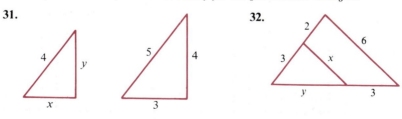

31.

32.

A.2 **Mathematical Induction**

Frequently a statement or proposition depending on the positive integers

$$N = \{1, 2, 3, \ldots\}$$

can be proved using the **principle of mathematical induction**.
  Suppose we can show:

a. a statement is true for the positive integer 1; and

b. whenever the statement is true for the positive integer k, then it is true for the next positive integer $k + 1$.

In other words, suppose we can demonstrate that the

statement
is
true for 1   (a)

and that the

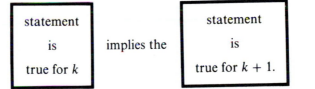

(b)

What can we conclude from this? From (a) we have that

the statement is true for the number 1,

and by (b),

the statement is true for the number 1 + 1 = 2.

In addition it now follows from (b) that

the statement is true for the number 2 + 1 = 3,

the statement is true for the number 3 + 1 = 4,

the statement is true for the number 4 + 1 = 5,

and so on. Symbolically we could represent this sequence of implications by

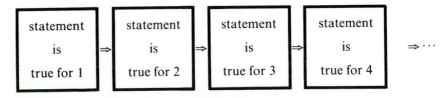

It seems clear that the statement must be true for *all* positive integers n. This is precisely the assertion of the following principle.

Principle of Mathematical Induction   *Let $S(n)$ be a statement involving a positive integer n such that*

a. *$S(1)$ is true, and*
b. *whenever $S(k)$ is true for a positive integer k, then $S(k + 1)$ is also true.*

Then $S(n)$ is true for every positive integer.

We now illustrate the use of induction with several examples.

Example 1   Prove by mathematical induction that

$$1 + 2 + 3 + \cdots + n = \frac{n(n + 1)}{2}$$

for every positive integer n.

Solution   Here the statement $S(n)$ is

$$1 + 2 + 3 + \cdots + n = \frac{n(n + 1)}{2}.$$

The first step is to show that $S(1)$ is true, where $S(1)$ is the statement

$$1 = \frac{1 \cdot 2}{2}.$$

Since this is clearly true, condition (a) is satisfied.

The next step is to verify condition (b). This requires that from the hypothesis "$S(k)$ is true" we prove that "$S(k + 1)$ is true." Thus we assume that the statement $S(k)$,

$$1 + 2 + 3 + \cdots + k = \frac{k(k + 1)}{2},$$

is true. If we add $k + 1$ to both sides of this equation, we have

$$1 + 2 + 3 + \cdots + k + (k + 1) = \frac{k(k + 1)}{2} + (k + 1)$$

$$= \frac{k(k + 1) + 2(k + 1)}{2}$$

$$= \frac{(k + 1)(k + 2)}{2}.$$

Thus we have shown that the statement $S(k + 1)$,

$$1 + 2 + 3 + \cdots + k + (k + 1) = \frac{(k + 1)[(k + 1) + 1]}{2},$$

is true. It now follows from the principle of mathematical induction that

$$1 + 2 + 3 + \cdots + n = \frac{n(n + 1)}{2}$$

is true for every positive integer n. ∎

In Chapter 0 we saw that

$$x - y = x - y$$
$$x^2 - y^2 = (x - y)(x + y)$$
$$x^3 - y^3 = (x - y)(x^2 + xy + y^2)$$

and
$$x^4 - y^4 = (x - y)(x + y)(x^2 + y^2).$$

From these factorizations we might conjecture that $x - y$ is always a factor of $x^n - y^n$ for any positive integer n. We now prove that this is so.

Example 2   Prove by mathematical induction that $x - y$ is a factor of $x^n - y^n$ for any positive integer n.

Solution   For the statement $S(n)$,

$$x - y \text{ is a factor of } x^n - y^n,$$

we must show that the two conditions of the principle of mathematical induction are satisfied.
   a. For $n = 1$ we have the true statement $S(1)$,

$$x - y \text{ is a factor of } x^1 - y^1.$$

   b. Assume that $S(k)$,

$$x - y \text{ is a factor of } x^k - y^k,$$

is true. Using this assumption, we must show that $x - y$ is a factor of $x^{k+1} - y^{k+1}$. Now subtracting and adding xy^k gives

$$x^{k+1} - y^{k+1} = x^{k+1} - xy^k + xy^k - y^{k+1},$$

or
$$x^{k+1} - y^{k+1} = x(x^k - y^k) + y^k(x - y).$$

But by hypothesis $x - y$ is a factor of $x^k - y^k$. Therefore $x - y$ is a factor of *each* term in the right-hand side of the above equation. It follows that $x - y$ is a factor of the right-hand side, and thus we have shown that the statement $S(k + 1)$,

$$x - y \text{ is a factor of } x^{k+1} - y^{k+1},$$

is true.

It now follows by mathematical induction that $x - y$ is a factor of $x^n - y^n$ for any positive integer n. ∎

Binomial Coefficient

If r is any integer such that $0 \leq r \leq n$, then

$$n(n - 1) \cdots (n - r + 1) = \frac{n(n - 1) \cdots (n - r + 1)(n - r)(n - r - 1) \cdots 3 \cdot 2 \cdot 1}{(n - r)(n - r - 1) \cdots 3 \cdot 2 \cdot 1}$$

$$= \frac{n!}{(n - r)!}.$$

Thus the **binomial coefficient**

$$\frac{n(n - 1) \cdots (n - r + 1)}{r!}$$

of $a^{n-r}b^r$ can be written as

$$\frac{n!}{r!(n - r)!}.$$

This latter quotient is usually denoted by $\binom{n}{r}$. Hence the binomial theorem can be written in the following alternative form.

Binomial Theorem Let a and b be real numbers and let n be a positive integer. Then

$$(a + b)^n = \binom{n}{0}a^n + \binom{n}{1}a^{n-1}b + \cdots + \binom{n}{r}a^{n-r}b^r + \cdots + \binom{n}{n}b^n.$$

In Chapter 0 we did not prove the binomial theorem. Now as an aid in proving this theorem we note that for $0 < r \leq n$,

$$\binom{n}{r - 1} + \binom{n}{r} = \frac{n!}{(r - 1)!(n - r + 1)!} + \frac{n!}{r!(n - r)!}$$

$$= \frac{n!r + n!(n - r + 1)}{r!(n - r + 1)!}$$

$$= \frac{n!(n + 1)}{r!(n + 1 - r)!}$$

$$= \frac{(n + 1)!}{r![(n + 1) - r]!} = \binom{n + 1}{r}.$$

Thus, we have shown that

$$\binom{n}{r-1} + \binom{n}{r} = \binom{n+1}{r}. \tag{1}$$

Proof of the Binomial Theorem

We use the principle of mathematical induction to prove this theorem.

a. For $n = 1$, we have the true statement

$$(a+b)^1 = \binom{1}{0}a^1 + \binom{1}{1}b^1,$$

since

$$\binom{1}{0} = \frac{1!}{0!1!} = 1 \quad \text{and} \quad \binom{1}{1} = \frac{1!}{1!0!} = 1.$$

b. Now assume that $S(k)$,

$$(a+b)^k = \binom{k}{0}a^k + \binom{k}{1}a^{k-1}b + \cdots + \binom{k}{r}a^{k-r}b^r + \cdots + \binom{k}{k}b^k,$$

is true. Multiplying both sides of this equation by $(a+b)$ gives

$$(a+b)(a+b)^k = (a+b)\left[\binom{k}{0}a^k + \binom{k}{1}a^{k-1}b + \cdots + \binom{k}{r}a^{k-r}b^r + \cdots + \binom{k}{k}b^k\right]$$

$$= \binom{k}{0}(a^{k+1} + a^k b) + \binom{k}{1}(a^k b + a^{k-1}b^2) + \cdots$$

$$+ \binom{k}{r}(a^{k-r+1}b^r + a^{k-r}b^{r+1}) + \cdots + \binom{k}{k}(ab^k + b^{k+1})$$

$$= \binom{k}{0}a^{k+1} + \left[\binom{k}{0} + \binom{k}{1}\right]a^k b + \left[\binom{k}{1} + \binom{k}{2}\right]a^{k-1}b^2 + \cdots$$

$$+ \left[\binom{k}{r-1} + \binom{k}{r}\right]a^{k-r+1}b^r + \cdots + \binom{k}{k}b^{k+1}.$$

From (1) and the facts that

$$\binom{k}{0} = 1 = \binom{k+1}{0} \quad \text{and} \quad \binom{k}{k} = 1 = \binom{k+1}{k+1}$$

we obtain

$$(a+b)^{k+1} = \binom{k+1}{0}a^{k+1} + \binom{k+1}{1}a^k b + \cdots$$

$$+ \binom{k+1}{r}a^{k+1-r}b^r + \cdots + \binom{k+1}{k+1}b^{k+1}.$$

which is the statement $S(k+1)$. Hence, by the principle of mathematical induction, the proof is complete.

Exercise A.2

In Problems 1–10, use mathematical induction to prove the given statement for any natural number n.

1. $2 + 4 + 6 + \cdots + 2n = n^2 + n$.

2. $1 + 3 + 5 + \cdots + (2n - 1) = n^2$.

3. $1^2 + 2^2 + 3^2 + \cdots + n^2 = \dfrac{n(n+1)(2n+1)}{6}$.

4. $1^3 + 2^3 + 3^3 + \cdots + n^3 = \dfrac{n^2(n+1)^2}{4}$.

5. $\dfrac{1}{1 \cdot 2} + \dfrac{1}{2 \cdot 3} + \dfrac{1}{3 \cdot 4} + \cdots + \dfrac{1}{n(n+1)} = \dfrac{n}{n+1}$.

6. $n^2 + n$ is divisible by 2.

7. $n^3 + 2n$ is divisible by 3.

8. $x + y$ is a factor of $x^{2n-1} + y^{2n-1}$.

9. For $a \geq -1$, $(1 + a)^n \geq 1 + na$.

10. $2n \leq 2^n$.

11. Assume that

$$2 + 4 + 6 + \cdots + 2n = n^2 + n + 1$$

is true for $n = k$, and show that the formula is true for $n = k + 1$. Show, however, that the formula itself is false. Explain why this does not violate the principle of mathematical induction.

A.3 Complex Numbers

A **complex number** is any number of the form

$$z = a + bi,$$

where a and b are real numbers and $i^2 = -1$. The real numbers a and b are called the **real** and **imaginary parts** of z, respectively. Choosing $b = 0$, we see that every real number is a complex number.

Since it is usual practice to write $i = \sqrt{-1}$, formal applications of the laws of exponents allow us to express square roots of negative numbers as multiples of i. For example,

$$\sqrt{-16} = \sqrt{16}\,\sqrt{-1} = 4i.$$

Example 1 We have already seen that use of the quadratic formula sometimes leads to complex numbers. Consider the additional example:

$$x^2 + 4x + 29 = 0.$$

It follows that

$$x = \frac{-4 \pm \sqrt{(4)^2 - 4(1)(29)}}{2}$$

$$= \frac{-4 \pm \sqrt{16 - 116}}{2}$$

$$= \frac{-4 \pm \sqrt{-100}}{2}$$

$$= \frac{-4 \pm 10i}{2}.$$

Hence, $x = -2 + 5i$ or $x = -2 - 5i$. ∎

Complex Conjugates

If $z = a + bi$, then $\bar{z} = a - bi$ is called the **complex conjugate** of z. To form the complex conjugate of z, we simply negate the imaginary part. For example, if $z = 3 - 6i$, then $\bar{z} = 3 - (-6i) = 3 + 6i$. Notice that solutions of the quadratic equation in the example above are complex conjugates. This is always the case when the coefficients of a quadratic equation are real numbers and the roots are complex numbers.

Properties

The sum, difference, product, and quotient of two complex numbers are also complex numbers. Suppose that $z_1 = a + bi$ and $z_2 = c + di$; we then define

1. $z_1 + z_2 = (a + c) + (b + d)i$

2. $z_1 - z_2 = (a - c) + (b - d)i$

3. $z_1 \cdot z_2 = (ac - bd) + (bc + ad)i$

4. $\dfrac{z_1}{z_2} = \dfrac{ac + bd}{c^2 + d^2} + \dfrac{bc - ad}{c^2 + d^2} i$

There is no need to memorize any of these rules. The first two formulas merely state that we add and subtract like quantities: the corresponding real and imaginary parts. The third formula is a consequence of the distributive law and the fact that $i^2 = -1$.

Example 2 If $z_1 = 2 - 3i$ and $z_2 = -5 + 7i$, evaluate:

    a. $z_1 + z_2$ **b.** $z_1 - z_2$ **c.** $z_1 \cdot z_2$

Solution   **a.** $z_1 + z_2 = (2 - 3i) + (-5 + 7i)$

$$= 2 - 5 - 3i + 7i$$

$$= -3 + 4i$$

  b. $z_1 - z_2 = (2 - 3i) - (-5 + 7i)$

$$= 2 + 5 - 3i - 7i = 7 - 10i$$

  c. $z_1 \cdot z_2 = (2 - 3i)(-5 + 7i)$

$$= (2 - 3i)(-5) + (2 - 3i)(7i)$$

$$= -10 + 15i + 14i - 21i^2$$

$$= -10 + 29i - 21(-1) = 11 + 29i$$ ∎

To divide complex numbers, we multiply numerator and denominator by the complex conjugate of the denominator. We also use the fact that if $z = c + di$ and $\bar{z} = c - di$, then

$$z \cdot \bar{z} = c^2 - cdi + cdi - d^2i^2$$

$$= c^2 + d^2.$$

That is, a complex number times its complex conjugate is equal to the sum of the squares of its real and imaginary parts. For example, $(2 + \sqrt{3}i)(2 - \sqrt{3}i) = (2)^2 + (\sqrt{3})^2 = 7$.

Example 3   Evaluate z_1/z_2 for $z_1 = 2 - 3i$, $z_2 = -5 + 7i$.

Solution

$$\frac{z_1}{z_2} = \frac{2 - 3i}{-5 + 7i}$$

$$= \frac{2 - 3i}{-5 + 7i} \cdot \frac{-5 - 7i}{-5 - 7i}$$

$$= \frac{(2 - 3i)(-5 - 7i)}{(-5)^2 + (7)^2}$$

$$= \frac{-10 - 14i + 15i + 21i^2}{74}$$

$$= \frac{-31 + i}{74}$$

$$= -\frac{31}{74} + \frac{1}{74}i.$$

■

Equality

Two complex numbers $z_1 = a + bi$ and $z_2 = c + di$ are equal if and only if $a = c$ and $b = d$.

Example 4   Find a complex number $z = x + yi$ for which $z^2 = i$.

Solution   Squaring $x + yi$ gives

$$x^2 - y^2 + 2xyi = 0 + i,$$

and thus by the definition of equality of two complex numbers we must have

$$x^2 - y^2 = 0$$

$$2xy = 1.$$

Since the first equation implies that $y = \pm x$, we can use this in the second to obtain

$$\pm 2x^2 = 1.$$

Because x is real, we must reject the minus sign, so

$$2x^2 = 1$$

and therefore
$$x = \pm \frac{1}{\sqrt{2}} = \pm \frac{\sqrt{2}}{2}.$$

It follows from $y = x$ that there are two complex numbers

$$z_1 = \frac{\sqrt{2}}{2} + \frac{\sqrt{2}}{2} i \qquad \text{and} \qquad z_2 = -\frac{\sqrt{2}}{2} - \frac{\sqrt{2}}{2} i$$

which satisfy $z^2 = i$.

Exercise A.3

In Problems 1–10, perform the indicated operations.

1. $\sqrt{-2}\sqrt{-6}$

2. i^3

3. $4(3 - i) - (2 + 6i) + 2(-1 - 5i)$

4. $(-5 + 6i)(2 - 8i)$

5. $7i(1 - i) - (1 + 3i)(4 + i)$

6. $1/i$

7. $\dfrac{1}{2 - 5i}$

8. $\dfrac{3 - 4i}{1 + 2i}$

9. $i\left(\dfrac{9 - 2i}{2 - 2i}\right)$

10. $\dfrac{1 + i}{1 - i} - i\left(\dfrac{-2 + i}{1 + i}\right)$

11. Find real numbers a and b for which

$$2(a + bi) - i(4 - i) = (3 - i)(2 + 4i).$$

12. Find real numbers a and b for which

$$(a + 2i)(1 + i) = (3 - bi)(2 - 4i).$$

13. Find real numbers a and b for which

$$(a + bi)^2 - 2i = 0.$$

A.4 Use of Tables

Although the availability of scientific calculators has reduced the need for tables of mathematical functions, there are still a number of functions used in scientific and business applications whose values are not readily obtainable from a calculator. In many instances there is still a need to be able to read a table of function values and apply the technique of linear interpolation.

Mantissa and Characteristic

Since, using scientific notation (see Section 0.2), we can write any positive real number as

$$N = n \times 10^c,$$

it follows that

$$\log_{10} N = \log_{10} n + \log_{10} 10^c$$
$$= \log_{10} n + c \log_{10} 10$$
$$= \log_{10} n + c$$
$$= m + c.$$

The number $m = \log_{10} n$ is called the **mantissa** of $\log_{10} N$ and the integer c is called its **characteristic**. Recall from Section 3.2 that $f(x) = \log_{10} x$ is an increasing function. Therefore, since $1 \leq n < 10$, we have $\log_{10} 1 \leq \log_{10} n < \log_{10} 10$, or

$$0 \leq \log_{10} n < 1.$$

Thus, the mantissa of $\log_{10} N$ is a decimal number satisfying $0 \leq m < 1$. This latter fact is very important and should be kept in mind whenever using a table of logarithms.

Use of Table I

Table I *contains only the mantissas of logarithms of numbers* $1 \leq n < 10$ which have at most two decimal places. To locate the mantissa of the logarithm of a particular number, note that the integer portion and the first decimal place locate the *row* in which the mantissa appears; the second decimal place determines the *column*. For example,

$$\text{if} \quad N = 1.18, \quad \text{then} \quad \log_{10} N \approx 0.0719$$

is obtained by locating the entry in the second row, opposite 1.1, and ninth column, below 8. Computations with logarithms are necessarily only approximations, since each mantissa is itself a rounded decimal. The number 0.0719 is a four-decimal-place approximation to the irrational number $\log_{10} 1.18$, which has a nonterminating and nonrepeating decimal part.

Example 1   Determine $\log_{10} N$ for

$$N = 6840.$$

Solution   First put the number into scientific notation,

$$N = 6.84 \times 10^3,$$

and then find its logarithm using Table I.

$$\log_{10} N = \log_{10} 6.84 + \log_{10} 10^3$$
$$= \log_{10} 6.84 + 3 \log_{10} 10$$
$$\approx 0.8351 + 3 = 3.8351.$$

■

Example 2 Determine $\log_{10}N$ for

$$N = 0.177.$$

Solution Since

$$N = 1.77 \times 10^{-1},$$

we have

$$\log_{10}N = \log_{10}1.77 + \log_{10}10^{-1}$$
$$= \log_{10}1.77 - 1 \cdot \log_{10}10$$
$$\approx 0.2480 - 1$$
$$= -0.7520.$$ ∎

Interpolation

The logarithm of a given real number may not appear in Table I. For example, the value of $\log_{10}2.346$ is not given in the table. However, an approximation to the value of a logarithm can be obtained by **linear interpolation**. This technique is demonstrated in the next example.

Example 3 Approximate

$$\log_{10}2.346.$$

Solution Since $f(x) = \log_{10}x$ is an increasing function, we know that $\log_{10}2.34 < \log_{10}2.346 < \log_{10}2.35$. Since 2.346 is $0.0006/0.010 = \frac{6}{10}$ of the way between 2.340 and 2.350, we reason that the value of $\log_{10}2.346$ should be close to $\frac{6}{10}$ of the way between $\log_{10}2.340$ and $\log_{10}2.350$. It is helpful to arrange the problem in the form:

$$0.0010 \left\{ 0.006 \left\{ \begin{matrix} \log_{10}2.340 \approx 0.3692 \\ \log_{10}2.346 \approx 0.3692 + d \end{matrix} \right\} d \\ \log_{10}2.350 \approx 0.3711 \end{matrix} \right\} 0.0019$$

Now we assume

$$\frac{0.006}{0.010} = \frac{d}{0.0019}$$

so that

$$d = 0.6(0.0019) = 0.00114 \approx 0.0011.$$

As is standard practice, we round to the same number of decimal places that appear in Table I. Thus we conclude

$$\log_{10}2.346 \approx 0.3692 + 0.0011$$
$$= 0.3703.$$ ∎

Antilogarithm

If we are given the logarithm of a real number N in the form (1), that is,

$$\log_{10} N = \log_{10} n + c, \qquad 1 \le n < 10,$$

then we may write

$$\log_{10} N = \log_{10} n + \log_{10} 10^c$$
$$= \log_{10}(n \cdot 10^c).$$

Since the logarithm function is one-to-one, we have

$$N = n \cdot 10^c.$$

The following examples illustrate the use of Table I in finding a number when its logarithm is given. The number N is called the **antilogarithm** of $\log_{10} N$.

Example 4   Find N if

$$\log_{10} N = 3.7118.$$

Solution   We first write the logarithm in the form of a mantissa plus a characteristic (that is, a nonnegative number less than 1 plus an integer)

$$\log_{10} N = 0.7118 + 3.$$

Now we can locate the mantissa in Table I:

$$\log_{10} 5.15 \approx 0.7118.$$

Thus

$$\log_{10} N \approx \log_{10} 5.15 + 3$$
$$N \approx (5.15)10^3$$
$$= 5150. \qquad \blacksquare$$

Example 5   Find N if

$$\log_{10} N = -2.3468.$$

Solution   To convert a negative logarithm into the correct form of mantissa plus characteristic, we should *add and subtract the smallest positive integer which will give a number in the form $m + c$, where $0 \le m < 1$.* Thus in the given problem we add and subtract 3:

$$\log_{10} N = -2.3468$$
$$= -2.3468 + \underbrace{3 - 3}_{\text{zero}}$$
$$= (-2.3468 + 3) - 3$$
$$= \underbrace{0.6532}_{m} - \underbrace{3.}_{c}$$

From Table I we find $\log_{10}4.50 \approx 0.6532$, so

$$N \approx (4.50)10^{-3}$$

$$= 0.0045.$$

[*Note:* A common *mistake* when working with negative logarithms is to rewrite -2.3468 as $0.3468 - 2$. This is *incorrect* since $0.3468 - 2 = -1.6532 \neq -2.3468$.] ∎

Example 6 Find N if

$$\log_{10}N = -0.1281.$$

Solution We add 1 to, and subtract 1 from, the given logarithm to find the mantissa and characteristic:

$$\log_{10}N = -0.1281 + 1 - 1$$

$$= 0.8719 - 1.$$

We then locate the mantissa 0.8719 between the entries 0.8716 and 0.8722 in Table I. Interpolating

$$0.01 \left\{ k \begin{cases} \log_{10}7.44 & \approx 0.8716 \\ \log_{10}(7.44 + k) \approx 0.8719 \end{cases} 0.0003 \\ \log_{10}7.45 & \approx 0.8722 \end{aligned} \right\} 0.0006$$

gives

$$\frac{k}{0.01} = \frac{0.0003}{0.0006},$$

or

$$k = 0.005.$$

Hence $7.44 + k = 7.445$ so that

$$\log_{10}N \approx \log_{10}7.445 - 1$$

and therefore

$$N \approx (7.445)10^{-1}$$

$$= 0.7445.$$ ∎

Use of Table IV

Table IV contains four-decimal-place approximations to $\sin\theta$, $\cos\theta$, $\tan\theta$, and $\cot\theta$ for values of the angle θ in the first quadrant. For values of θ between 0 and $45°$, we use the *left-hand* column together with the headings at the *top* of the table. For an angle θ between $45°$ and $90°$ we use the *right-hand* column with the headings at the *bottom* of the table. The table can be arranged in this fashion, since from formulas (8) in

Section 4.2, we have that

$$\sin(90° - \theta) = \cos \theta$$

and

$$\cos(90° - \theta) = \sin \theta,$$

and therefore

$$\tan(90° - \theta) = \frac{\sin(90° - \theta)}{\cos(90° - \theta)} = \frac{\cos \theta}{\sin \theta} = \cot \theta.$$

For example, using the top and left-hand headings on page 405, we find sin 37° to be 0.6018. The cosine of 53° = 90° − 37° is also found to be 0.6018 using the bottom and right-hand headings. Thus the table is read *downward* for angles between 0° and 45° and read *upward* for angles between 45° and 90°.

Example 7   Evaluate
a. sin 27°50′;   **b.** tan 69°10′.

Solution   **a.** Reading down the column labeled *sin* in Table IV we find

$$\sin 27°50′ \approx 0.4669.$$

b. Since the specified angle is between 45° and 90°, we read up the column labeled *tan* at the bottom of the table. From page 404 we find that

$$\tan 69°10′ \approx 2.628.$$

∎

Table V gives the values of sin t, cos t, tan t, and cot t for angles between 0 and 1.60 radians in increments of 0.01 radian. Since $\pi/2 \approx 1.57$, Table V contains values of these four functions for angles throughout the first quadrant.

Example 8   Find the value of
a. sin 0.39;   **b.** cos 1.47.

Solution   **a.** From Table V, page 406, we find that

$$\sin 0.39 \approx 0.3802.$$

b. From Table V we find that

$$\cos 1.47 \approx 0.1006.$$

∎

As shown in the next example, values of the secant and cosecant may be obtained by taking reciprocals of the cosine and sine, respectively.

Example 9 Evaluate
a. sec 78°20′;   **b.** csc 0.40.

Solution **a.** Since

$$\sec 78°20' = \frac{1}{\cos 78°20'},$$

we find from Table IV that

$$\sec 78°20' \approx \frac{1}{0.2022} \approx 4.946.$$

b. From Table V we have

$$\csc 0.40 = \frac{1}{\sin 0.40} \approx \frac{1}{0.3894} \approx 2.568. \qquad ■$$

Exercise A.4

In Problems 1–6, find the mantissa and characteristic of the logarithm of the given number.

1. 39.4

2. 0.000301

3. 45×10^6

4. $(6.3)^2$

5. $\sqrt{487}$

6. 1750

In Problems 7–12, use linear interpolation to find the common logarithm of the given number.

7. 4.327

8. 7.962

9. 326.8

10. 0.004986

11. $\sqrt{17.25}$

12. $(3783)^3$

In Problems 13–18, use Table I to find the approximate value of N if $\log_{10} N$ is as given.

13. 2.4609

14. 4.6130

15. -1.6271

16. -3.0072

17. $1.9305 + 2$

18. 3.8920

In Problems 19–30, use Table IV and linear interpolation when necessary to find the indicated value.

19. cos 82°

20. cos 25°30′

21. sin 142°20′

22. sec 75°10′

23. csc 79°40′

24. cot 72°50′

25. tan 15°12′

26. sin 83°25′

27. cos 8°7′

28. cot 30°35′

29. csc 18°52′

30. sec 52°19′

In Problems 31–36, use Table V to find the indicated value.

31. sin(−0.89)

32. cos(−0.54)

33. cos 1.3

34. tan 1.60

35. sec(−1.0)

36. csc(−0.5)

Tables

Table I Common Logarithms

x	0	1	2	3	4	5	6	7	8	9
1.0	0.0000	0.0043	0.0086	0.0128	0.0170	0.0212	0.0253	0.0294	0.0334	0.0374
1.1	0.0414	0.0453	0.0492	0.0531	0.0569	0.0607	0.0645	0.0682	0.0719	0.0755
1.2	0.0792	0.0828	0.0864	0.0899	0.0934	0.0969	0.1004	0.1038	0.1072	0.1106
1.3	0.1139	0.1173	0.1206	0.1239	0.1271	0.1303	0.1335	0.1367	0.1399	0.1430
1.4	0.1461	0.1492	0.1523	0.1553	0.1584	0.1614	0.1644	0.1673	0.1703	0.1732
1.5	0.1761	0.1790	0.1818	0.1847	0.1875	0.1903	0.1931	0.1959	0.1987	0.2014
1.6	0.2041	0.2068	0.2095	0.2122	0.2148	0.2175	0.2201	0.2227	0.2253	0.2279
1.7	0.2304	0.2330	0.2355	0.2380	0.2405	0.2430	0.2455	0.2480	0.2504	0.2529
1.8	0.2553	0.2577	0.2601	0.2625	0.2648	0.2672	0.2695	0.2718	0.2742	0.2765
1.9	0.2788	0.2810	0.2833	0.2856	0.2878	0.2900	0.2923	0.2945	0.2967	0.2989
2.0	0.3010	0.3032	0.3054	0.3075	0.3096	0.3118	0.3139	0.3160	0.3181	0.3201
2.1	0.3222	0.3243	0.3263	0.3284	0.3304	0.3324	0.3345	0.3365	0.3385	0.3404
2.2	0.3424	0.3444	0.3464	0.3483	0.3502	0.3522	0.3541	0.3560	0.3579	0.3598
2.3	0.3617	0.3636	0.3655	0.3674	0.3692	0.3711	0.3729	0.3747	0.3766	0.3784
2.4	0.3802	0.3820	0.3838	0.3856	0.3874	0.3892	0.3909	0.3927	0.3945	0.3962
2.5	0.3979	0.3997	0.4014	0.4031	0.4048	0.4065	0.4082	0.4099	0.4116	0.4133
2.6	0.4150	0.4166	0.4183	0.4200	0.4216	0.4232	0.4249	0.4265	0.4281	0.4298
2.7	0.4314	0.4330	0.4346	0.4362	0.4378	0.4393	0.4409	0.4425	0.4440	0.4456
2.8	0.4472	0.4487	0.4502	0.4518	0.4533	0.4548	0.4564	0.4579	0.4594	0.4609
2.9	0.4624	0.4639	0.4654	0.4669	0.4683	0.4698	0.4713	0.4728	0.4742	0.4757
3.0	0.4771	0.4786	0.4800	0.4814	0.4829	0.4843	0.4857	0.4871	0.4886	0.4900
3.1	0.4914	0.4928	0.4942	0.4955	0.4969	0.4983	0.4997	0.5011	0.5024	0.5038
3.2	0.5051	0.5065	0.5079	0.5092	0.5105	0.5119	0.5132	0.5145	0.5159	0.5172
3.3	0.5185	0.5198	0.5211	0.5224	0.5237	0.5250	0.5263	0.5276	0.5289	0.5302
3.4	0.5315	0.5328	0.5340	0.5353	0.5366	0.5378	0.5391	0.5403	0.5416	0.5428
3.5	0.5441	0.5453	0.5465	0.5478	0.5490	0.5502	0.5514	0.5527	0.5539	0.5551
3.6	0.5563	0.5575	0.5587	0.5599	0.5611	0.5623	0.5635	0.5647	0.5658	0.5670
3.7	0.5682	0.5694	0.5705	0.5717	0.5729	0.5740	0.5752	0.5763	0.5775	0.5786
3.8	0.5798	0.5809	0.5821	0.5832	0.5843	0.5855	0.5866	0.5877	0.5888	0.5899
3.9	0.5911	0.5922	0.5933	0.5944	0.5955	0.5966	0.5977	0.5988	0.5999	0.6010
4.0	0.6021	0.6031	0.6042	0.6053	0.6064	0.6075	0.6085	0.6096	0.6107	0.6117
4.1	0.6128	0.6138	0.6149	0.6160	0.6170	0.6180	0.6191	0.6201	0.6212	0.6222
4.2	0.6232	0.6243	0.6253	0.6263	0.6274	0.6284	0.6294	0.6304	0.6314	0.6325
4.3	0.6335	0.6345	0.6355	0.6365	0.6375	0.6385	0.6395	0.6405	0.6415	0.6425
4.4	0.6435	0.6444	0.6454	0.6464	0.6474	0.6484	0.6493	0.6503	0.6513	0.6522
4.5	0.6532	0.6542	0.6551	0.6561	0.6571	0.6580	0.6590	0.6599	0.6609	0.6618
4.6	0.6628	0.6637	0.6646	0.6656	0.6665	0.6675	0.6684	0.6693	0.6702	0.6712
4.7	0.6721	0.6730	0.6739	0.6749	0.6758	0.6767	0.6776	0.6785	0.6794	0.6803
4.8	0.6812	0.6821	0.6830	0.6839	0.6848	0.6857	0.6866	0.6875	0.6884	0.6893
4.9	0.6902	0.6911	0.6920	0.6928	0.6937	0.6946	0.6955	0.6964	0.6972	0.6981
5.0	0.6990	0.6998	0.7007	0.7016	0.7024	0.7033	0.7042	0.7050	0.7059	0.7067
5.1	0.7076	0.7084	0.7093	0.7101	0.7110	0.7118	0.7126	0.7135	0.7143	0.7152
5.2	0.7160	0.7168	0.7177	0.7185	0.7193	0.7202	0.7210	0.7218	0.7226	0.7235
5.3	0.7243	0.7251	0.7259	0.7267	0.7275	0.7284	0.7292	0.7300	0.7308	0.7316
5.4	0.7324	0.7332	0.7340	0.7348	0.7356	0.7364	0.7372	0.7380	0.7388	0.7396
x	0	1	2	3	4	5	6	7	8	9

Table I   *(continued)*

x	0	1	2	3	4	5	6	7	8	9
5.5	0.7404	0.7412	0.7419	0.7427	0.7435	0.7443	0.7451	0.7459	0.7466	0.7474
5.6	0.7482	0.7490	0.7497	0.7505	0.7513	0.7520	0.7528	0.7536	0.7543	0.7551
5.7	0.7559	0.7566	0.7574	0.7582	0.7589	0.7597	0.7604	0.7612	0.7619	0.7627
5.8	0.7634	0.7642	0.7649	0.7657	0.7664	0.7672	0.7679	0.7686	0.7694	0.7701
5.9	0.7709	0.7716	0.7723	0.7731	0.7738	0.7745	0.7752	0.7760	0.7767	0.7774
6.0	0.7782	0.7789	0.7796	0.7803	0.7810	0.7818	0.7825	0.7832	0.7839	0.7846
6.1	0.7853	0.7860	0.7868	0.7875	0.7882	0.7889	0.7896	0.7903	0.7910	0.7917
6.2	0.7924	0.7931	0.7938	0.7945	0.7952	0.7959	0.7966	0.7973	0.7980	0.7987
6.3	0.7993	0.8000	0.8007	0.8014	0.8021	0.8028	0.8035	0.8041	0.8048	0.8055
6.4	0.8062	0.8069	0.8075	0.8082	0.8089	0.8096	0.8102	0.8109	0.8116	0.8122
6.5	0.8129	0.8136	0.8142	0.8149	0.8156	0.8162	0.8169	0.8176	0.8182	0.8189
6.6	0.8195	0.8202	0.8209	0.8215	0.8222	0.8228	0.8235	0.8241	0.8248	0.8254
6.7	0.8261	0.8267	0.8274	0.8280	0.8287	0.8293	0.8299	0.8306	0.8312	0.8319
6.8	0.8325	0.8331	0.8338	0.8344	0.8351	0.8357	0.8363	0.8370	0.8376	0.8382
6.9	0.8388	0.8395	0.8401	0.8407	0.8414	0.8420	0.8426	0.8432	0.8439	0.8445
7.0	0.8451	0.8457	0.8463	0.8470	0.8476	0.8482	0.8488	0.8494	0.8500	0.8506
7.1	0.8513	0.8519	0.8525	0.8531	0.8537	0.8543	0.8549	0.8555	0.8561	0.8567
7.2	0.8573	0.8579	0.8585	0.8591	0.8597	0.8603	0.8609	0.8615	0.8621	0.8627
7.3	0.8633	0.8639	0.8645	0.8651	0.8657	0.8663	0.8669	0.8675	0.8681	0.8686
7.4	0.8692	0.8698	0.8704	0.8710	0.8716	0.8722	0.8727	0.8733	0.8739	0.8745
7.5	0.8751	0.8756	0.8762	0.8768	0.8774	0.8779	0.8785	0.8791	0.8797	0.8802
7.6	0.8808	0.8814	0.8820	0.8825	0.8831	0.8837	0.8842	0.8848	0.8854	0.8859
7.7	0.8865	0.8871	0.8876	0.8882	0.8887	0.8893	0.8899	0.8904	0.8910	0.8915
7.8	0.8921	0.8927	0.8932	0.8938	0.8943	0.8949	0.8954	0.8960	0.8965	0.8971
7.9	0.8976	0.8992	0.8987	0.8993	0.8998	0.9004	0.9009	0.9015	0.9020	0.9025
8.0	0.9031	0.9036	0.9042	0.9047	0.9053	0.9058	0.9063	0.9069	0.9074	0.9079
8.1	0.9085	0.9090	0.9096	0.9101	0.9106	0.9112	0.9117	0.9122	0.9128	0.9133
8.2	0.9138	0.9143	0.9149	0.9154	0.9159	0.9165	0.9170	0.9175	0.9180	0.9186
8.3	0.9191	0.9196	0.9201	0.9206	0.9212	0.9217	0.9222	0.9227	0.9232	0.9238
8.4	0.9243	0.9248	0.9253	0.9258	0.9263	0.9269	0.9274	0.9279	0.9284	0.9289
8.5	0.9294	0.9299	0.9304	0.9309	0.9315	0.9320	0.9325	0.9330	0.9335	0.9340
8.6	0.9345	0.9350	0.9355	0.9360	0.9365	0.9370	0.9375	0.9380	0.9385	0.9390
8.7	0.9395	0.9400	0.9405	0.9410	0.9415	0.9420	0.9425	0.9430	0.9435	0.9440
8.8	0.9445	0.9450	0.9455	0.9460	0.9465	0.9469	0.9474	0.9479	0.9484	0.9489
8.9	0.9494	0.9499	0.9504	0.9509	0.9513	0.9518	0.9523	0.9528	0.9533	0.9538
9.0	0.9542	0.9547	0.9552	0.9557	0.9562	0.9566	0.9571	0.9576	0.9581	0.9586
9.1	0.9590	0.9595	0.9600	0.9605	0.9609	0.9614	0.9619	0.9624	0.9628	0.9633
9.2	0.9638	0.9643	0.9647	0.9652	0.9657	0.9661	0.9666	0.9671	0.9675	0.9680
9.3	0.9685	0.9689	0.9694	0.9699	0.9703	0.9708	0.9713	0.9717	0.9722	0.9727
9.4	0.9731	0.9736	0.9741	0.9745	0.9750	0.9754	0.9759	0.9763	0.9768	0.9773
9.5	0.9777	0.9782	0.9786	0.9791	0.9795	0.9800	0.9805	0.9809	0.9814	0.9818
9.6	0.9823	0.9827	0.9832	0.9836	0.9841	0.9845	0.9850	0.9854	0.9859	0.9863
9.7	0.9868	0.9872	0.9877	0.9881	0.9886	0.9890	0.9894	0.9899	0.9903	0.9908
9.8	0.9912	0.9917	0.9921	0.9926	0.9930	0.9934	0.9939	0.9943	0.9948	0.9952
9.9	0.9956	0.9961	0.9965	0.9969	0.9974	0.9978	0.9983	0.9987	0.9991	0.9996
x	0	1	2	3	4	5	6	7	8	9

Table II Natural Logarithms

x	$\ln x$	x	$\ln x$	x	$\ln x$
		4.5	1.5041	9.0	2.1972
0.1	−2.3026	4.6	1.5261	9.1	2.2083
0.2	−1.6094	4.7	1.5476	9.2	2.2192
0.3	−1.2040	4.8	1.5686	9.3	2.2300
0.4	−0.9163	4.9	1.5892	9.4	2.2407
0.5	−0.6931	5.0	1.6094	9.5	2.2513
0.6	−0.5108	5.1	1.6292	9.6	2.2618
0.7	−0.3567	5.2	1.6487	9.7	2.2721
0.8	−0.2231	5.3	1.6677	9.8	2.2824
0.9	−0.1054	5.4	1.6864	9.9	2.2925
1.0	0.0000	5.5	1.7047	10	2.3026
1.1	0.0953	5.6	1.7228	11	2.3979
1.2	0.1823	5.7	1.7405	12	2.4849
1.3	0.2624	5.8	1.7579	13	2.5649
1.4	0.3365	5.9	1.7750	14	2.6391
1.5	0.4055	6.0	1.7918	15	2.7081
1.6	0.4700	6.1	1.8083	16	2.7726
1.7	0.5306	6.2	1.8245	17	2.8332
1.8	0.5878	6.3	1.8405	18	2.8904
1.9	0.6419	6.4	1.8563	19	2.9444
2.0	0.6931	6.5	1.8718	20	2.9957
2.1	0.7419	6.6	1.8871	25	3.2189
2.2	0.7885	6.7	1.9021	30	3.4012
2.3	0.8329	6.8	1.9169	35	3.5553
2.4	0.8755	6.9	1.9315	40	3.6889
2.5	0.9163	7.0	1.9459	45	3.8067
2.6	0.9555	7.1	1.9601	50	3.9120
2.7	0.9933	7.2	1.9741	55	4.0073
2.8	1.0296	7.3	1.9879	60	4.0943
2.9	1.0647	7.4	2.0015	65	4.1744
3.0	1.0986	7.5	2.0149	70	4.2485
3.1	1.1314	7.6	2.0281	75	4.3175
3.2	1.1632	7.7	2.0412	80	4.3820
3.3	1.1939	7.8	2.0541	85	4.4427
3.4	1.2238	7.9	2.0669	90	4.4998
3.5	1.2528	8.0	2.0794	100	4.6052
3.6	1.2809	8.1	2.0919	110	4.7005
3.7	1.3083	8.2	2.1041	120	4.7875
3.8	1.3350	8.3	2.1163	130	4.8676
3.9	1.3610	8.4	2.1282	140	4.9416
4.0	1.3863	8.5	2.1401	150	5.0106
4.1	1.4110	8.6	2.1518	160	5.0752
4.2	1.4351	8.7	2.1633	170	5.1358
4.3	1.4586	8.8	2.1748	180	5.1930
4.4	1.4816	8.9	2.1861	190	5.2470

Table III Exponential Functions

x	e^x	e^{-x}	x	e^x	e^{-x}
0.00	1.0000	1.0000	1.5	4.4817	0.2231
0.01	1.0101	0.9901	1.6	4.9530	0.2019
0.02	1.0202	0.9802	1.7	5.4739	0.1827
0.03	1.0305	0.9705	1.8	6.0496	0.1653
0.04	1.0408	0.9608	1.9	6.6859	0.1496
0.05	1.0513	0.9512	2.0	7.3891	0.1353
0.06	1.0618	0.9418	2.1	8.1662	0.1225
0.07	1.0725	0.9324	2.2	9.0250	0.1108
0.08	1.0833	0.9331	2.3	9.9742	0.1003
0.09	1.0942	0.9139	2.4	11.023	0.0907
0.10	1.1052	0.9048	2.5	12.182	0.0821
0.11	1.1163	0.8958	2.6	13.464	0.0743
0.12	1.1275	0.8869	2.7	14.880	0.0672
0.13	1.1388	0.8781	2.8	16.445	0.0608
0.14	1.1503	0.8694	2.9	18.174	0.0550
0.15	1.1618	0.8607	3.0	20.086	0.0498
0.16	1.1735	0.8521	3.1	22.198	0.0450
0.17	1.1853	0.8437	3.2	24.533	0.0408
0.18	1.1972	0.8353	3.3	27.113	0.0369
0.19	1.2092	0.8270	3.4	29.964	0.0334
0.20	1.2214	0.8187	3.5	33.115	0.0302
0.21	1.2337	0.8106	3.6	36.598	0.0273
0.22	1.2461	0.8025	3.7	40.447	0.0247
0.23	1.2586	0.7945	3.8	44.701	0.0224
0.24	1.2712	0.7866	3.9	49.402	0.0202
0.25	1.2840	0.7788	4.0	54.598	0.0183
0.30	1.3499	0.7408	4.1	60.340	0.0166
0.35	1.4191	0.7047	4.2	66.686	0.0150
0.40	1.4918	0.6703	4.3	73.700	0.0136
0.45	1.5683	0.6376	4.4	81.451	0.0123
0.50	1.6487	0.6065	4.5	90.017	0.0111
0.55	1.7333	0.5769	4.6	99.484	0.0101
0.60	1.8221	0.5488	4.7	109.95	0.0091
0.65	1.9155	0.5220	4.8	121.51	0.0082
0.70	2.0138	0.4966	4.9	134.29	0.0074
0.75	2.1170	0.4724	5.0	148.41	0.0067
0.80	2.2255	0.4493	5.5	244.69	0.0041
0.85	2.3396	0.4274	6.0	403.43	0.0025
0.90	2.4596	0.4066	6.5	665.14	0.0015
0.95	2.5857	0.3867	7.0	1096.6	0.0009
1.0	2.7183	0.3679	7.5	1808.0	0.0006
1.1	3.0042	0.3329	8.0	2981.0	0.0003
1.2	3.3201	0.3012	8.5	4914.8	0.0002
1.3	3.6693	0.2725	9.0	8103.1	0.0001
1.4	4.0552	0.2466	10.0	22026	0.00005

Table IV Trigonometric Functions (Degrees)

θ	sin	cos	tan	cot		θ	sin	cos	tan	cot	
0°00′	0.0000	1.000	0.0000	—	90°00′	8°00′	0.1392	0.9903	0.1405	7.115	82°00′
10′	0.0029	1.000	0.0029	343.8	89°50′	10′	0.1421	0.9899	0.1435	6.968	81°50′
20′	0.0058	1.000	0.0058	171.9	40′	20′	0.1449	0.9894	0.1465	6.827	40′
30′	0.0087	1.000	0.0087	114.6	30′	30′	0.1478	0.9890	0.1495	6.691	30′
40′	0.0116	0.9999	0.0116	85.94	20′	40′	0.1507	0.9886	0.1524	6.561	20′
0°50′	0.0145	0.9999	0.0145	68.75	10′	8°50′	0.1536	0.9881	0.1554	6.435	10′
1°00′	0.0175	0.9998	0.0175	57.29	89°00′	9°00′	0.1564	0.9877	0.1584	6.314	81°00′
10′	0.0204	0.9998	0.0204	49.10	88°50′	10′	0.1593	0.9872	0.1614	6.197	80°50′
20′	0.0233	0.9997	0.0233	42.96	40′	20′	0.1622	0.9868	0.1644	6.084	40′
30′	0.0262	0.9997	0.0262	38.19	30′	30′	0.1650	0.9863	0.1673	5.976	30′
40′	0.0291	0.9996	0.0291	34.37	20′	40′	0.1679	0.9858	0.1703	5.871	20′
1°50′	0.0320	0.9995	0.0320	31.24	10′	9°50′	0.1708	0.9853	0.1733	5.769	10′
2°00′	0.0349	0.9994	0.0349	28.64	88°00′	10°00′	0.1736	0.9848	0.1763	5.671	80°00′
10′	0.0378	0.9993	0.0378	26.43	87°50′	10′	0.1765	0.9843	0.1793	5.576	79°50′
20′	0.0407	0.9992	0.0407	24.54	40′	20′	0.1794	0.9838	0.1823	5.485	40′
30′	0.0436	0.9990	0.0437	22.90	30′	30′	0.1822	0.9833	0.1853	5.396	30′
40′	0.0465	0.9989	0.0466	21.47	20′	40′	0.1851	0.9827	0.1883	5.309	20′
2°50′	0.0494	0.9988	0.0495	20.21	10′	10°50′	0.1880	0.9822	0.1914	5.226	10′
3°00′	0.0523	0.9986	0.0524	19.08	87°00′	11°00′	0.1908	0.9816	0.1944	5.145	79°00′
10′	0.0552	0.9985	0.0553	18.07	86°50′	10′	0.1937	0.9811	0.1974	5.066	78°50′
20′	0.0581	0.9983	0.0582	17.17	40′	20′	0.1965	0.9805	0.2004	4.989	40′
30′	0.0610	0.9981	0.0612	16.35	30′	30′	0.1994	0.9799	0.2035	4.915	30′
40′	0.0640	0.9980	0.0641	15.60	20′	40′	0.2022	0.9793	0.2065	4.843	20′
3°50′	0.0669	0.9978	0.0670	14.92	10′	11°50′	0.2051	0.9787	0.2095	4.773	10′
4°00′	0.0698	0.9976	0.0699	14.30	86°00′	12°00′	0.2079	0.9781	0.2126	4.705	78°00′
10′	0.0727	0.9974	0.0729	13.73	85°50′	10′	0.2108	0.9775	0.2156	4.638	77°50′
20′	0.0756	0.9971	0.0758	13.20	40′	20′	0.2136	0.9769	0.2186	4.574	40′
30′	0.0785	0.9969	0.0787	12.71	30′	30′	0.2164	0.9763	0.2217	4.511	30′
40′	0.0814	0.9967	0.0816	12.25	20′	40′	0.2193	0.9757	0.2247	4.449	20′
4°50′	0.0843	0.9964	0.0846	11.83	10′	12°50′	0.2221	0.9750	0.2278	4.390	10′
5°00′	0.0872	0.9962	0.0875	11.43	85°00′	13°00′	0.2250	0.9744	0.2309	4.331	77°00′
10′	0.0901	0.9959	0.0904	11.06	84°50′	10′	0.2278	0.9737	0.2339	4.275	76°50′
20′	0.0929	0.9957	0.0934	10.71	40′	20′	0.2306	0.9730	0.2370	4.219	40′
30′	0.0958	0.9954	0.0963	10.39	30′	30′	0.2334	0.9724	0.2401	4.165	30′
40′	0.0987	0.9951	0.0992	10.08	20′	40′	0.2363	0.9717	0.2432	4.113	20′
5°50′	0.1016	0.9948	0.1022	9.788	10′	13°50′	0.2391	0.9710	0.2462	4.061	10′
6°00′	0.1045	0.9945	0.1051	9.514	84°00′	14°00′	0.2419	0.9703	0.2493	4.011	76°00′
10′	0.1074	0.9942	0.1080	9.255	83°50′	10′	0.2447	0.9696	0.2524	3.962	75°50′
20′	0.1103	0.9939	0.1110	9.010	40′	20′	0.2476	0.9689	0.2555	3.914	40′
30′	0.1132	0.9936	0.1139	8.777	30′	30′	0.2504	0.9681	0.2586	3.867	30′
40′	0.1161	0.9932	0.1169	8.556	20′	40′	0.2532	0.9674	0.2617	3.821	20′
6°50′	0.1190	0.9929	0.1198	8.345	10′	14°50′	0.2560	0.9667	0.2648	3.776	10′
7°00′	0.1219	0.9925	0.1228	8.144	83°00′	15°00′	0.2588	0.9659	0.2679	3.732	75°00′
10′	0.1248	0.9922	0.1257	7.953	82°50′	10′	0.2616	0.9652	0.2711	3.689	74°50′
20′	0.1276	0.9918	0.1287	7.770	40′	20′	0.2644	0.9644	0.2742	3.647	40′
30′	0.1305	0.9914	0.1317	7.596	30′	30′	0.2672	0.9636	0.2773	3.606	30′
40′	0.1334	0.9911	0.1346	7.429	20′	40′	0.2700	0.9628	0.2805	3.566	20′
7°50′	0.1363	0.9907	0.1376	7.269	10′	15°50′	0.2728	0.9621	0.2836	3.526	10′
	cos	sin	cot	tan	θ		cos	sin	cot	tan	θ

Table IV *(continued)*

θ	sin	cos	tan	cot	
16°00′	0.2756	0.9613	0.2867	3.487	74°00′
10′	0.2784	0.9605	0.2899	3.450	73°50′
20′	0.2812	0.9596	0.2931	3.412	40′
30′	0.2840	0.9588	0.2962	3.376	30′
40′	0.2868	0.9580	0.2994	3.340	20′
16°50′	0.2896	0.9572	0.3026	3.305	10′
17°00′	0.2924	0.9563	0.3057	3.271	73°00′
10′	0.2952	0.9555	0.3089	3.237	72°50′
20′	0.2979	0.9546	0.3121	3.204	40′
30′	0.3007	0.9537	0.3153	3.172	30′
40′	0.3035	0.9528	0.3185	3.140	20′
17°50′	0.3062	0.9520	0.3217	3.108	10′
18°00′	0.3090	0.9511	0.3249	3.078	72°00′
10′	0.3118	0.9502	0.3281	3.047	71°50′
20′	0.3145	0.9492	0.3314	3.018	40′
30′	0.3173	0.9483	0.3346	2.989	30′
40′	0.3201	0.9474	0.3378	2.960	20′
18°50′	0.3228	0.9465	0.3411	2.932	10′
19°00′	0.3256	0.9455	0.3443	2.904	71°00′
10′	0.3283	0.9446	0.3476	2.877	70°50′
20′	0.3311	0.9436	0.3508	2.850	40′
30′	0.3338	0.9426	0.3541	2.824	30′
40′	0.3365	0.9417	0.3574	2.798	20′
19°50′	0.3393	0.9407	0.3607	2.773	10′
20°00′	0.3420	0.9397	0.3640	2.747	70°00′
10′	0.3448	0.9387	0.3673	2.723	69°50′
20′	0.3475	0.9377	0.3706	2.699	40′
30′	0.3502	0.9367	0.3739	2.675	30′
40′	0.3529	0.9356	0.3772	2.651	20′
20°50′	0.3557	0.9346	0.3805	2.628	10′
21°00′	0.3584	0.9336	0.3839	2.605	69°00′
10′	0.3611	0.9325	0.3872	2.583	68°50′
20′	0.3638	0.9315	0.3906	2.560	40′
30′	0.3665	0.9304	0.3939	2.539	30′
40′	0.3692	0.9293	0.3973	2.517	20′
21°50′	0.3719	0.9283	0.4006	2.496	10′
22°00′	0.3746	0.9272	0.4040	2.475	68°00′
10′	0.3773	0.9261	0.4074	2.455	67°50′
20′	0.3800	0.9250	0.4108	2.434	40′
30′	0.3827	0.9239	0.4142	2.414	30′
40′	0.3854	0.9228	0.4176	2.394	20′
22°50′	0.3881	0.9216	0.4210	2.375	10′
23°00′	0.3907	0.9205	0.4245	2.356	67°00′
10′	0.3934	0.9194	0.4279	2.337	66°50′
20′	0.3961	0.9182	0.4314	2.318	40′
30′	0.3987	0.9171	0.4348	2.300	30′
40′	0.4014	0.9159	0.4383	2.282	20′
23°50′	0.4041	0.9147	0.4417	2.264	10′
	cos	sin	cot	tan	θ

θ	sin	cos	tan	cot	
24°00′	0.4067	0.9135	0.4452	2.246	66°00′
10′	0.4094	0.9124	0.4487	2.229	65°50′
20′	0.4120	0.9112	0.4522	2.211	40′
30′	0.4147	0.9100	0.4557	2.194	30′
40′	0.4173	0.9088	0.4592	2.177	20′
24°50′	0.4200	0.9075	0.4628	2.161	10′
25°00′	0.4226	0.9063	0.4663	2.145	65°00′
10′	0.4253	0.9051	0.4699	2.128	64°50′
20′	0.4279	0.9038	0.4734	2.112	40′
30′	0.4305	0.9026	0.4770	2.097	30′
40′	0.4331	0.9013	0.4806	2.081	20′
25°50′	0.4358	0.9001	0.4841	2.066	10′
26°00′	0.4384	0.8988	0.4877	2.050	64°00′
10′	0.4410	0.8975	0.4913	2.035	63°50′
20′	0.4436	0.8962	0.4950	2.020	40′
30′	0.4462	0.8949	0.4986	2.006	30′
40′	0.4488	0.8936	0.5022	1.991	20′
26°50′	0.4514	0.8923	0.5059	1.977	10′
27°00′	0.4540	0.8910	0.5095	1.963	63°00′
10′	0.4566	0.8897	0.5132	1.949	62°50′
20′	0.4592	0.8884	0.5169	1.935	40′
30′	0.4617	0.8870	0.5206	1.921	30′
40′	0.4643	0.8857	0.5243	1.907	20′
27°50′	0.4669	0.8843	0.5280	1.894	10′
28°00′	0.4695	0.8829	0.5317	1.881	62°00′
10′	0.4720	0.8816	0.5354	1.868	61°50′
20′	0.4746	0.8802	0.5392	1.855	40′
30′	0.4772	0.8788	0.5430	1.842	30′
40′	0.4797	0.8774	0.5467	1.829	20′
28°50′	0.4823	0.8760	0.5505	1.816	10′
29°00′	0.4848	0.8746	0.5543	1.804	61°00′
10′	0.4878	0.8732	0.5581	1.792	60°50′
20′	0.4899	0.8718	0.5619	1.780	40′
30′	0.4924	0.8704	0.5658	1.767	30′
40′	0.4950	0.8689	0.5696	1.756	20′
29°50′	0.4975	0.8675	0.5735	1.744	10′
30°00′	0.5000	0.8660	0.5774	1.732	60°00′
10′	0.5025	0.8646	0.5812	1.720	59°50′
20′	0.5050	0.8631	0.5851	1.709	40′
30′	0.5075	0.8616	0.5890	1.698	30′
40′	0.5100	0.8601	0.5930	1.686	20′
30°50′	0.5125	0.8587	0.5969	1.675	10′
31°00′	0.5150	0.8572	0.6009	1.664	59°00′
10′	0.5175	0.8557	0.6048	1.653	58°50′
20′	0.5200	0.8542	0.6088	1.643	40′
30′	0.5225	0.8526	0.6128	1.632	30′
40′	0.5250	0.8511	0.6168	1.621	20′
31°50′	0.5275	0.8496	0.6208	1.611	10′
	cos	sin	cot	tan	θ

Table IV *(continued)*

θ	sin	cos	tan	cot		θ	sin	cos	tan	cot	
32°00′	0.5299	0.8480	0.6249	1.600	58°00′	39°00′	0.6293	0.7771	0.8098	1.235	51°00′
10′	0.5324	0.8465	0.6289	1.590	57°50′	10′	0.6316	0.7753	0.8146	1.228	50°50′
20′	0.5348	0.8450	0.6330	1.580	40′	20′	0.6338	0.7735	0.8195	1.220	40′
30′	0.5373	0.8434	0.6371	1.570	30′	30′	0.6361	0.7716	0.8243	1.213	30′
40′	0.5398	0.8418	0.6412	1.560	20′	40′	0.6383	0.7698	0.8292	1.206	20′
32°50′	0.5422	0.8403	0.6453	1.550	10′	39°50′	0.6406	0.7679	0.8342	1.199	10′
33°00′	0.5446	0.8387	0.6496	1.540	57°00′	40°00′	0.6428	0.7660	0.8391	1.192	50°00′
10′	0.5471	0.8371	0.6536	1.530	56°50′	10′	0.6450	0.7642	0.8441	1.185	49°50′
20′	0.5495	0.8355	0.6577	1.520	40′	20′	0.6472	0.7623	0.8491	1.178	40′
30′	0.5519	0.8339	0.6619	1.511	30′	30′	0.6494	0.7604	0.8541	1.171	30′
40′	0.5544	0.8323	0.6661	1.501	20′	40′	0.6517	0.7585	0.8591	1.164	20′
33°50′	0.5568	0.8307	0.6703	1.492	10′	40°50′	0.6539	0.7566	0.8642	1.157	10′
34°00′	0.5592	0.8290	0.6745	1.483	56°00′	41°00′	0.6561	0.7547	0.8693	1.150	49°00′
10′	0.5616	0.8274	0.6787	1.473	55°50′	10′	0.6583	0.7528	0.8744	1.144	48°50′
20′	0.5640	0.8258	0.6830	1.464	40′	20′	0.6604	0.7509	0.8796	1.137	40′
30′	0.5664	0.8241	0.6873	1.455	30′	30′	0.6626	0.7490	0.8847	1.130	30′
40′	0.5688	0.8225	0.6916	1.446	20′	40′	0.6648	0.7470	0.8899	1.124	20′
34°50′	0.5712	0.8208	0.6959	1.437	10′	41°50′	0.6670	0.7451	0.8952	1.117	10′
35°00′	0.5736	0.8192	0.7002	1.428	55°00′	42°00′	0.6691	0.7431	0.9004	1.111	48°00′
10′	0.5760	0.8175	0.7046	1.419	54°50′	10′	0.6713	0.7412	0.9057	1.104	47°50′
20′	0.5783	0.8158	0.7089	1.411	40′	20′	0.6734	0.7392	0.9110	1.098	40′
30′	0.5807	0.8141	0.7133	1.402	30′	30′	0.6756	0.7373	0.9163	1.091	30′
40′	0.5831	0.8124	0.7177	1.393	20′	40′	0.6777	0.7353	0.9217	1.085	20′
35°50′	0.5854	0.8107	0.7221	1.385	10′	42°50′	0.6799	0.7333	0.9271	1.079	10′
36°00′	0.5878	0.8090	0.7265	1.376	54°00′	43°00′	0.6820	0.7314	0.9325	1.072	47°00′
10′	0.5901	0.8073	0.7310	1.368	53°50′	10′	0.6841	0.7294	0.9380	1.066	46°50′
20′	0.5925	0.8056	0.7355	1.360	40′	20′	0.6862	0.7274	0.9435	1.060	40′
30′	0.5948	0.8039	0.7400	1.351	30′	30′	0.6884	0.7254	0.9490	1.054	30′
40′	0.5972	0.8021	0.7445	1.343	20′	40′	0.6905	0.7234	0.9545	1.048	20′
36°50′	0.5995	0.8004	0.7490	1.335	10′	43°50′	0.6926	0.7214	0.9601	1.042	10′
37°00′	0.6018	0.7986	0.7536	1.327	53°00′	44°00′	0.6947	0.7193	0.9657	1.036	46°00′
10′	0.6041	0.7969	0.7581	1.319	52°50′	10′	0.6967	0.7173	0.9713	1.030	45°50′
20′	0.6065	0.7951	0.7627	1.311	40′	20′	0.6988	0.7153	0.9770	1.024	40′
30′	0.6088	0.7934	0.7673	1.303	30′	30′	0.7009	0.7133	0.9827	1.018	30′
40′	0.6111	0.7916	0.7720	1.295	20′	40′	0.7030	0.7112	0.9884	1.012	20′
37°50′	0.6134	0.7898	0.7766	1.288	10′	44°50′	0.7050	0.7092	0.9942	1.006	10′
38°00′	0.6157	0.7880	0.7813	1.280	52°00′	45°00′	0.7071	0.7071	1.000	1.000	45°00′
10′	0.6180	0.7862	0.7860	1.272	51°50′						
20′	0.6202	0.7844	0.7907	1.265	40′						
30′	0.6225	0.7826	0.7954	1.257	30′						
40′	0.6248	0.7808	0.8002	1.250	20′						
38°50′	0.6271	0.7790	0.8050	1.242	10′						
	cos	sin	cot	tan	θ		cos	sin	cot	tan	θ

Table V Trigonometric Functions (Radians)

t	sin	cos	tan	cot
0.00	0.0000	1.0000	0.0000	—
0.01	0.0100	1.0000	0.0100	99.997
0.02	0.0200	0.9998	0.0200	49.993
0.03	0.0300	0.9996	0.0300	33.323
0.04	0.0400	0.9992	0.0400	24.987
0.05	0.0500	0.9988	0.0500	19.983
0.06	0.0600	0.9982	0.0601	16.647
0.07	0.0699	0.9976	0.0701	14.262
0.08	0.0799	0.9968	0.0802	12.473
0.09	0.0899	0.9960	0.0902	11.081
0.10	0.0998	0.9950	0.1003	9.967
0.11	0.1098	0.9940	0.1104	9.054
0.12	0.1197	0.9928	0.1206	8.293
0.13	0.1296	0.9916	0.1307	7.649
0.14	0.1395	0.9902	0.1409	7.096
0.15	0.1494	0.9888	0.1511	6.617
0.16	0.1593	0.9872	0.1614	6.197
0.17	0.1692	0.9856	0.1717	5.826
0.18	0.1790	0.9838	0.1820	5.495
0.19	0.1889	0.9820	0.1923	5.200
0.20	0.1987	0.9801	0.2027	4.933
0.21	0.2085	0.9780	0.2131	4.692
0.22	0.2182	0.9759	0.2236	4.472
0.23	0.2280	0.9737	0.2341	4.271
0.24	0.2377	0.9713	0.2447	4.086
0.25	0.2474	0.9689	0.2553	3.916
0.26	0.2571	0.9664	0.2660	3.759
0.27	0.2667	0.9638	0.2768	3.613
0.28	0.2764	0.9611	0.2876	3.478
0.29	0.2860	0.9582	0.2984	3.351
0.30	0.2955	0.9553	0.3093	3.233
0.31	0.3051	0.9523	0.3203	3.122
0.32	0.3146	0.9492	0.3314	3.018
0.33	0.3240	0.9460	0.3425	2.920
0.34	0.3335	0.9428	0.3537	2.827
0.35	0.3429	0.9394	0.3650	2.740
0.36	0.3523	0.9359	0.3764	2.657
0.37	0.3616	0.9323	0.3879	2.578
0.38	0.3709	0.9287	0.3994	2.504
0.39	0.3802	0.9249	0.4111	2.433
0.40	0.3894	0.9211	0.4228	2.365
0.41	0.3986	0.9171	0.4346	2.301
0.42	0.4078	0.9131	0.4466	2.239
0.43	0.4169	0.9090	0.4586	2.180
0.44	0.4259	0.9048	0.4708	2.124

t	sin	cos	tan	cot
0.45	0.4350	0.9004	0.4831	2.070
0.46	0.4439	0.8961	0.4954	2.018
0.47	0.4529	0.8916	0.5080	1.969
0.48	0.4618	0.8870	0.5206	1.921
0.49	0.4706	0.8823	0.5334	1.875
0.50	0.4794	0.8776	0.5463	1.830
0.51	0.4882	0.8727	0.5594	1.788
0.52	0.4969	0.8678	0.5726	1.747
0.53	0.5055	0.8628	0.5859	1.707
0.54	0.5141	0.8577	0.5994	1.668
0.55	0.5227	0.8525	0.6131	1.631
0.56	0.5312	0.8473	0.6269	1.595
0.57	0.5396	0.8419	0.6410	1.560
0.58	0.5480	0.8365	0.6552	1.526
0.59	0.5564	0.8309	0.6696	1.494
0.60	0.5646	0.8253	0.6841	1.462
0.61	0.5729	0.8196	0.6989	1.431
0.62	0.5810	0.8139	0.7139	1.401
0.63	0.5891	0.8080	0.7291	1.372
0.64	0.5972	0.8021	0.7445	1.343
0.65	0.6052	0.7961	0.7602	1.315
0.66	0.6131	0.7900	0.7761	1.288
0.67	0.6210	0.7838	0.7923	1.262
0.68	0.6288	0.7776	0.8087	1.237
0.69	0.6365	0.7712	0.8253	1.212
0.70	0.6442	0.7648	0.8423	1.187
0.71	0.6518	0.7584	0.8595	1.163
0.72	0.6594	0.7518	0.8771	1.140
0.73	0.6669	0.7452	0.8949	1.117
0.74	0.6743	0.7385	0.9131	1.095
0.75	0.6816	0.7317	0.9316	1.073
0.76	0.6889	0.7248	0.9505	1.052
0.77	0.6961	0.7179	0.9697	1.031
0.78	0.7033	0.7109	0.9893	1.011
0.79	0.7104	0.7038	1.009	0.9908
0.80	0.7174	0.6967	1.030	0.9712
0.81	0.7243	0.6895	1.050	0.9520
0.82	0.7311	0.6822	1.072	0.9331
0.83	0.7379	0.6749	1.093	0.9146
0.84	0.7446	0.6675	1.116	0.8964
0.85	0.7513	0.6600	1.138	0.8785
0.86	0.7578	0.6524	1.162	0.8609
0.87	0.7643	0.6448	1.185	0.8437
0.88	0.7707	0.6372	1.210	0.8267
0.89	0.7771	0.6294	1.235	0.8100

Table V *(continued)*

t	sin	cos	tan	cot	t	sin	cos	tan	cot
0.90	0.7833	0.6216	1.260	0.7936	1.25	0.9490	0.3153	3.010	0.3323
0.91	0.7895	0.6137	1.286	0.7774	1.26	0.9521	0.3058	3.113	0.3212
0.92	0.7956	0.6058	1.313	0.7615	1.27	0.9551	0.2963	3.224	0.3102
0.93	0.8016	0.5978	1.341	0.7458	1.28	0.9580	0.2867	3.341	0.2993
0.94	0.8076	0.5898	1.369	0.7303	1.29	0.9608	0.2771	3.467	0.2884
0.95	0.8134	0.5817	1.398	0.7151	1.30	0.9636	0.2675	3.602	0.2776
0.96	0.8192	0.5735	1.428	0.7001	1.31	0.9662	0.2579	3.747	0.2669
0.97	0.8249	0.5653	1.459	0.6853	1.32	0.9687	0.2482	3.903	0.2562
0.98	0.8305	0.5570	1.491	0.6707	1.33	0.9711	0.2385	4.072	0.2456
0.99	0.8360	0.5487	1.524	0.6563	1.34	0.9735	0.2288	4.256	0.2350
1.00	0.8415	0.5403	1.557	0.6421	1.35	0.9757	0.2190	4.455	0.2245
1.01	0.8468	0.5319	1.592	0.6281	1.36	0.9779	0.2092	4.673	0.2140
1.02	0.8521	0.5234	1.628	0.6142	1.37	0.9799	0.1994	4.913	0.2035
1.03	0.8573	0.5148	1.665	0.6005	1.38	0.9819	0.1896	5.177	0.1931
1.04	0.8624	0.5062	1.704	0.5870	1.39	0.9837	0.1798	5.471	0.1828
1.05	0.8674	0.4976	1.743	0.5736	1.40	0.9854	0.1700	5.798	0.1725
1.06	0.8724	0.4889	1.784	0.5604	1.41	0.9871	0.1601	6.165	0.1622
1.07	0.8772	0.4801	1.827	0.5473	1.42	0.9887	0.1502	6.581	0.1519
1.08	0.8820	0.4713	1.871	0.5344	1.43	0.9901	0.1403	7.055	0.1417
1.09	0.8866	0.4625	1.917	0.5216	1.44	0.9915	0.1304	7.602	0.1315
1.10	0.8912	0.4536	1.965	0.5090	1.45	0.9927	0.1205	8.238	0.1214
1.11	0.8957	0.4447	2.014	0.4964	1.46	0.9939	0.1106	8.989	0.1113
1.12	0.9001	0.4357	2.066	0.4840	1.47	0.9949	0.1006	9.887	0.1011
1.13	0.9044	0.4267	2.120	0.4718	1.48	0.9959	0.0907	10.983	0.0910
1.14	0.9086	0.4176	2.176	0.4596	1.49	0.9967	0.0807	12.350	0.0810
1.15	0.9128	0.4085	2.234	0.4475	1.50	0.9975	0.0707	14.101	0.0709
1.16	0.9168	0.3993	2.296	0.4356	1.51	0.9982	0.0608	16.428	0.0609
1.17	0.9208	0.3902	2.360	0.4237	1.52	0.9987	0.0508	19.670	0.0508
1.18	0.9246	0.3809	2.427	0.4120	1.53	0.9992	0.0408	24.498	0.0408
1.19	0.9284	0.3717	2.498	0.4003	1.54	0.9995	0.0308	32.461	0.0308
1.20	0.9320	0.3624	2.572	0.3888	1.55	0.9998	0.0208	48.078	0.0208
1.21	0.9356	0.3530	2.650	0.3773	1.56	0.9999	0.0108	92.620	0.0108
1.22	0.9391	0.3436	2.733	0.3659	1.57	1.0000	0.0008	1,255.8	0.0008
1.23	0.9425	0.3342	2.820	0.3546	1.58	1.0000	−0.0092	−108.65	−0.0092
1.24	0.9458	0.3248	2.912	0.3434	1.59	0.9998	−0.0192	−52.067	−0.0192
					1.60	0.9996	−0.0292	−34.233	−0.0292

Answers to Odd-Numbered Problems

1.

3. $1.16666\ldots$

5. $0.714285714285\ldots$

7. $0.416666\ldots$

9. 3.375

11. $\frac{2}{11}$

13. $\frac{5}{4}$

15. $\frac{939}{125}$

17. $\frac{173}{250}$

19. 7

21. 22

23. x

25. 0.13

27. 3

29. 4

31. $3 - \sqrt{5}$

33. -5

35. $b - a$

37. a. 3 **b.** 3.5

39. a. 355 **b.** 77.5

41. a. 10.5 **b.** 0.75

43. a. 2 **b.** $\frac{3}{4}$

45. $a = 3, b = 9$

47. $a = -3, b = 5$

49. $b = 20, d(a, b) = 12$

51. $b = 6 + 2\sqrt{2}, m = 6 + \sqrt{2}$

53. $\frac{22}{7} - \pi$

55. $2\pi - 6.28$

57. No, no

59. Let x be a rational number and y be an irrational number. Assume the sum $x + y = r$ is rational. We have $y = r - x$. Since the difference of two rational numbers is rational, y is rational. This is a contradiction.

1. $\left(\frac{1}{8}\right)^3$

3. $(2y)^4$

5. $\left(\frac{y}{x}\right)^{-2}$

7. x^{-3}

9. a. 81 **b.** $\frac{1}{81}$

11. a. 49 **b.** $\frac{1}{49}$

13. a. 1 **b.** 1

15. $-\frac{3}{2}$

17. $\frac{1}{5}$

19. x^4

21. $-21x^6$

23. 32

25. 0.00000000001

27. $25x^2$

29. $5^6 = 15{,}625$

31. $\frac{64x^6}{y^3}$

33. x

35. $49ab$

37. $\frac{9y^9}{x}$

39. $a^6 b^{10}$

41. $-x^2 y^4 z^6$

43. 2.371×10^3

45. 2.453×10^6

47. 6.75×10^5

49. 6.67×10^{-11}

51. -5

53. 0.1

55. $\frac{1}{2}$

57. $x^2 y$

59. $\frac{2\sqrt{3}}{3}$

61. $\frac{\sqrt[3]{100}}{10}$

63. $\frac{2\sqrt{10} - 7}{3}$

65. $\frac{x - 2\sqrt{xy} + y}{x - y}$

67. $\frac{-1}{2 - \sqrt{10}}$

69. $\frac{1}{\sqrt{x + h} + \sqrt{x}}$

71. a. 7 **b.** $\frac{1}{7}$

73. a. 0.0000128 **b.** $\frac{1}{0.0000128}$

75. a. $\frac{1}{3}$ **b.** $\frac{4}{9}$

77. $x^{1/2}$

79. $q^{1/2} b$

81. $125 x^{1/2} y^{3/2}$

83. $\frac{64x^3 z}{y^4}$

85. $\frac{9a^2}{b^4}$

87. $\frac{w^4}{2}$

89. $\sqrt[6]{500}$

91. 2

93. $\sqrt[4]{x^3}$

95. $-\frac{41}{55}$

97. True

99. $x + \dfrac{h}{2}$

Exercise 0.3 [Page 26]

1. $x^2 + 2xy + y^2$

3. $4x^2 + 4x + 1$

5. $4a^4 - 12a^2 + 9$

7. $r^3 - 3r^2s + 3rs^2 - s^3$

9. $x^6 - 3x^4y^2 + 3x^2y^4 - y^6$

11. $x^{-4} + 4x^{-3} + 6x^{-2} + 4x^{-1} + 1$

13. $x^{10} + 5x^8y^2 + 10x^6y^4 + 10x^4y^6$
$+ 5x^2y^8 + y^{10}$

15. $x^2 + 4xy + 2x + 4y^2 + 4y + 1$

17. 6

19. $\frac{1}{20}$

21. 12

23. 10

25. $6ab^5$

27. $-20x^6y^6$

29. $35(4)^3x^4 = 2240x^4$

31. $2002x^5y^9$

33. $-144y^7$

35. $15xy^2$

37. 1 8 28 56 70 56 28 8 1

39. $2x + h$

41. $3x^2 + 3xh + h^2$

43. 0.970299

45. $\sqrt[24]{x}$

47. $x^2(x^2 + y^2)^{1/2}$

49. False

Exercise 0.4 [Page 30]

1. $x^2 + x - 2$

3. $2x^6 - 13x^3 - 7$

5. $10x^2 + 26x - 56$

7. $24x - 2\sqrt{x} - 2$

9. $3.0x^2 + 7.63x + 1.47$

11. $x^2 - \frac{13}{24}x - \frac{1}{12}$

13. $7 + 37x + 10x^2$

15. $a^2 - b^2$

17. $20x^2 + 51x + 28$

19. $y^{-2} + 5xy^{-1} + 6x^2$

21. $1 - x^2$

23. $a^3 - 27$

25. $x^3 + z^6$

27. $3x^4 + 3x^3 + 5x^2 + 2x + 2$

29. $3y^5 - 15y^3 + 7y^2 + y - 5$
$- 3y^{-1} + 2y^{-2}$

31. $625x^4 - y^4$

33. $\frac{1}{25}x^2 - 25y^2$

35. $x^{4/3} - x^{2/3}$

37. $-4x^2 + 12xy - 9y^2$

39. $16x^4 - 80x^2 + 100$

43. $\frac{3}{2}$

45. True

47. $120t^3s^{21}$

Exercise 0.5 [Page 37]

1. $2x(6x^2 + x + 3)$

3. $(y + 3)(2y - z)$

5. $(6x - 5)(6x + 5)$

7. $(2xy - 1)(2xy + 1)$

9. Does not factor with integer coefficients.

11. $(x - y)(x + y)(x^2 + y^2)(x^4 + y^4)$

13. $(2xy + 3)(4x^2y^2 - 6xy + 9)$

15. $(y + 1)(y - 1)(y^2 - y + 1)$
$\cdot (y^2 + y + 1)$

17. $(x - 2)(x - 3)$

19. $(x + 2)(x + 5)$

21. $(x - 2)(x + 2)(x^2 + 1)$

23. Does not factor with integer coefficients.

25. $(x - 2y)(x + y)$

27. $(x + 5)^2$

29. $(x - 4y)^2$

31. $(2x + 5)(x + 1)$

33. $(6x^2 - 5)(x^2 + 3)$

35. $(2x - y)(x - 3y)$

37. $(x^2 + y^2) \cdot$
$\cdot (x^4 - x^2y^2 + y^4 + 3x^2 - 3y^2 + 3)$

39. $(x - y)^2$

41. $(y - x)(y + x)$
$\cdot (x^4 + y^4 + x^2y^2 - 3x^2 - 3y^2 + 3)$

43. $(1 - 2x)(1 + 2x)(1 + 4x^2)$
$\cdot (1 + 16x^4)$

45. $(x - 1)(x^2 + x + 1)(x + 2)$
$\cdot (x^2 - 2x + 4)$

47. a. $25z^6 - 4y^4$
b. $a^2 - b^2 - 2b - 1$

49. $3^{-1} - 6^{-1} = 6^{-1}$

51. a. 8 **b.** 25

Exercise 0.6 [Page 43]

1. $\{x \mid x \neq \pm 2\}$

3. $\{x \mid x \neq -1, 0\}$

5. $\dfrac{x + 1}{x + 4}$

7. $\dfrac{x - 3}{x^2 - 3x + 9}$

9. $(x + 2)(x - 1)$

11. $x^2(x + 3)(x - 2)(x - 6)$

13. 1

15. $\dfrac{7x + 1}{7x - 1}$

17. $\dfrac{2x^2 - 2x + 5}{x^2 - 1}$

19. $\dfrac{y^2 + x^2}{y^2 - x^2}$

21. $\dfrac{x^2 - 4x + 2}{x^2 - x - 12}$

23. $\dfrac{-8x + 2}{2x^2 + 3x - 2}$

25. $\dfrac{x^2 + x - 20}{x^2 + x - 6}$

27. $\dfrac{x^2 - 1}{x^2 + x + 1}$

29. $\dfrac{1}{3x - 6}$

31. $\dfrac{x + 7}{x + 2}$

33. $\dfrac{x^2}{x^2 + 9x + 20}$

35. $\dfrac{x + 1}{x - 4}$

37. $\dfrac{x - 3}{x + 5}$

39. $\dfrac{1 - x^3}{1 + x^3}$

41. $\dfrac{x}{2}$

43. $\dfrac{xy}{x - y}$

45. $\dfrac{ab}{b - a}$

47. $\dfrac{a}{a - 1}$

49. $\dfrac{t^2}{st^2 + 1}$

51. $x + a$

53. $\dfrac{4}{(3x + 4)(3a + 4)}$

55. a. $(r - 5)(s - t)$
b. $(u + v)(u - v)(r + s)(r - s)$
$\cdot (r^2 + s^2)$

57. $\left(\dfrac{35}{2}\right)\dfrac{y^{12}}{x^6}$

59. $-\frac{4}{3}$

Exercise 0.7 [Page 52]

1. -7
3. $-\frac{1}{5}$
5. 3
7. $2, -3$
9. ± 4
11. $\frac{1}{3}, 4$
13. $\frac{5}{4}, \frac{3}{2}$
15. $0, \pm\frac{1}{4}$
17. $0, \pm\frac{5}{2}$
19. $-1, \frac{1}{2}$
21. $\dfrac{2 \pm 2\sqrt{6}}{5}$
23. No real roots
25. $\frac{1}{3}, 2$
27. $\dfrac{-5 \pm \sqrt{105}}{-20}$
29. $\frac{3}{2}$
31. $\pm\sqrt{3 + \sqrt{2}}, \pm\sqrt{3 - \sqrt{2}}$
33. $-\frac{1}{2}, \pm 1$
35. -1
37. $\dfrac{3}{2}, \dfrac{-3 \pm \sqrt{17}}{4}$
39. $\frac{1}{4}, \frac{2}{3}, -\frac{3}{2}$
41. -2
43. $2, 18$
45. 4
47. $248°F$
49. $250 \text{ ft} \times 250 \text{ ft}$
51. (approximately) $12.72 \text{ in} \times 7.86 \text{ in}$
53. $\dfrac{2x - 2\sqrt{a}}{x^2 - a}$
55. $\dfrac{x^2 + 2x + h + xh}{(x + 1)(x + h + 1)}$
57. $2(x - 2)^3(3x^2 - 2x + 2)$

Exercise 0.8 [Page 58]

1. $-\frac{1}{2}$
3. The solution set is empty.
5. There are no real solutions.
7. $-\frac{7}{3}, -\frac{3}{2}$
9. $-\frac{4}{5}, -3$
11. $-1, 8^5 = 32,768$
13. 256
15. $-\frac{191}{64}, -\frac{82}{27}$
17. $\pm\sqrt{\frac{23}{3}}$
19. 63

21. The solution set is empty.
23. 0
25. $3, 7$
27. 0
29. 1
31. $-\frac{1}{4}, \frac{3}{4}$
33. $-\frac{1}{2}, \frac{5}{6}$
35. $\frac{15}{4}$
37. The solution set is empty.
39. $\pm\sqrt{6}$
41. $\sqrt{2x^5 y^3}$
43. The fourth term in the expansion is $35x^4(-y)^3$, so the coefficient is 35.

Exercise 0.9 [Page 63]

1. $-3 < 15$
3. $1.33 < \frac{4}{3}$
5. $2.5 \geq \frac{5}{2}$ or $\frac{5}{2} \geq 2.5$
7. $22 \geq 7\pi$
9. If $a \geq b$ and c is any real number, then $a + c \geq b + c$.

11.
$$x + 3 > -2$$
$$x + 3 - 3 > -2 - 3 \quad \text{(rule I)}$$
$$x > -5$$

13.
$$\tfrac{3}{2}x + 4 \leq 10$$
$$\tfrac{3}{2}x + 4 - 4 \leq 10 - 4 \quad \text{(rule I)}$$
$$\tfrac{3}{2}x \leq 6$$
$$\tfrac{2}{3}(\tfrac{3}{2}x) \leq \tfrac{2}{3}(6) \quad \text{(rule II)}$$
$$x \leq 4$$

15.
$$-7 < x - 2 < 1$$
$$-7 + 2 < x - 2 + 2 < 1 + 2 \quad \text{(rule I)}$$
$$-5 < x < 3$$

17.
$$7 < 3 - \tfrac{1}{2}x \leq 8$$
$$-3 + 7 < -3 + 3 - \tfrac{1}{2}x$$
$$\leq -3 + 8 \quad \text{(rule I)}$$
$$4 < -\tfrac{1}{2}x \leq 5$$
$$-2(4) > -2(-\tfrac{1}{2}x) \geq -2(5)$$
$$\text{(rule III)}$$
$$-8 > x \geq -10$$
$$-10 \leq x < -8$$

19.
$$-x < x + 1 < 3 - x$$
$$-x + x < x + 1 + x$$
$$< 3 - x + x \quad \text{(rule I)}$$
$$0 < 2x + 1 < 3$$
$$0 - 1 < 2x + 1 - 1 < 3 - 1 \text{ (rule I)}$$
$$-1 < 2x < 2$$
$$\tfrac{1}{2}(-1) < \tfrac{1}{2}(2x) < \tfrac{1}{2}(2) \quad \text{(rule II)}$$
$$-\tfrac{1}{2} < x < 1$$

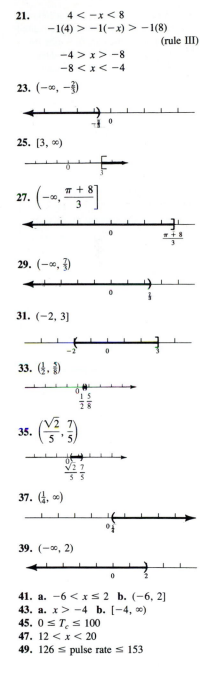

21.
$$4 < -x < 8$$
$$-1(4) > -1(-x) > -1(8)$$
$$\text{(rule III)}$$
$$-4 > x > -8$$
$$-8 < x < -4$$

23. $\left(-\infty, -\frac{2}{3}\right)$

25. $[3, \infty)$

27. $\left(-\infty, \dfrac{\pi + 8}{3}\right]$

29. $\left(-\infty, \frac{7}{3}\right)$

31. $(-2, 3]$

33. $\left(\frac{1}{2}, \frac{5}{8}\right)$

35. $\left(\dfrac{\sqrt{2}}{5}, \dfrac{7}{5}\right)$

37. $\left(\frac{1}{4}, \infty\right)$

39. $(-\infty, 2)$

41. **a.** $-6 < x \leq 2$ **b.** $(-6, 2]$
43. **a.** $x > -4$ **b.** $[-4, \infty)$
45. $0 \leq T_c \leq 100$
47. $12 < x < 20$
49. $126 \leq \text{pulse rate} \leq 153$

51. Assume that $a < b$; then it follows that $b - a$ is positive. If we multiply the positive number $b - a$ by the negative number c, the product $(b - a)c = bc - ac$ is negative. Hence, we have $ac > bc$.

53. $\frac{9}{8}$

55. 5

57. $\dfrac{1}{\sqrt{2 + h} + \sqrt{2}}$

Exercise 0.10 [Page 68]

1. $(-5, 5)$

3. $(-5, 9)$

5. $(-\infty, -4) \cup (4, \infty)$

7. $[-5, 13]$

9. $(-\infty, 3] \cup [4, \infty)$

11. $(-\infty, -1 - \sqrt{2}]$ $\cup [1 - \sqrt{2}, \infty)$

13. The solution set is empty.

15. $\frac{6}{5}$

17. $(-\infty, -1] \cup [\frac{11}{5}, \infty)$

19. $|x - 12| < 5;$ $(7, 17)$

21. $|x - 3.5| < 0.5;$ $(3.0, 4.0)$

23. $|x - \pi| < \sqrt{3};$ $(\pi - \sqrt{3}, \pi + \sqrt{3})$

25. $|x - 10| \le 7$

27. $|x + 1| < 4$

29. $|x - 14| < 7$

31. $|x| \ge 3$

33. True

35. $\dfrac{-x^3 + 3x^2 - 3x + 2}{(x - 1)^3(x - 2)}$

37. $\dfrac{x + 1}{(2x + 1)^{3/2}}$

39. $\frac{2}{7}, 2$

Chapter 0
Review Exercises [Page 71]

1. True

3. True

5. False

7. False

9. False

11. False

13. True

15. False

17. $5 - 2\sqrt{6}$

19. $8x^3 + 36x^2y + 54xy^2 + 27y^3$

21. $-20x^3y^3$

23. $y^2 - x^2$

25. $x^3 + 3x^2 + 3x + 1$

27. $(x^2 - 4y)(x^2 + 4y)$

29. $(x - 7)(x - 2)$

31. $(2x - 1)(2x - 3)$

33. $\dfrac{3x - 3}{x^2 - 4}$

35. -2

37. $-\frac{1}{4}, 1$

39. $-\frac{3}{2}, 1$

41. $\pm\dfrac{\sqrt{-1 + \sqrt{17}}}{2}$

43. $\frac{1}{2}, -1 \pm \sqrt{3}$

45. 4

47. -5

49. There are no solutions.

51. $-\frac{1}{5}, \frac{7}{5}$

53. $(-\infty, \frac{7}{4})$

55. $[-\frac{5}{4}, \frac{3}{4}]$

57. $[-5, 1]$

59. $|x - \frac{1}{2}| > 6$

61. $|x - \frac{9}{2}| < 4$

Exercise 1.1 [Page 81]

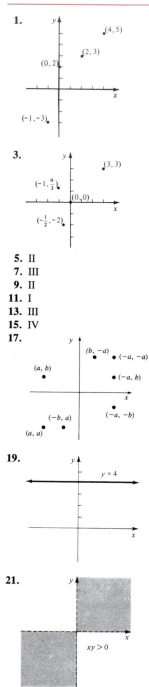

1.

3.

5. II

7. III

9. II

11. I

13. III

15. IV

17.

19.

21.

23.

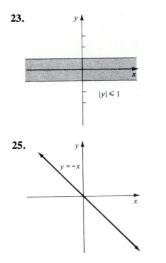

$|y| \leq 1$

25.

$y = -x$

27. Symmetric with respect to the y axis.

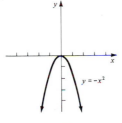

$y = -x^2$

29. Symmetric with respect to the y axis.

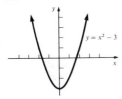

$y = x^2 - 3$

31. No symmetry.

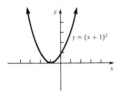

$y = (x + 1)^2$

33. No symmetry.

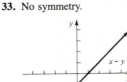

$x - y = 1$

35. Symmetric with respect to the origin.

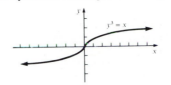

$y^3 = x$

37. Symmetric with respect to the y axis.

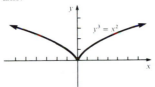

$y^3 = x^2$

39. Symmetric with respect to the y axis.

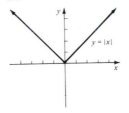

$y = |x|$

41. Symmetric with respect to the origin.

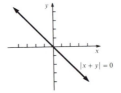

$|x + y| = 0$

43. Symmetric with respect to the origin.

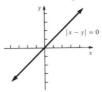

$|x - y| = 0$

45. Symmetric with respect to the y axis.

$y = |x|^3$

47. A graph is symmetric with respect to the line $y = 1$ if, whenever the point (x, y) is on the graph, $(x, -y + 2)$ is also on the graph.

49. A graph is symmetric with respect to the line $x = -3$ if, whenever the point (x, y) is on the graph, $(-x - 6, y)$ is also on the graph.

51. A graph is symmetric with respect to the line $y = k$ if, whenever the point (x, y) is on the graph, $(x, -y + 2k)$ is also on the graph.

53. A graph is symmetric with respect to the point $(1, -2)$ if, whenever the point (x, y) is on the graph, $(-x + 2, -y - 4)$ is also on the graph.

55. We need $(-x + 8, y)$ to be on the graph whenever (x, y) is. Since $(-x + 8 - 4)^2 + 6 = (-x + 4)^2 + 6 = (x - 4)^2 + 6$, the graph is symmetric with respect to the line $x = 4$.

Exercise 1.2 [Page 88]

1. $\sqrt{20}$
3. 10
5. 5
7. $d(A, B) = \sqrt{125},$
$d(A, C) = \sqrt{20},$
$d(B, C) = \sqrt{205},$
$[d(A, B)]^2 + [d(A, C)]^2$
$= 125 + 20 = 145$
$\neq 205 = [d(B, C)]^2.$ Thus, ABC is *not* a right triangle.
9. $d(A, B) = \sqrt{125},$
$d(A, C) = 5, d(B, C) = 10,$
$[d(A, C)]^2 + [d(B, C)]^2$
$= 25 + 100 = 125$
$= [d(A, B)]^2.$ Thus, ABC *is* a right triangle.
11. $d(A, C) = \sqrt{65},$
$d(B, C) = \sqrt{65}$

13. $(1, \frac{5}{2})$

15. $(\frac{3}{2}, -1)$

17. $(5, -1)$

19. $(\frac{38}{21}, -\frac{13}{4})$

21. Center $(1, 3)$, $r = 7$

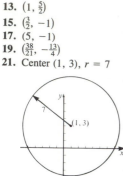

23. Center $(\frac{1}{2}, \frac{3}{2})$, $r = \sqrt{5}$

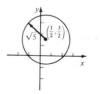

25. $x^2 + (y + 4)^2 = 16$; center $(0, -4)$, $r = 4$

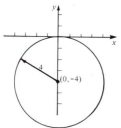

27. $(x - 9)^2 + (y - 3)^2 = 100$; center $(9, 3)$, $r = 10$

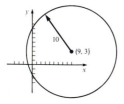

29. $(x + 1)^2 + (y + 4)^2 = 22$; center $(-1, -4)$, $r = \sqrt{22}$

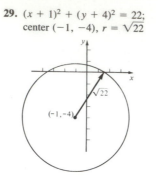

31. $(x - \frac{5}{6})^2 + (y + \frac{7}{6})^2 = \frac{157}{18}$ center $(\frac{5}{6}, -\frac{7}{6})$, radius $\frac{1}{3}\sqrt{\frac{157}{2}}$

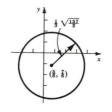

33. Completing the square, we have
$$x^2 + y^2 + 2y + 1 = -9 + 1$$
$$x^2 + (y + 1)^2 = -8.$$
Since the sum of squares cannot be negative, this equation has no graph, and thus does not describe a circle.

35. $x^2 + y^2 = 1$

37. $x^2 + (y - 2)^2 = 2$

39. $(x - 1)^2 + (y - 6)^2 = 8$

41. $x^2 + y^2 = 5$

43. $(x - 5)^2 + (y - 6)^2 = 36$

45. $(3 \pm \sqrt{13}, 0)$, $(-6 \pm 2\sqrt{10})$

47. Yes

49. $4 + \sqrt{5}$ ft

51. Tank: $x^2 + (y - 1)^2 = 6.25$ Hole: $x^2 + y^2 = \frac{1}{4}$

53. a. III **b.** II **c.** IV **d.** I

55. Symmetric with respect to the x axis, y axis, and origin.

57. Symmetric with respect to $x = 1$, $y = 2$, and $(1, 2)$.

Exercise 1.3 [Page 98]

1. $\Delta x = 1$, $\Delta y = -1$

3. $\Delta x = 4$, $\Delta y = 6$

5. $-\frac{7}{2}$

7. -1

9. $\frac{5}{3}$

11. $-\frac{6}{5}$

13. $\frac{b}{a}$

15.

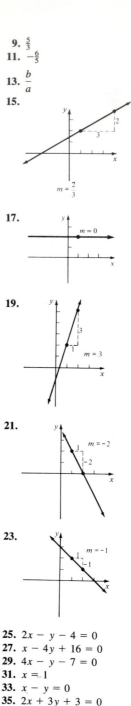

17.

19.

21.

23.

25. $2x - y - 4 = 0$

27. $x - 4y + 16 = 0$

29. $4x - y - 7 = 0$

31. $x = 1$

33. $x - y = 0$

35. $2x + 3y + 3 = 0$

37. $7x + y + 3 = 0$

39. $y = mx$

41. $y = -4$

43. $y = 22x$

45. $2x + 5y - 10 = 0$
47. $m = \frac{1}{2}, b = -\frac{7}{4}$
49. $m = \frac{1}{6}, b = \frac{2}{3}$
51. $m = -2, b = 0$
53.

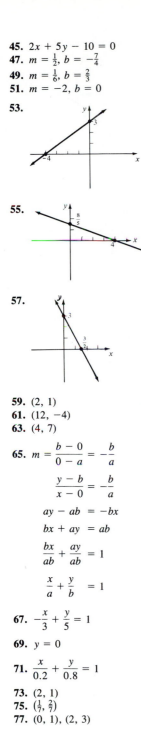

55.

57.

59. $(2, 1)$
61. $(12, -4)$
63. $(4, 7)$
65. $m = \dfrac{b - 0}{0 - a} = -\dfrac{b}{a}$

$\dfrac{y - b}{x - 0} = -\dfrac{b}{a}$

$ay - ab = -bx$

$bx + ay = ab$

$\dfrac{bx}{ab} + \dfrac{ay}{ab} = 1$

$\dfrac{x}{a} + \dfrac{y}{b} = 1$

67. $-\dfrac{x}{3} + \dfrac{y}{5} = 1$
69. $y = 0$
71. $\dfrac{x}{0.2} + \dfrac{y}{0.8} = 1$
73. $(2, 1)$
75. $\left(\frac{1}{7}, \frac{2}{7}\right)$
77. $(0, 1), (2, 3)$

79. $\left(\dfrac{9\sqrt{10}}{10}, \dfrac{10 - 3\sqrt{10}}{10}\right),$

$\left(-\dfrac{9\sqrt{10}}{10}, \dfrac{10 + 3\sqrt{10}}{10}\right)$

81. $\left(\frac{1}{5}, \frac{17}{5}\right)$
83. No
87. $(1 + \sqrt{1/101}, 2 + 10\sqrt{1/101}),$
$(1 - \sqrt{1/101}, 2 - 10\sqrt{1/101})$
89. $6y + 7x - 8 = 0$
91. $y = \sqrt{3}x$
93. $y = -4x - 64$
95. $(x + 2)^2 + (y - 1)^2 = 20$
97. Yes, it does.
99. a. $\left(\frac{5}{2}, 0\right)$ **b.** $\dfrac{\sqrt{41}}{2}$

Exercise 1.4 [Page 107]

1. $3x + y + 2 = 0$
3. $y = -4$
5. $y = 3$
7. $4x + y - 28 = 0$
9. $2x - 5y - 17 = 0$
11. Yes
13. No
15. No
17. No
19. Yes
21. Parallel lines: a and c, b and e;
perpendicular lines: a and b, a and
e, b and c, c and e, d and f
23. Perpendicular lines: a and c, b and
d, e and f
25. $3x + y - 3 = 0$
27. $y = 0$
29. $x + 3y - 4 = 0$
31. $7x - 3y = 0$
33. $x = 1$
35. Not a right triangle
37. Not a right triangle
39. A right triangle
41. $x + 2y - 5 = 0$
43. $x = 3$
45. $7, -3, 2 \pm \sqrt{5}$
47. If both lines have slope 0, they are
both horizontal and hence parallel.
Otherwise, use the fact that the
slopes are equal to prove that the
two right triangles shown are
congruent. Then the indicated angles
are equal. Now use the fact that if
two lines are intersected by a third

line in equal angles, the two lines
are parallel.

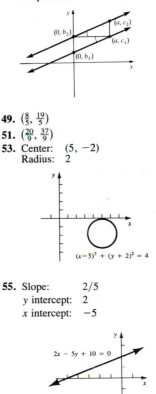

49. $\left(\frac{8}{5}, \frac{19}{5}\right)$
51. $\left(\frac{20}{9}, \frac{37}{9}\right)$
53. Center: $(5, -2)$
Radius: 2

$(x-5)^2 + (y + 2)^2 = 4$

55. Slope: $2/5$
y intercept: 2
x intercept: -5

$2x - 5y + 10 = 0$

57. $b = 0$

Chapter 1
Review Exercises [Page 110]

1. True
3. False
5. False
7. True
9. False
11.

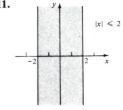

$|x| \leqslant 2$

13. Symmetric with respect to the y axis

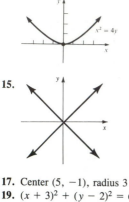

15.

17. Center $(5, -1)$, radius 3
19. $(x + 3)^2 + (y - 2)^2 = 68$
21. $2x - 3y + 17 = 0$
23. $x - 2y - 6 = 0$
25. Slope: $5/7$
   y intercept: $1/7$
   x intercept: $-1/5$
27. $x - 2y - 7 = 0$
29. $x + 3y = 14$
31. $\sqrt{3}$
33. $(5, 7)$
35. $(0, 0)$, $\left(-\frac{32}{17}, -\frac{8}{17}\right)$
37. $(0, -4)$
39. a. $c = 0$ **b.** $a = 0$ **c.** $b = 0$

Exercise 2.1 [Page 121]

1. This correspondence is not a
   function: $1 \to 2$ and $1 \to 5$.
3. This correspondence is a function.
5. This correspondence is a function.
7. $-1; 0; 1; 3$
9. $1; 2; 0; \sqrt{6}$
11. $0; \frac{3}{2}; \sqrt{2}; -\frac{3}{2}$
13. $3a^3 - a; 3(a + 1)^3 - (a + 1)$
   $= 3a^3 + 9a^2 + 8a + 2;$
   $3(a^2)^3 - (a^2) = 3a^6 - a^2;$
   $3(1/a)^3 - (1/a) = 3/a^3 - 1/a;$
   $3(-a)^3 - (-a) = -3a^3 + a$
15. $3; 0; 2 + 2\sqrt{2}; 8$
17. 3
19. $2x + h$
21. $-1/x(x + h)$
23. 2
25. $\dfrac{\sqrt{x} - \sqrt{a}}{x - a}$

27. $\{x \mid x \neq 1\}$
29. All real numbers
31. $\{x \mid x \geq -1,\ x \neq 0\}$
33. Domain: all real numbers
   Range: $\{6\}$
35. Domain: all real numbers
   Range: $\{y \mid y \geq 0\}$
37. Domain: $\{x \mid x \geq 5\}$
   Range: $\{y \mid y \geq 0\}$
39. Domain: $\{x \mid x \geq 0\}$
   Range: $\{y \mid y \geq 2\}$

41. $20; 4$
43. $-2; 2$
45. Domain: all real numbers
   Range: $\{3\}$

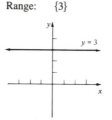

47. Domain: all real numbers
   Range: all real numbers

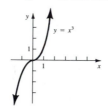

49. Domain: all real numbers
   Range: $\{y \mid y \neq -1\}$

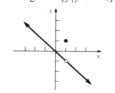

51. Domain: all real numbers
   Range: $\{y \mid -3 \leq y \leq 3\}$

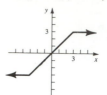

53. Domain: all real numbers
   Range: all real numbers

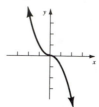

55. a. $\sqrt{8}$ **b.** $\sqrt{5}$ **c.** $\sqrt{a + 5}$,
   provided that $a \geq -5$

57.

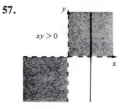

This is not the graph of a function.

59.

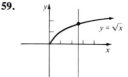

This is the graph of a function.

61.

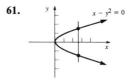

This is not the graph of a function.

63. y is a function of x.
   Domain: all real numbers
   Range: $\{y \mid -1 \leq y \leq 1\}$

65. y is a function of x.
 Domain: $\{x \mid -1 < x < 1\}$
 Range: $\{-1, 1\}$

67. $A = \dfrac{\pi d^2}{4}$

 $A = \dfrac{C^2}{4\pi}$

69. Let w be the width of the box and l the length.

 $A = 10w^3; \quad A = \dfrac{5l^3}{4}$

71. $S = 2x^2 + \dfrac{256{,}000}{x}$

 Domain: $\{x \mid 0 < x < 80\sqrt{10}\}$

73. Domain: $\{x \mid x \geq 0\}$
 Range: $\{y \mid y = 0.22 + k(0.17),\ k$ a nonnegative integer$\}$

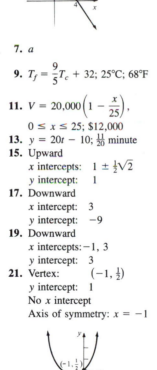

75. $A = 2x(4 - x^2)$
77. $A = (x + 4)\sqrt{16 - x^2}$

79. $m = \dfrac{f(1 + h) - f(1)}{h}$

81. $A(-1, 2), B(\frac{1}{2}, \frac{5}{4}), C(2, 5),$
 $D(a, a^2 + 1)$
83. $1.27 \times 10^{30}, 2.66 \times 10^{-5}, 2.70$
85. $1.15, 1.02, -1.62$

Exercise 2.2 [Page 132]

1. x intercept: 4
 y intercept: -8

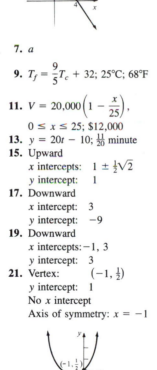

3. x intercept: 2
 y intercept: 6

5. x intercept: 4
 y intercept: 6

7. a

9. $T_f = \dfrac{9}{5}T_c + 32$; 25°C; 68°F

11. $V = 20{,}000\left(1 - \dfrac{x}{25}\right),$
 $0 \leq x \leq 25$; \$12,000
13. $y = 20t - 10$; $\frac{11}{20}$ minute
15. Upward
 x intercepts: $1 \pm \frac{1}{2}\sqrt{2}$
 y intercept: 1
17. Downward
 x intercept: 3
 y intercept: -9
19. Downward
 x intercepts: $-1, 3$
 y intercept: 3
21. Vertex: $\left(-1, \frac{1}{2}\right)$
 y intercept: 1
 No x intercept
 Axis of symmetry: $x = -1$

23. Vertex: $(1, -8)$
 y intercept: -7
 x intercepts: $1 \pm 2\sqrt{2}$
 Axis of symmetry: $x = 1$

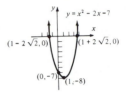

25. Vertex: $(0, -4)$
 y intercept: -4
 x intercepts: $\pm\dfrac{2\sqrt{3}}{3}$
 Axis of symmetry: $x = 0$

27. Yes, the two numbers are 15 and 15; no.
29. $(1, 2)$; $\sqrt{20}$
31. $\frac{5}{2}$ in
33. If the sides are x and y, then the area of the rectangle is
 $A(x) = x(20 - x) = -xc^2 + 20x.$
 The vertex of the graph is at
 $x = -20/2(-1) = 10$. Thus, the rectangle has greatest area when it is a square, 10 inches on each side.
35. 3 seconds
37. Linear

39. Quadratic

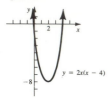

$y = 2x(x - 4)$

41. Quadratic

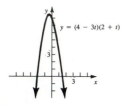

$y = (4 - 3t)(2 + t)$

43. Linear

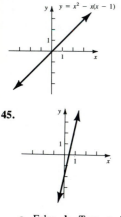

$y = x^2 - x(x - 1)$

45.

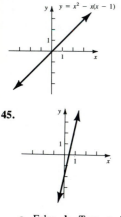

a. False b. True c. True
d. False
47. $f(x) = 4x + 2$
49. $f(x) = \frac{5}{12}x + \frac{11}{3}$
51. $f(x) = 2x^2 + 3x + 5$
53. $f(x) = -2x^2 - 4x + 8$
55. $S = \$1737.50$
$t = 0.61$ years
57. $t = 1.27$
59. $D(2) = 1385$
$D(3) = 868$
$D(4) = 2553$
Minimum when $x = 2.73$. The
graph of $D(x)$ is a parabola. For

$x \geq 2.73$ the demand increases as
the price increases.

61.

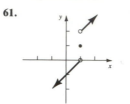

Exercise 2.3 [Page 143]

1.

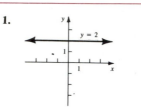

$y = 2$

Constant everywhere

3.

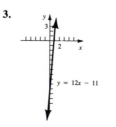

$y = 12x - 11$

Increasing everywhere

5.

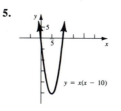

$y = x(x - 10)$

Decreasing on $(-\infty, 5)$.
Increasing on $(5, \infty)$.

7.

$y = -2x^2 - 6x + 3$

Increasing on $\left(-\infty, -\frac{3}{2}\right)$.
Decreasing on $\left(-\frac{3}{2}, \infty\right)$.
9. Odd
11. Neither
13. Even
15. Odd

17.

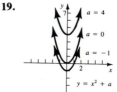

$a = 1$
$a = 0$
$a = -2$

$y = x + a$

19.

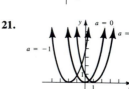

$a = 4$
$a = 0$
$a = -1$

$y = x^2 + a$

21.

$a = 0$
$a = 2$
$a = -1$

$y = (x + a)^2$

23.

$a = 4$
$a = 1$

$y = ax^2$

$a = -1$

25.

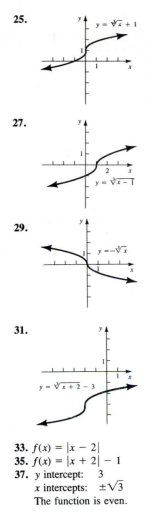

$y = \sqrt[3]{x} + 1$

27.

$y = \sqrt[3]{x - 1}$

29.

$y = -\sqrt[3]{x}$

31.

$y = \sqrt[3]{x + 2} - 3$

33. $f(x) = |x - 2|$
35. $f(x) = |x + 2| - 1$
37. y intercept: 3
    x intercepts: $\pm\sqrt{3}$
    The function is even.

$y = 3 - x^2$

Increasing on $(-\infty, 0)$.
Decreasing on $(0, \infty)$.

39. y intercept: 4
    x intercepts: ± 2
    The function is even.

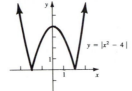

$y = |x^2 - 4|$

Increasing on $(-2, 0)$ and $(2, \infty)$.
Decreasing on $(-\infty, -2)$ and $(0, 2)$.

41. y intercept: 4
    x intercepts: none
    The function is even.

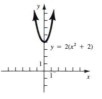

$y = 2(x^2 + 2)$

Increasing on $(0, \infty)$.
Decreasing on $(-\infty, 0)$.

43. y intercept: 5
    x intercepts: $\pm 1, \pm\sqrt{5}$
45. y intercept: -18
    x intercepts: $-1, \frac{2}{3}, \frac{3}{2}, 3$

47. $d = a$ (stretch by a factor of d)
    $h = b/2a$ (horizontal shift of h
    units)
    $k = c - b^2/4a$ (vertical shift of k
    units)

49. a.

$y = \llbracket x - 2 \rrbracket$

b.

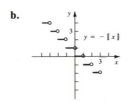

$y = -\llbracket x \rrbracket$

c.

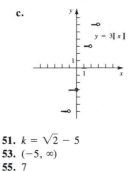

$y = 3\llbracket x \rrbracket$

51. $k = \sqrt{2} - 5$
53. $(-5, \infty)$
55. 7

Exercise 2.4 [Page 153]

1.

$y = -x^3$

x	$-x^3$
2	-8
1	-1
$\frac{1}{2}$	$-\frac{1}{8}$
0	0
$-\frac{1}{2}$	$\frac{1}{8}$
-1	1
-2	8

3.

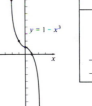

$y = 1 - x^3$

x	$1 - x^3$
2	-7
1	0
0	1
-1	2
-2	9

5.

$y = x^4$

x	x^4
3	81
2	16
1	1
0	0
-1	1
-2	16
-3	81

7. No intercepts
  Asymptotes: $x = 0$; $y = 0$

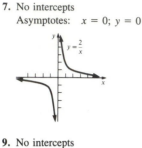

9. No intercepts
  Asymptotes: $x = 0$; $y = 0$

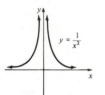

11. x intercept: $\frac{9}{4}$
   y intercept: -3
   Asymptotes: $x = -\frac{3}{2}$; $y = 2$

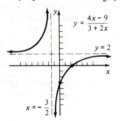

13. x intercepts: ± 1
   y intercept: none
   Asymptotes: $x = 0$; $y = -1$

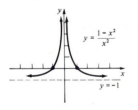

15. x intercept: 0
   y intercept: 0
   Asymptotes: $x = \pm 1$; $y = 0$

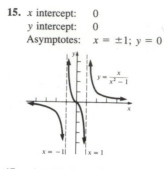

17. x intercepts: ± 3
   y intercept: none
   Asymptote: $x = 0$

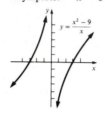

19. No

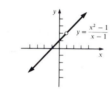

21.

23.

25.

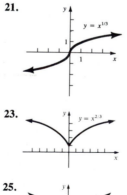

27.

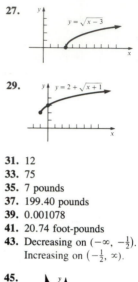

29.

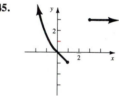

31. 12
33. 75
35. 7 pounds
37. 199.40 pounds
39. 0.001078
41. 20.74 foot-pounds
43. Decreasing on $\left(-\infty, -\frac{1}{2}\right)$.
   Increasing on $\left(-\frac{1}{2}, \infty\right)$.

45.

Exercise 2.5 [Page 161]

1. $(f + g)(x) = 3x^2 - x + 4$, domain:
   all real numbers;
   $(fg)(x) = 2x^4 - x^3 + 5x^2 - x + 3$,
   domain: all real numbers
3. $(f + g)(x) = 4x$, $\{x | x > 0\}$;
   $(fg)(x) = 4x^2 - 1/x$, $\{x | x > 0\}$
5. $(fg)(x) = x^3 + 2x^2 - 4x - 8$,
   all real numbers;
   $\left(\dfrac{f}{g}\right)(x) = x - 2$, $\{x | x \neq -2\}$
7. $(f + g)(x) = \sqrt{1 - x} + \sqrt{x + 2}$,
   $\{x | -2 \leq x \leq 1\}$;
   $(fg)(x) = \sqrt{1 - x}\sqrt{x + 2}$,
   $\{x | -2 \leq x \leq 1\}$
9. $(f + g)(x) = \begin{cases} 3x + 2, & x < 0 \\ \sqrt{x} + x + 3, \\ & 0 \leq x \leq 2 \\ 2\sqrt{x} - 4, & x > 2 \end{cases}$

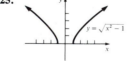

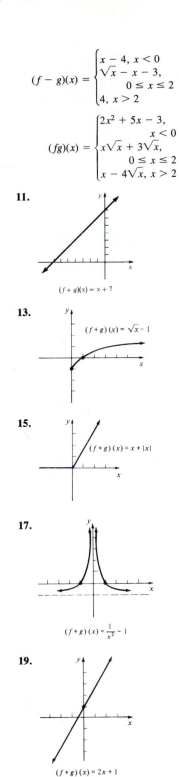

$$(f - g)(x) = \begin{cases} x - 4, \ x < 0 \\ \sqrt{x} - x - 3, \\ \qquad 0 \le x \le 2 \\ 4, \ x > 2 \end{cases}$$

$$(fg)(x) = \begin{cases} 2x^2 + 5x - 3, \\ \qquad\qquad x < 0 \\ x\sqrt{x} + 3\sqrt{x}, \\ \qquad 0 \le x \le 2 \\ x - 4\sqrt{x}, \ x > 2 \end{cases}$$

11.

$(f + g)(x) = x + 7$

13.

$(f + g)(x) = \sqrt{x} - 1$

15.

$(f + g)(x) = x + |x|$

17.

$(f + g)(x) = \dfrac{1}{x^2} - 1$

19.

$(f + g)(x) = 2x + 1$

21.

$y = x - |x|$

23.

$y = x^2 - 2|x - 2| - 3$

25.

27. $(f \circ g)(x) = x;$
$(g \circ f)(x) = |x|$

29. $(f \circ g)(x) = 1/(2x^2 + 1);$
$(g \circ f)(x) = (4x^2 - 4x + 2)/$
$(4x^2 - 4x + 1)$

31. $(f \circ g)(x) = x; (g \circ f)(x) = x$

33. $(f \circ g)(x) = \dfrac{1}{x} + x^2;$

$(g \circ f)(x) = \dfrac{x^2}{x^3 + 1}$

35. $(f \circ g)(x) = x + \sqrt{x - 1} + 1;$
$(g \circ f)(x) = x + 1 + \sqrt{x}$

37. $(f \circ g)(x) = 2$
$(g \circ f)(x) = 3$

39. $(f \circ g)(x) = \begin{cases} -1, & x = 0 \\ x^2 + 1, & x \neq 0 \end{cases}$

$(g \circ f)(x) = \begin{cases} (x - 1)^2, & x \le 0 \\ (x + 1)^2, & x > 0 \end{cases}$

41. $(f \circ g)(x) =$
$$\begin{cases} x^4 - 3x^2 + 2, & 0 \le |x| < 1 \\ x^2 - 1, & 1 \le |x| \le \sqrt{2} \\ -x^4 + 4x^2 - 3, & |x| > \sqrt{2} \end{cases}$$

$(g \circ f)(x) =$
$$\begin{cases} x^4 - 2x^3 + x^2 - 1, & x < 0 \\ x^2 - 1, & 0 \le x \le 1 \\ x^4 - 4x^3 + 4x^2 - 1, & x > 1 \end{cases}$$

43. $(f \circ f)(x) = \sqrt[4]{x}$
$(f \circ 1/f)(x) = 1/\sqrt[4]{x}$

45. $(f \circ f)(x) = x^4 + 2x^2 + 2$

$(f \circ 1/f)(x) = \dfrac{x^4 + 2x^2 + 2}{x^4 + 2x^2 + 1}$

47. **a.** 1025 **b.** 578 **c.** 5
49. **a.** -12 **b.** 12 **c.** -6
51. $(f \circ g \circ h)(x) = \sqrt{10x - 11}$
53. $f(x) = x^4$ and $g(x) = 3x - 5$ are
possible choices for $f(x)$ and $g(x)$.

55. $f(x) = \sqrt{x}$ and $g(x) = x + x^{-1}$ are
possible choices for $f(x)$ and $g(x)$.
57. $f(x) = \sqrt{x} - 4$ and $g(x) = |x|$ are
possible choices for $f(x)$ and $g(x)$.
59. $(f \circ g)(x) = x^4 + 2x^2$
$(g \circ f)(x) = x^4 - 2x^2 + 2$

61. False
63. $d = \sqrt{3}s$

65. $A(t) = \dfrac{16\pi t^4}{(t^2 + 1)^2}$

67. $v = 16$
69. $a = -1, 0, 1$

71. $\dfrac{\sqrt{1 + h} - 1}{h}$

Exercise 2.6 [Page 172]

1. One-to-one
3. Not one-to-one
5. One-to-one
7. If $f(x_1) = f(x_2)$, then
$$3x_1 - 5 = 3x_2 - 5$$
$$3x_1 = 3x_2$$
$$x_1 = x_2.$$
9. If $f(x_1) = f(x_2)$, then
$$\frac{1}{x_1} = \frac{1}{x_2}$$
$$x_1 = x_2.$$

11. If $f(x_1) = f(x_2)$, then
$$\sqrt{x_1} = \sqrt{x_2}$$
$$(\sqrt{x_1})^2 = (\sqrt{x_2})^2$$
$$x_1 = x_2.$$

13. Not one-to-one

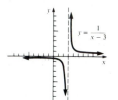

15. One-to-one

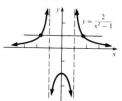

17. Not one-to-one

19. $a \neq 0$

21. If $x_1 \neq x_2$, then either $x_1 < x_2$ or $x_2 < x_1$. In the first case we have $f(x_1) < f(x_2)$, since f is increasing. Hence, $f(x_1) \neq f(x_2)$ and so f is one-to-one.

23. $f^{-1}(x) = \dfrac{x + 9}{3}$

25. $f^{-1}(x) = \dfrac{1}{x} + 3$

27. $f^{-1}(x) = \dfrac{1}{x}$

29. $f^{-1}(x) = \dfrac{x}{2x - 3}$

31. $f^{-1}(x) = \sqrt[3]{x - 2}$

33. $f^{-1}(x) = x^3 - 1$

35. Since $f^{-1}(v) =$ that number u such that $f(u) = v$, we have $(f \circ f^{-1})(v) = f(f^{-1}(v)) = f(u) = v.$

37.

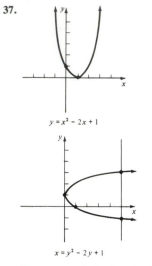

$$y = x^2 - 2x + 1$$

$$x = y^2 - 2y + 1$$

The inverse relation is not a function $y = f(x)$.

39.

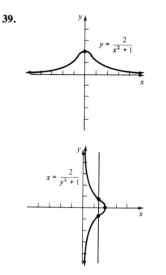

The inverse relation is not a function $y = f(x)$.

41.

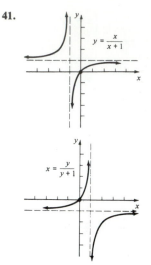

The inverse relation is a function $y = f(x)$.

43. $f^{-1}(x) = \frac{1}{2} - \frac{1}{4}x$

45. $f^{-1}(x) = \dfrac{5 - x}{2x}$

47. $f^{-1}(x) = \dfrac{x}{x - 1}$

49. $f^{-1}(x) = x^3 + 1$

51. $f^{-1}(x) = \sqrt[5]{\dfrac{x + 2}{3}}$

53. $f(0) = 32 = f(8)$, so f is not one-to-one.

55.

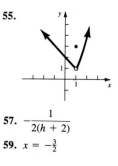

57. $-\dfrac{1}{2(h + 2)}$

59. $x = -\frac{3}{2}$

Chapter 2
Review Exercises [Page 176]

1. True
3. False
5. True
7. True
9. False
11. $-2; 0; -6; -\frac{13}{8}$
13. Domain: $\{x \mid x \leq 2\}$
    Range: $\{y \mid y \geq 0\}$
15. 2
17. a.

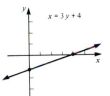

$$y = \begin{cases} -x, & x < -1 \\ x^2, & x \geq -1 \end{cases}$$

b. $-3, \sqrt{3}$
19. This is the graph of a function
    $y = f(x)$.

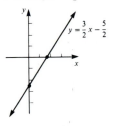

$x = 3y + 4$

21. x intercept: $\frac{5}{3}$
    y intercept: $-\frac{5}{2}$

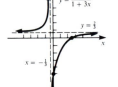

$y = \frac{3}{2}x - \frac{5}{2}$

23. x intercept: 2
    y intercept: -4
    Vertex: $(2, 0)$

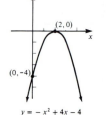

$(2, 0)$
$(0, -4)$
$y = -x^2 + 4x - 4$

25. a. 64 feet   **b.** $t = 2$ seconds
    c. $t = 4$ seconds
27. Odd

29.

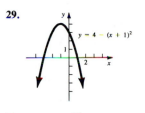

$y = 4 - (x + 1)^2$

31.

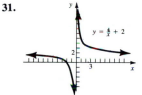

$y = \frac{4}{x} + 2$

33. $g(x) = x^{2/3} + 1$
35. $r(x) = -(x - 2)^{2/3}$
37. x intercept: $\frac{5}{2}$
    y intercept: -5
    Asymptotes: $x = -\frac{1}{3}, y = \frac{2}{3}$

$y = \frac{2x - 5}{1 + 3x}$
$y = \frac{2}{3}$
$x = -\frac{1}{3}$

39. x intercept: 0
    y intercept: 0
    Asymptotes: $x = \pm 2, y = 0$

$y = \frac{x}{x^2 - 4}$
$x = -2$ $x = 2$

41. $(f + g)(x) = x^3 + 1$
    $+ \dfrac{1}{x - 1}, \{x \mid x \neq 1\};$
    $(f - g)(x) = x^3 + 1$
    $- \dfrac{1}{x - 1}, \{x \mid x \neq 1\};$
    $(fg)(x) = \dfrac{x^3 + 1}{x - 1}, \{x \mid x \neq 1\};$
    $\left(\dfrac{f}{g}\right)(x) = (x^3 + 1)(x - 1),$
    $\{x \mid x \neq 1\};$
    $(f \circ g)(x) = \left(\dfrac{1}{x - 1}\right)^3 + 1$
    $= \dfrac{x^3 - 3x^2 + 3x}{x^3 - 3x^2 + 3x - 1},$
    $\{x \mid x \neq 1\};$
    $(g \circ f)(x) = \dfrac{1}{(x^3 + 1) - 1}$
    $= \dfrac{1}{x^3}, \{x \mid x \neq 0\}$

43. a.

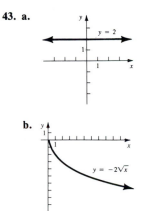

$y = 2$

b.

$y = -2\sqrt{x}$

45. $f(x_1) = f(x_2)$

$$\frac{5}{x_1 - 2} = \frac{5}{x_2 - 2}$$

$$x_2 - 2 = x_1 - 2$$

$$x_2 = x_1$$

$$f^{-1}(x) = \frac{5 + 2x}{x}$$

47.

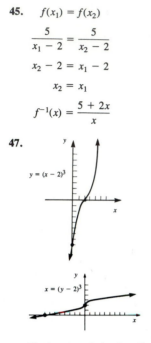

The inverse relation is a function $y = f(x)$.

49. $a = -\frac{1}{4}, 0, 8$

51. $f(x) = \frac{7}{4}x^2 + 7x + 7$

53. $f(x) = -\frac{4}{25}x^2 + 9$

Exercise 3.1 [Page 184]

1.

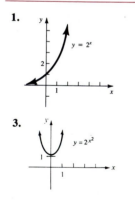

3.

$y = 2^{x^2}$

5.

$y = 2^{|x|}$

7.

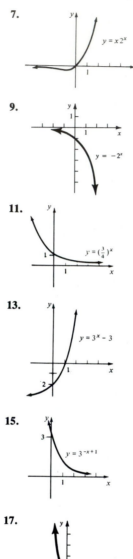

$y = x \, 2^x$

9.

$y = -2^x$

11.

$y = (\frac{3}{4})^x$

13.

$y = 3^x - 3$

15.

$y = 3^{-x+1}$

17.

$y = e^{-2x}$

19.

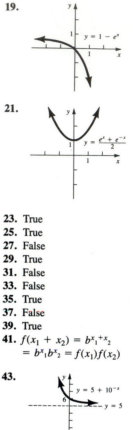

$y = 1 - e^x$

21.

$y = \frac{e^x + e^{-x}}{2}$

23. True
25. True
27. False
29. True
31. False
33. False
35. True
37. False
39. True
41. $f(x_1 + x_2) = b^{x_1 + x_2}$
$= b^{x_1} b^{x_2} = f(x_1) f(x_2)$

43.

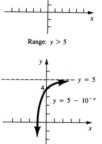

$y = 5 + 10^{-x}$

$y = 5$

Range: $y > 5$

$y = 5$

$y = 5 - 10^{-x}$

Range: $y < 5$

45. $(1, 2)$
47. a. $(100)2^{-t}$ **b.** 0.78 grams, 0.006 grams

49.

h	$f(h)$
1	2
0.1	2.59374
0.01	2.70481
0.001	2.71692
0.0001	2.71815
0.00001	2.71827
0.000001	2.71828
0.0000001	2.71828

51. e

53. **a.** 0.375 **b.** 16.375 **c.** 1.6484

Exercise 3.2 [Page 192]

1. $\log_4 \frac{1}{2} = -\frac{1}{2}$
3. $\log_9 1 = 0$
5. $\log_{10} x = y$
7. $\log_{1/64} 8 = -\frac{1}{2}$
9. $\log_{36} \frac{1}{216} = -\frac{3}{2}$
11. $\log_6 \frac{1}{3} = -0.6131$
13. $3^4 = 81$
15. $10^1 = 10$
17. $5^{-2} = \frac{1}{25}$
19. $16^{1/4} = 2$
21. $b^2 = b^2$
23. $b^2 = 9$
25. -1
27. 3
29. $-\frac{5}{6}$
31. -4
33. -3
35. -3
37. $b = 5$
39. $y = 3$
41. $N = \frac{1}{32}$
43. $N = 3$
45. $k = -3$
47. $r = \frac{4}{3}$
49. 0.3011
51. 1.8063
53. 0.8011
55. 0.2330
57. -0.0969
59. -1.2042
61. $\log_{10} 10 = 1$
63. $\log_{10}(x^2 - 2)$
65. $\log_2 1 = 0$

67.
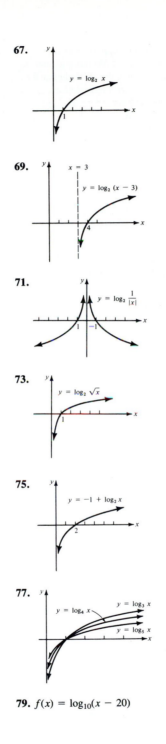
$y = \log_2 x$

69.
$x = 3$
$y = \log_2 (x - 3)$

71.
$y = \log_2 \frac{1}{|x|}$

73.
$y = \log_2 \sqrt{x}$

75.
$y = -1 + \log_2 x$

77.
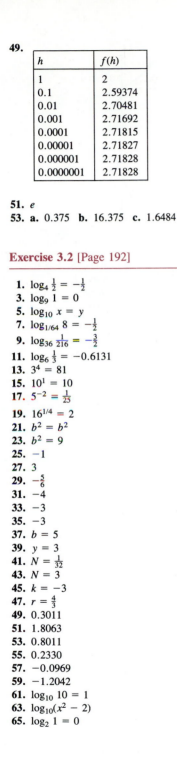
$y = \log_4 x$ $y = \log_3 x$ $y = \log_5 x$

79. $f(x) = \log_{10}(x - 20)$

81.
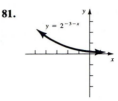
$y = 2^{-3-x}$

83. 7.3928, 7.3887

Exercise 3.3 [Page 197]

1. $\log_{10}(0.001) = -3$
3. $\log_{10} \frac{1}{5} = -0.6990$
5. $\ln 95 = 4.5539$
7. -6
9. 5
11. $\frac{3}{4}$
13. -2.7806
15. 1.3444
17. 1.2535
19. 158.4893
21. 1.9953
23. 78.2571
25. 2.3219
27. 2.7375
29. -1.0959

31.

h	$f(h)$
0.1	0.48790
0.01	0.49875
0.001	0.49988
0.0001	0.49999
0.00001	0.50000

33. **a.** Since $x = e^y$ we have
$$\log_{10} x = \log_{10} e^y$$
$$= y \log_{10} e$$
and so $y = \dfrac{\log_{10} x}{\log_{10} e}$.

Now use the fact that $y = \ln x$.

b. $\dfrac{1}{\log_{10} e} \approx 2.3026$

35. $u^v = w$

37. $\ln \dfrac{x^4}{\sqrt{x^2 - 25}}$

Exercise 3.4 [Page 202]

1. $x = 2$

3. $x = 2$

5. $x = \log_2\left(\dfrac{3 \pm \sqrt{5}}{2}\right)$ or

$x \approx \pm 1.3885$

7. $x = 3$

9. $x = -\frac{2}{3}$

11. $x = 3$

13. $x = -1$

15. $x = \pm 4$

17. $x = -\frac{5}{2}$

19. $x = -1 + \log_{10} 21 \approx 0.3222$

21. $x = 2$

23. $x = 1, x = 5$

25. $x = 5$

27. $x = 16$

29. $x = \pm \frac{1}{10}$

31. $x = \pm 8$

33. $x = 81$

35. $x = 60$

37. $x = 1$

39. $x = -\frac{1}{3}$

41. $x = 4$

43. $x = 2$

45. $t = \frac{11}{3}$

47. $x = e^{\pm 4}$ or $x \approx 54.5982$,

$x \approx 0.0183$

49. $x = 12, y = 9$

51. 8, −4, 12

53. $e^0 = 1$

55. $t = -\dfrac{1}{b} \ln\left[-\dfrac{1}{c}\left(-\dfrac{a}{b} + \ln P\right)\right]$

Exercise 3.5 [Page 210]

1. 6

3. 7.6

5. 6.2

7. 5×10^{-4}

9. 2.5×10^{-7}

11. 7.9×10^{-9}

13. 10 times as acidic

15. 90 dB

17. 60 dB

19. 74.8 dB

21. $b = 10[\log_{10} I - \log_{10} 10^{-16}]$

$= 10[\log_{10} I$

$\quad - (-16)\log_{10} 10]$

$= 160 + 10 \log_{10} I$

23. 4×10^{-5}

25. 28 dB

27. The 1906 earthquake was approximately 200 times greater in intensity than the 1979 earthquake. The 1979 San Fernando Valley earthquake was about 45 times greater in intensity.

29. $1,419

31. $10,272

33. $2,466

35. $2,308

37. Approximately 14 years

39. The solution to $2P_0 = P_0 e^{kt}$ is $t = (\ln 2)/k$.

41. Approximately 1660 years

43. 2.7% of I_0

45. $t = -\dfrac{1}{a} \ln \dfrac{(a - bP)P_0}{(a - bP_0)P}$

47. $t = -\dfrac{1}{b} \ln \dfrac{a}{2(a - bC_0)}$

49. 0.2877

51. $x = \ln(y \pm \sqrt{y^2 - 1})$

53. $\dfrac{1 + \sqrt{13}}{2}$

55. 4.6985

Chapter 3
Review Exercises [Page 214]

1. True

3. True

5. True

7. False

9. True

11. True

13. True

15. $\log_9 3$

17. 2

19. 5

21. 2

23. 1000

25. $f(-2) = -1, f(-\frac{3}{2}) = -2$, $f(-1) = -4, f(0) = -16$, $f(\frac{1}{2}) = -32, f(2) = -256$

27.

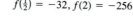
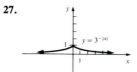

29.

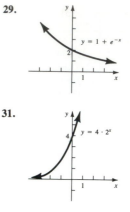

31.

33. a. All real numbers **b.** $x \neq 0$
c. $x \neq 0$ **d.** $x \geq -1$
e. All real numbers
f. All real numbers

35. $\log_3 \frac{32}{27}$

37.

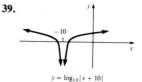

39.

41. $x = -3$

43. $x = 5$

45. $x = \frac{1}{2}\log_5 3 \approx 0.3413$

47. 1.5131

49. Approximately 55 hours

51. $\dfrac{3^{1-h} - 3}{h}$

53. $f(x) = 5e^{-0.2682x}$ (approximately)

55. $a = 3$

57. $a = 5$

Exercise 4.1 [Page 223]

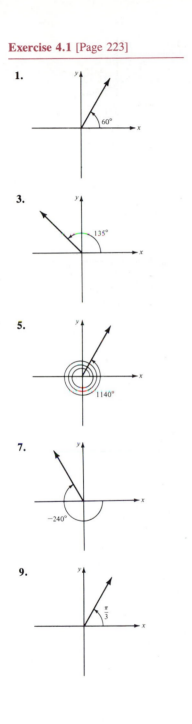

1.

3.

5.

7.

9.

11.

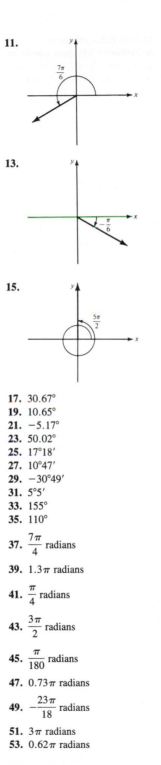

13.

15.

17. 30.67°
19. 10.65°
21. −5.17°
23. 50.02°
25. 17°18′
27. 10°47′
29. −30°49′
31. 5°5′
33. 155°
35. 110°

37. $\dfrac{7\pi}{4}$ radians

39. 1.3π radians

41. $\dfrac{\pi}{4}$ radians

43. $\dfrac{3\pi}{2}$ radians

45. $\dfrac{\pi}{180}$ radians

47. 0.73π radians

49. $-\dfrac{23\pi}{18}$ radians

51. 3π radians
53. 0.62π radians

55. 0.71π radians
57. 120°
59. 30°
61. 225°
63. 177.62°
65. 687.55°
67. −143.24°

Exercise 4.2 [Page 233]

1.

t	$\cos t$	$\sin t$
0	1	0
$\dfrac{\pi}{6}$	$\dfrac{\sqrt{3}}{2}$	$\dfrac{1}{2}$
$\dfrac{\pi}{4}$	$\dfrac{\sqrt{2}}{2}$	$\dfrac{\sqrt{2}}{2}$
$\dfrac{\pi}{3}$	$\dfrac{1}{2}$	$\dfrac{\sqrt{3}}{2}$
$\dfrac{\pi}{2}$	0	1
$\dfrac{2\pi}{3}$	$-\dfrac{1}{2}$	$\dfrac{\sqrt{3}}{2}$
$\dfrac{3\pi}{4}$	$-\dfrac{\sqrt{2}}{2}$	$\dfrac{\sqrt{2}}{2}$
$\dfrac{5\pi}{6}$	$-\dfrac{\sqrt{3}}{2}$	$\dfrac{1}{2}$
π	−1	0
$\dfrac{7\pi}{6}$	$-\dfrac{\sqrt{3}}{2}$	$-\dfrac{1}{2}$
$\dfrac{5\pi}{4}$	$-\dfrac{\sqrt{2}}{2}$	$-\dfrac{\sqrt{2}}{2}$
$\dfrac{4\pi}{3}$	$-\dfrac{1}{2}$	$-\dfrac{\sqrt{3}}{2}$
$\dfrac{3\pi}{2}$	0	−1
$\dfrac{5\pi}{3}$	$\dfrac{1}{2}$	$-\dfrac{\sqrt{3}}{2}$
$\dfrac{7\pi}{4}$	$\dfrac{\sqrt{2}}{2}$	$-\dfrac{\sqrt{2}}{2}$
$\dfrac{11\pi}{6}$	$\dfrac{\sqrt{3}}{2}$	$-\dfrac{1}{2}$
2π	1	0

3. $\dfrac{\sqrt{21}}{5}$

5. $-\dfrac{\sqrt{5}}{3}$

7. $\pm\dfrac{3\sqrt{5}}{7}$

9. ± 0.98

11. $\sin t = \dfrac{2}{\sqrt{13}}$, $\cos t = \dfrac{3}{\sqrt{13}}$

or $\sin t = -\dfrac{2}{\sqrt{13}}$, $\cos t = -\dfrac{3}{\sqrt{13}}$

13. $\sin t = -\dfrac{3\sqrt{10}}{10}$,

$\cos t = -\dfrac{\sqrt{10}}{10}$

15. -1

17. $\frac{1}{2}$

19. $\dfrac{\sqrt{2}}{2}$

21. 0

23. $-\frac{1}{2}$

25. $\dfrac{\sqrt{3}}{2}$

27. $\dfrac{2\pi}{3} + 2n\pi$ and

$\dfrac{4\pi}{3} + 2n\pi$ for n an integer

29. $\cos n\pi$

$= \begin{cases} 1, & n \text{ even} \\ -1, & n \text{ odd} \end{cases}$

$\sin n\pi = 0$

35. The lengths of the hypotenuses are equal. In each triangle the angles at the origin are $\pi/6$ radians. Since they are both right triangles, the other two acute angles are equal. Hence the triangles are congruent.

37. $\dfrac{3\pi}{4}$ radians

39. $405° = \dfrac{9\pi}{4}$ radians,

$\sin \dfrac{9\pi}{4} = \cos \dfrac{9\pi}{4} = \dfrac{\sqrt{2}}{2}$

Exercise 4.3 [Page 241]

1.

t	$\tan t$	$\cot t$	$\sec t$	$\csc t$
0	0	—	1	—
$\dfrac{\pi}{6}$	$\dfrac{\sqrt{3}}{3}$	$\sqrt{3}$	$\dfrac{2\sqrt{3}}{3}$	2
$\dfrac{\pi}{4}$	1	1	$\sqrt{2}$	$\sqrt{2}$
$\dfrac{\pi}{3}$	$\sqrt{3}$	$\dfrac{\sqrt{3}}{3}$	2	$\dfrac{2\sqrt{3}}{3}$
$\dfrac{\pi}{2}$	—	0	—	1
$\dfrac{2\pi}{3}$	$-\sqrt{3}$	$-\dfrac{\sqrt{3}}{3}$	-2	$\dfrac{2\sqrt{3}}{3}$
$\dfrac{3\pi}{4}$	-1	-1	$-\sqrt{2}$	$\sqrt{2}$
$\dfrac{5\pi}{6}$	$-\dfrac{\sqrt{3}}{3}$	$-\sqrt{3}$	$-\dfrac{2\sqrt{3}}{3}$	2
π	0	—	-1	—
$\dfrac{7\pi}{6}$	$\dfrac{\sqrt{3}}{3}$	$\sqrt{3}$	$-\dfrac{2\sqrt{3}}{3}$	-2
$\dfrac{5\pi}{4}$	1	1	$-\sqrt{2}$	$-\sqrt{2}$
$\dfrac{4\pi}{3}$	$\sqrt{3}$	$\dfrac{\sqrt{3}}{3}$	-2	$-\dfrac{2\sqrt{3}}{3}$
$\dfrac{3\pi}{2}$	—	0	—	-1
$\dfrac{5\pi}{3}$	$-\sqrt{3}$	$-\dfrac{\sqrt{3}}{3}$	2	$-\dfrac{2\sqrt{3}}{3}$
$\dfrac{7\pi}{4}$	-1	-1	$\sqrt{2}$	$-\sqrt{2}$
$\dfrac{11\pi}{6}$	$-\dfrac{\sqrt{3}}{3}$	$-\sqrt{3}$	$\dfrac{2\sqrt{3}}{3}$	-2
2π	0	—	1	—

3. $\sqrt{3}$

5. Undefined

7. -2

9. -1

11. -2

13. Undefined

15. -1

17. $\frac{1}{2}$

19. -2

21. $\sqrt{2}$

23. 0.64278761

25. -1.23606798

27. 2.30476487

29. 0.58103516

31. 1255.765990

33. -0.37460659

35. 1.21309700

37. 1.02494482

39. $-\sqrt{5}$

41. $\pm\frac{5}{4}$

43. $\sin t = -\dfrac{3\sqrt{11}}{10}$

$\tan t = -3\sqrt{11}$

$\cot t = -\dfrac{\sqrt{11}}{33}$

$\sec t = 10$

$\csc t = -\dfrac{10\sqrt{11}}{33}$

45. $\sin t = -\frac{1}{2}$

$\cos t = \dfrac{\sqrt{3}}{2}$

$\tan t = -\dfrac{1}{\sqrt{3}}$

$\cot t = -\sqrt{3}$

$\sec t = \dfrac{2}{\sqrt{3}}$

$\csc t = -2$

or

$\sin t = -\frac{1}{2}$

$\cos t = -\dfrac{\sqrt{3}}{2}$

$\tan t = \dfrac{1}{\sqrt{3}}$

$\cot t = \sqrt{3}$

$\sec t = -\dfrac{2}{\sqrt{3}}$

$\csc t = -2$

47. $t = \dfrac{\pi}{3} + n\pi$, $n = 0, \pm1, \pm2, \dots$

49. $\tan t$ is undefined, $\cot t = 0$, $\sec t$ is undefined,

$\csc t = \begin{cases} 1, & n \text{ even} \\ -1, & n \text{ odd} \end{cases}$

51. From 0 to ∞

53. From 1 to ∞

57. $4y^2 - x^2 = 4$

59. $y^2 - 25x^2 = 25$

61. a. 0.841470984

0.954929658

0.900316316

b. $n = 1: 0.998334167$

$n = 2: 0.999983334$

$n = 3: 0.999999833$

$n = 4: 0.999999998$

$n = 5: 1.000000000$

63. a. $585°$

b. $143.81°$

65. $-\frac{4}{5}$

Exercise 4.4 [Page 250]

1. $\dfrac{\sqrt{2}}{4}(1 + \sqrt{3})$

3. $\frac{1}{2}\sqrt{2 + \sqrt{2}}$

5. $\dfrac{\sqrt{2}}{4}(1 + \sqrt{3})$

7. $2 + \sqrt{3}$

9. $\frac{1}{2}\sqrt{2 - \sqrt{2}}$

11. $\dfrac{\sqrt{2}}{4}(\sqrt{3} - 1)$

13. $-\frac{1}{2}\sqrt{2 + \sqrt{2}}$

15. $-\frac{1}{2}\sqrt{2 - \sqrt{2}}$

17. $-\dfrac{\sqrt{2}}{4}(1 + \sqrt{3})$

19. $-2 + \sqrt{3}$

21. $\dfrac{\sqrt{2}}{4}(1 - \sqrt{3})$

23. $\dfrac{\sqrt{2}}{4}(\sqrt{3} - 1)$

25. $\sin \dfrac{t}{2} = \sqrt{\dfrac{3 + \sqrt{7}}{6}}$

$\cos \dfrac{t}{2} = \sqrt{\dfrac{3 - \sqrt{7}}{6}}$

$\sin 2t = -\dfrac{2\sqrt{14}}{9}$

$\cos 2t = \frac{5}{9}$

27. $\sin \dfrac{t}{2} = \sqrt{\dfrac{3 + 2\sqrt{2}}{6}}$

$\cos \dfrac{t}{2} = -\sqrt{\dfrac{3 - 2\sqrt{2}}{6}}$

$\sin 2t = \dfrac{4\sqrt{2}}{9}$

$\cos 2t = \frac{7}{9}$

53. $\dfrac{\sqrt{2}}{2 + \sqrt{2}} = \sqrt{2} - 1$

55. $\dfrac{\sqrt{2} - 2}{\sqrt{2}} = 1 - \sqrt{2}$

57. $-2 - \sqrt{3}$

59. $\dfrac{2}{\sqrt{2} - \sqrt{2}}$

61. a. $-\frac{1}{2}$

b. $-\dfrac{\sqrt{3}}{3}$

c. 1

63. $4x^2 + 9y^2 = 36$

Exercise 4.5 [Page 255]

1. $\dfrac{\sin t}{\csc t} = \dfrac{\sin t}{\dfrac{1}{\sin t}} = \sin^2 t$

$1 - \dfrac{\cos t}{\sec t} = 1 - \dfrac{\cos t}{\dfrac{1}{\cos t}}$

$= 1 - \cos^2 t = \sin^2 t$

3. $1 - \cos^4 t$
$= (1 + \cos^2 t)(1 - \cos^2 t)$
$= (1 + 1 - \sin^2 t)\sin^2 t$
$= (2 - \sin^2 t)\sin^2 t$

5. $1 - 2\sin^2 t$
$= 1 - \sin^2 t - \sin^2 t$
$= \cos^2 t - \sin^2 t$
$= \cos^2 t - (1 - \cos^2 t)$
$= 2\cos^2 t - 1$

7. $\dfrac{\sec t - \csc t}{\sec t + \csc t} = \dfrac{\dfrac{1}{\cos t} - \dfrac{1}{\sin t}}{\dfrac{1}{\cos t} + \dfrac{1}{\sin t}}$

$= \dfrac{\sin t\left(\dfrac{1}{\cos t} - \dfrac{1}{\sin t}\right)}{\sin t\left(\dfrac{1}{\cos t} + \dfrac{1}{\sin t}\right)}$

$= \dfrac{\tan t - 1}{\tan t + 1}$

9. $\dfrac{\sec^4 t - \tan^4 t}{1 + 2\tan^2 t}$

$= \dfrac{(\sec^2 t - \tan^2 t)(\sec^2 t + \tan^2 t)}{1 + \tan^2 t + \tan^2 t}$

$= \dfrac{1(\sec^2 t + \tan^2 t)}{\sec^2 t + \tan^2 t}$

$= 1$

11. $\sin^2 t \cot^2 t + \cos^2 t \tan^2 t$

$= \sin^2 t \dfrac{\cos^2 t}{\sin^2 t} + \cos^2 t \dfrac{\sin^2 t}{\cos^2 t}$

$= \cos^2 t + \sin^2 t$

$= 1$

13. $\sec t - \dfrac{\cos t}{1 + \sin t}$

$= \dfrac{1}{\cos t} - \dfrac{\cos t}{1 + \sin t}$

$= \dfrac{1 + \sin t - \cos^2 t}{\cos t(1 + \sin t)}$

$= \dfrac{1 + \sin t - (1 - \sin^2 t)}{\cos t(1 + \sin t)}$

$= \dfrac{\sin t + \sin^2 t}{\cos t(1 + \sin t)}$

$= \dfrac{\sin t(1 + \sin t)}{\cos t(1 + \sin t)}$

$= \tan t$

15. $\dfrac{\tan^2 t}{1 + \cos t} = \dfrac{\dfrac{\sin^2 t}{\cos^2 t}}{1 + \cos t}$

$= \dfrac{\dfrac{1 - \cos^2 t}{\cos^2 t}}{1 + \cos t}$

$= \dfrac{(1 - \cos t)(1 + \cos t)}{\cos^2 t(1 + \cos t)}$

$= \dfrac{1 - \cos t}{\cos^2 t}$

$\dfrac{\sec t - 1}{\cos t} = \dfrac{\dfrac{1}{\cos t} - 1}{\cos t}$

$= \dfrac{\dfrac{1 - \cos t}{\cos t}}{\cos t} = \dfrac{1 - \cos t}{\cos^2 t}$

17. $(\csc t - \cot t)^2$

$= \left(\dfrac{1}{\sin t} - \dfrac{\cos t}{\sin t}\right)^2$

$= \left(\dfrac{1 - \cos t}{\sin t}\right)^2$

$= \dfrac{(1 - \cos t)^2}{\sin^2 t} = \dfrac{(1 - \cos t)^2}{1 - \cos^2 t}$

$= \dfrac{(1 - \cos t)(1 - \cos t)}{(1 - \cos t)(1 + \cos t)}$

$= \dfrac{1 - \cos t}{1 + \cos t}$

19. $\dfrac{\tan^2 t}{\sec t - 1} = \dfrac{\sec^2 t - 1}{\sec t - 1}$

$= \dfrac{(\sec t - 1)(\sec t + 1)}{\sec t - 1}$

$= \sec t + 1 = 1 + \dfrac{1}{\cos t}$

21. $\dfrac{\cot t - \tan t}{\cot t + \tan t} = \dfrac{\dfrac{\cos t}{\sin t} - \dfrac{\sin t}{\cos t}}{\dfrac{\cos t}{\sin t} + \dfrac{\sin t}{\cos t}}$

$= \dfrac{\cos^2 t - \sin^2 t}{\cos^2 t + \sin^2 t}$

$= 1 - \sin^2 t - \sin^2 t$

$= 1 - 2\sin^2 t$

23. $\dfrac{1 - \tan^2 t}{1 + \tan^2 t} = \dfrac{1 - \tan^2 t}{\sec^2 t}$

$= \dfrac{1 - \dfrac{\sin^2 t}{\cos^2 t}}{\dfrac{1}{\cos^2 t}} = \cos^2 t - \sin^2 t$

$= \cos^2 t - (1 - \cos^2 t)$

$= 2 \cos^2 t - 1$

25. $\dfrac{\cos t}{\sin t} + \dfrac{\sin t}{\cos t}$

$= \dfrac{\cos^2 t + \sin^2 t}{\sin t \cos t}$

$= \dfrac{1}{\sin t \dfrac{1}{\sec t}} = \dfrac{\sec t}{\sin t}$

27. $\dfrac{1 - \cos 2t}{\sin 2t}$

$= \dfrac{1 - (\cos^2 t - \sin^2 t)}{2 \sin t \cos t}$

$= \dfrac{1 - \cos^2 t + \sin^2 t}{2 \sin t \cos t}$

$= \dfrac{\sin^2 t + \sin^2 t}{2 \sin t \cos t}$

$= \dfrac{2 \sin^2 t}{2 \sin t \cos t}$

$= \dfrac{\sin t}{\cos t}$

$= \tan t$

29. $\dfrac{1 - \tan^2 t}{\cos 2t} = \dfrac{1 - \dfrac{\sin^2 t}{\cos^2 t}}{\cos^2 t - \sin^2 t}$

$= \dfrac{\dfrac{\cos^2 t - \sin^2 t}{\cos^2 t}}{\cos^2 t - \sin^2 t}$

$= \dfrac{1}{\cos^2 t}$

$\dfrac{2 \tan t}{\sin 2t} = \dfrac{2 \dfrac{\sin t}{\cos t}}{2 \sin t \cos t}$

$= \dfrac{\sin t}{\sin t \cos^2 t}$

$= \dfrac{1}{\cos^2 t}$

31. $\dfrac{\sin^2 2t}{(1 + \cos 2t)^2}$

$= \dfrac{(2 \sin t \cos t)^2}{(1 + 2 \cos^2 t - 1)^2}$

$= \dfrac{(2 \sin t \cos t)^2}{(2 \cos^2 t)^2}$

$= \dfrac{4 \sin^2 t \cos^2 t}{4 \cos^4 t}$

$= \dfrac{\sin^2 t}{\cos^2 t}$

$= \tan^2 t$

$= \sec^2 t - 1$

33. $\sin (u + v) \sin (u - v)$
$= (\sin u \cos v + \cos u \sin v)$
$\quad \cdot (\sin u \cos v - \cos u \sin v)$
$= \sin^2 u \cos^2 v$
$\quad - \sin u \cos v \cos u \sin v$
$\quad + \cos u \sin v \sin u \cos v$
$\quad - \cos^2 u \sin^2 v$
$= \sin^2 u \cos^2 v - \cos^2 u \sin^2 v$
$= \sin^2 u(1 - \sin^2 v)$
$\quad - (1 - \sin^2 u) \sin^2 v$
$= \sin^2 u - \sin^2 u \sin^2 v - \sin^2 v$
$\quad + \sin^2 u \sin^2 v$
$= \sin^2 u - \sin^2 v$

35. $y = A \sin (\omega t - kx)$
$\quad - A \sin (\omega t + kx)$
$= A [\sin \omega t \cos kx$
$\quad - \sin kx \cos \omega t]$
$\quad - A [\sin \omega t \cos kx$
$\quad + \sin kx \cos \omega t]$
$= A \sin \omega t \cos kx$
$\quad - A \sin kx \cos \omega t$
$\quad - A \sin \omega t \cos kx$
$\quad - A \sin kx \cos \omega t$
$= -2A \sin kx \cos \omega t$
$= -2A \cos \omega t \sin kx$

37. $\sin t = 1$
$\cos t = 0$
$\tan t$ is undefined

39. $\dfrac{\sqrt{2} - \sqrt{6}}{4}$

41. $\sin 4t = 2 \sin 2t \cos 2t$
$= 2(2 \sin t \cos t)$
$\quad \cdot (\cos^2 t - \sin^2 t)$
$= 4 \sin t \cos t(2 \cos^2 t - 1)$
$= 8 \sin t \cos^3 t$
$\quad - 4 \sin t \cos t$

Exercise 4.6 [Page 259]

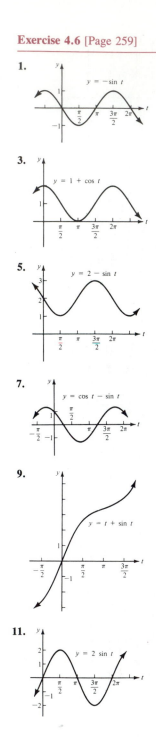

1. $y = -\sin t$

3. $y = 1 + \cos t$

5. $y = 2 - \sin t$

7. $y = \cos t - \sin t$

9. $y = t + \sin t$

11. $y = 2 \sin t$

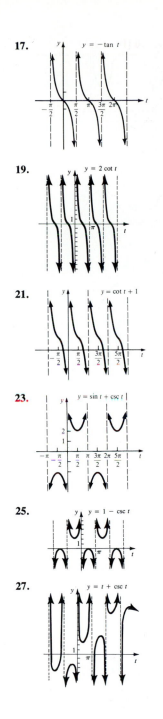

17.

19.

21.

23.

25.

27.

29. a. π **b.** 6 **c.** $\dfrac{\pi}{2}$ **d.** 3π

31. $\tan\left(t + \dfrac{\pi}{4}\right) = \dfrac{\tan t + \tan\dfrac{\pi}{4}}{1 - \tan t \tan\dfrac{\pi}{4}}$

$= \dfrac{\tan t + 1}{1 - \tan t}$

33. $\dfrac{1 + \csc t}{\sec t} = \dfrac{1 + \dfrac{1}{\sin t}}{\dfrac{1}{\cos t}}$

$= \cos t\left(1 + \dfrac{1}{\sin t}\right)$

$= \cos t + \cot t$

Exercise 4.7 [Page 266]

1.

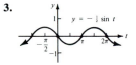

Amplitude: 4
Period: 2π
y intercept: 4
t intercepts: $\dfrac{\pi}{2}, \dfrac{3\pi}{2}, \dfrac{5\pi}{2}$

3.

Amplitude: $\dfrac{1}{2}$
Period: 2π
y intercept: 0
t intercepts: $\pi, 2\pi, 3\pi$

5.

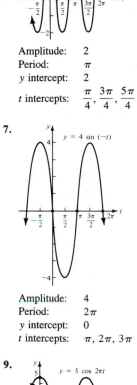

Amplitude: 2
Period: π
y intercept: 2
t intercepts: $\dfrac{\pi}{4}, \dfrac{3\pi}{4}, \dfrac{5\pi}{4}$

7.

Amplitude: 4
Period: 2π
y intercept: 0
t intercepts: $\pi, 2\pi, 3\pi$

9.

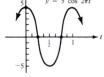

Amplitude: 5
Period: 1
y intercept: 5
t intercepts: $\dfrac{1}{4}, \dfrac{3}{4}, \dfrac{5}{4}$

11.

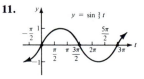

Amplitude: 1
Period: 3π
y intercept: 0
t intercepts: $\dfrac{3\pi}{2}, 3\pi, \dfrac{9\pi}{2}$

13.

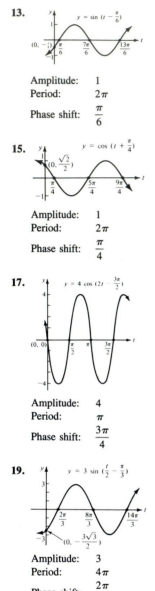

Amplitude:	1
Period:	2π
Phase shift:	$\dfrac{\pi}{6}$

15.

Amplitude: 1
Period: 2π
Phase shift: $\dfrac{\pi}{4}$

17.

Amplitude: 4
Period: π
Phase shift: $\dfrac{3\pi}{4}$

19.

Amplitude: 3
Period: 4π
Phase shift: $\dfrac{2\pi}{3}$

21.

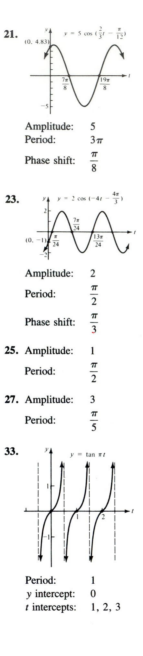

Amplitude: 5
Period: 3π
Phase shift: $\dfrac{\pi}{8}$

23.

Amplitude: 2
Period: $\dfrac{\pi}{2}$
Phase shift: $\dfrac{\pi}{3}$

25. Amplitude: 1
     Period: $\dfrac{\pi}{2}$

27. Amplitude: 3
     Period: $\dfrac{\pi}{5}$

33.

Period: 1
y intercept: 0
t intercepts: 1, 2, 3

35.

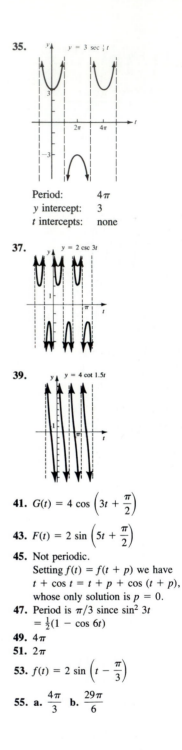

Period: 4π
y intercept: 3
t intercepts: none

37.

39.

41. $G(t) = 4 \cos\left(3t + \dfrac{\pi}{2}\right)$

43. $F(t) = 2 \sin\left(5t + \dfrac{\pi}{2}\right)$

45. Not periodic.
Setting $f(t) = f(t + p)$ we have
$t + \cos t = t + p + \cos (t + p)$,
whose only solution is $p = 0$.

47. Period is $\pi/3$ since $\sin^2 3t$
$= \frac{1}{2}(1 - \cos 6t)$

49. 4π

51. 2π

53. $f(t) = 2 \sin\left(t - \dfrac{\pi}{3}\right)$

55. a. $\dfrac{4\pi}{3}$ **b.** $\dfrac{29\pi}{6}$

57. $\dfrac{1 + \cos^2 t \csc^2 t}{1 + \cot^2 t} = \dfrac{1 + \dfrac{\cos^2 t}{\sin^2 t}}{1 + \cot^2 t}$

$= \dfrac{1 + \cot^2 t}{1 + \cot^2 t} = 1$

59.

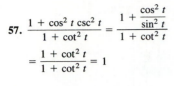

$y = 1 - \sin(t - \frac{\pi}{4})$

Exercise 4.8 [Page 272]

1. $x = n\pi,\ n = 0, \pm 1, \pm 2, \ldots$

3. $x = \dfrac{3\pi}{4} + 2n\pi,$ or

$x = \dfrac{5\pi}{4} + 2n\pi,$

where $n = 0, \pm 1, \pm 2, \ldots$

5. $x = 2n\pi,$ or $x = (2n + 1)\pi,$
where $n = 0, \pm 1, \pm 2, \ldots$

7. $x = \dfrac{2\pi}{3} + 2n\pi,$ or

$x = \dfrac{4\pi}{3} + 2n\pi,$

where $n = 0, \pm 1, \pm 2, \ldots$

9. No solutions

11. $x = n\pi,\ x = \dfrac{2\pi}{3} + 2n\pi,$ or

$x = \dfrac{4\pi}{3} + 2n\pi,$ where $n = 0,$

$\pm 1, \pm 2, \ldots$

13. $x = \dfrac{\pi}{6} + 2n\pi,\ x = \dfrac{5\pi}{6} + 2n\pi,$

or $x = \dfrac{3\pi}{2} + 2n\pi,$ where $n = 0,$

$\pm 1, \pm 2, \ldots$

15. $x = n\pi,$ where $n = 0, \pm 1, \pm 2, \ldots$

17. $x = \dfrac{\pi}{6} + 2n\pi,$ or

$x = \dfrac{5\pi}{6} + 2n\pi,$ where $n = 0,$

$\pm 1, \pm 2, \ldots$

19. $x = (2n + 1)\dfrac{\pi}{2},$ or $x = 2n\pi,$

where $n = 0, \pm 1, \pm 2, \ldots$

21. $x = 0, \dfrac{3\pi}{2}$

23. $x = 0, \dfrac{\pi}{3}, \pi$

25. No solutions

27. $x = \dfrac{\pi}{4}, \dfrac{3\pi}{4}, \dfrac{5\pi}{4}, \dfrac{7\pi}{4}$

29. $x = \dfrac{\pi}{3}, \dfrac{2\pi}{3}, \dfrac{4\pi}{3}, \dfrac{5\pi}{3}$

31. Since $\sin x \le 1$ and $\cos x \le 1$ for all x, $\sin x + \cos x \le 2$ for all x.

33. Since $\sec^2 x \ge 1$ and $\csc^2 x \ge 1$ for all x, $\sec^2 x + \csc^2 x \ge 2$ for all x.

35.

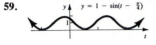

$y = \tan x$, $y = x$

The equation has infinitely many solutions.

37. a. $\dfrac{9\pi}{8}$ **b.** $\dfrac{15\pi}{4}$

c. $\dfrac{3\pi}{8}$ **d.** $\dfrac{451\pi}{80}$

39. Since $\sin(\pi/2 - t) = \cos t$,
$\sin^2 t + \sin^2(\pi/2 - t)$
$= \sin^2 t + \cos^2 t = 1$

41. 4π

Exercise 4.9 [Page 280]

1. 0

3. π

5. $\dfrac{\pi}{3}$

7. $\dfrac{\pi}{4}$

9. $\dfrac{\pi}{4}$

11. $\dfrac{2\pi}{3}$

13. $-\dfrac{\pi}{4}$

15. $\dfrac{\pi}{6}$

17. $-\dfrac{\sqrt{5}}{2}$

19. $\dfrac{\sqrt{3}}{2}$

21. $\dfrac{5}{3}$

23. $\dfrac{2\sqrt{21}}{21}$

25. $\dfrac{1}{3}$

27. $\dfrac{\pi}{5}$

29. $-\dfrac{7\pi}{36}$

31. $\dfrac{\pi}{4}$

33. 0.7800

35. 1.4000

37. -1.2947

39. 0.6816

41. 0.5899

43. $\dfrac{x}{\sqrt{1 + x^2}}$

45. $\dfrac{x}{\sqrt{1 - x^2}}$

47. $\dfrac{\sqrt{1 - x^2}}{x}$

49. $\dfrac{1}{x}$

51. Domain: $(-\infty, -1] \cup [1, \infty)$

Range: $\left[0, \dfrac{\pi}{2}\right) \cup \left(\dfrac{\pi}{2}, \pi\right]$

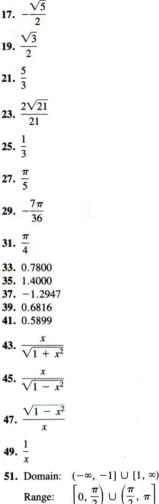

$y = \text{arcsec } x$

55.

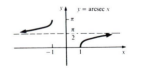

$\cot^{-1} x = \begin{cases} \pi + \tan^{-1}\left(\dfrac{1}{x}\right) & \text{if } x < 0 \\[2mm] \dfrac{\pi}{2} & \text{if } x = 0 \\[2mm] \tan^{-1}\left(\dfrac{1}{x}\right) & \text{if } x > 0 \end{cases}$

57. 0.9273

59. 2.5536

61. 2.5559

63. 1.0385

65. 1.3258, 4.4674

67. $\dfrac{\pi}{4}, \dfrac{5\pi}{4}$, 2.6779, 5.8195

69. 2.0344, 5.1760

71. $\dfrac{\pi}{6}, \dfrac{5\pi}{6}$, 2.3005, 3.9827

73.

75.

77. a. Odd **b.** Even **c.** Odd
 d. Even **e.** Odd **f.** Even

79. $\tan 2t = \tan (t + t) = \dfrac{2 \tan t}{1 - \tan^2 t}$

$\tan 3t = \tan (2t + t)$

$= \dfrac{\dfrac{2 \tan t}{1 - \tan^2 t} + \tan t}{1 - \dfrac{2 \tan t}{1 - \tan^2 t} \tan t}$

$= \dfrac{3 \tan t - \tan^3 t}{1 - 3 \tan^2 t}$

81. Amplitude: 4
 Period: 6π

Chapter 4
Review Exercises [Page 283]

1. True

3. True

5. True

7. False

9. False

11. $-\dfrac{4\pi}{3}$

13. 210°

15. -1

17. $-\dfrac{1}{2}$

19. $\sin t = \dfrac{4\sqrt{17}}{17}$

$\cos t = -\dfrac{\sqrt{17}}{17}$

$\cot t = -\dfrac{1}{4}$

$\sec t = -\sqrt{17}$

$\csc t = \dfrac{\sqrt{17}}{4}$

21. $\dfrac{\sqrt{2 + \sqrt{2}}}{2}$

23. $\cos \left(t + \dfrac{3\pi}{2}\right)$

$= \cos t \cos \dfrac{3\pi}{2} - \sin t \sin \dfrac{3\pi}{2}$

$= -(\sin t)(-1) = \sin t$

25. $\tan t + \cot t = \dfrac{\sin t}{\cos t} + \dfrac{\cos t}{\sin t}$

$= \dfrac{\sin^2 t + \cos^2 t}{\cos t \sin t}$

$= \dfrac{1}{\cos t \sin t}$

$= \dfrac{1}{\cos t} \dfrac{1}{\sin t}$

$= \sec t \csc t$

27. Consider $u = v = \dfrac{\pi}{2}$

29.

y intercept: $\frac{1}{4}$

t intercepts: $\frac{1}{2}, \frac{3}{2}, \frac{5}{2}$

31. $x = n\pi, n = \pm 1, \pm 3, \pm 5, \ldots$

---

33. 0.4115, 2.7301

35. $\dfrac{\sqrt{7}}{3}$

37. -0.4257

39. 1.8993

Exercise 5.1 [Page 289]

1. $b = 2.04, r = 4.49$

3. $a = 11.71, r = 14.18$

5. $b = 3.30, r = 6.85$

7. $\alpha = 60°, \beta = 30°, a = 2.60$

9. $\alpha = 21.8°, \beta = 68.2°, r = 10.77$

11. $\alpha = 48.59°, \beta = 41.41°, b = 7.94$

13. $a = 8.49, r = 21.73$

15. $a = 4.63, b = 11.07$

17. $\alpha = 36.2°, a = 3.15, r = 5.33$

19. $\beta = 66°10', a = 2.06, b = 4.67$

21. $\beta = 74°55', a = 4.58, r = 17.61$

23. $\beta = 42°35', b = 14.43, r = 21.32$

25. $\alpha = 36.87°, \beta = 53.13°, b = 40$

27. $\alpha = 86.5°, b = 0.29, r = 4.81$

Exercise 5.2 [Page 294]

1. 52.1 meters

3. 66.35 feet

5. 409.7 feet

7. Height: 15.54 feet;
 Distance: 12.59 feet

9. $\theta = 8.71°$

11. 34,157 feet $\approx$ 6.47 miles

13. 13.47 feet

15. 66.04°

17. 44.43°

19. 170.5 miles

21. 15.96

23. 22.62°, 67.38°

Exercise 5.3 [Page 303]

1. $\gamma = 80°, a = 20.16, c = 20.16$

3. $\alpha = 92°, b = 3.01, c = 3.89$

5. $\alpha = 79.6°, \gamma = 28.4°, a = 12.41$

7. No solution

9. $\alpha = 24.46°, \beta = 140.54°,$
 $b = 12.28$
 $\alpha = 155.54°, \beta = 9.46°, b = 3.18$

11. No solution

13. $\alpha = 64°40'$, $a = 35.27$, $c = 38.17$

15. No solution

17. $\alpha = 45.58°$, $\gamma = 104.42°$,
$c = 13.56$
$\alpha = 134.42°$, $\gamma = 15.58°$, $c = 3.76$

19. $\alpha = 70.5°$, $\beta = 29.5°$, $a = 7.66$

21. $\alpha = 26.31°$, $\gamma = 53.69°$, $c = 16.37$

23. No solution

25. $\alpha = 34.7°$, $\beta = 50.3°$, $b = 27.03$

27. 15.80 feet

29. 9.1 meters

31. 8.81°

33. 11.82°

35. 290.3 meters

39. If the lengths of the sides are a, b, and c, then a triangle will be determined if and only if $a + b > c$, $a + c > b$, and $b + c > a$.

41. a. $b = \frac{5}{2}\sqrt{2}$ **b.** $b < \frac{5}{2}\sqrt{2}$
c. $\frac{5}{2}\sqrt{2} < b < 5$
d. $b = \frac{5}{2}\sqrt{2}$ or $b \geq 5$

43. a. $b \leq c$ **b.** $b > c$

45. 16.79

47. $A = ab \sin \alpha$

Exercise 5.4 [Page 309]

1. $\alpha = 37.59°$, $\beta = 77.41°$, $c = 7.43$

3. $\alpha = 52.62°$, $\beta = 83.33°$,
$\gamma = 44.05°$

5. $\alpha = 36.87°$, $\beta = 53.13°$, $\gamma = 90°$

7. $\alpha = 76.45°$, $\beta = 57.10°$,
$\gamma = 46.45°$

9. $\alpha = 27.66°$, $\beta = 40.54°$,
$\gamma = 111.8°$

11. $\alpha = 25°$, $\beta = 57°40'$, $c = 7.04$

13. $\alpha = 26.38°$, $\beta = 36.33°$,
$\gamma = 117.29°$

15. $\beta = 10.24°$, $\gamma = 147.76°$, $a = 6.32$

17. Front: 18.88°; back: 35.12°

19. 39.42 feet

21. 66.68°

25. 30

27. 6.19

29. 24.15°, 30.75°, 125.10°

31. 19.88

Chapter 5
Review Exercises [Page 312]

1. True

3. False

5. False

7. True

9. False

11. $\alpha = 36.87°$, $\beta = 53.13°$, $r = 50$

13. $a = 3.37$, $r = 6.03$

15. $a = 6.65$, $b = 7.47$

17. $\gamma = 80°$, $a = 5.32$, $c = 10.48$

19. $\alpha = 81.09°$, $\gamma = 53.91°$, $a = 9.78$
$\alpha = 8.91°$, $\gamma = 126.09°$, $a = 1.53$

21. $\alpha = 123.65°$, $\beta = 31.35°$, $c = 4.06$

23. 93.82°, 29.93°, 56.25°

25. 40.6°

27. Horizontal distance: 21,447 feet.
Straight-line distance: 29,326 feet.

29. Angle at (0, 0): 36.86°
Angle at (5, 0): 71.57°
Angle at (4, 3): 71.57°

31. 6.96 centimeters and 14.20 centimeters

33. Area $= \frac{1}{2}ac \sin \beta$

35. $A = \dfrac{r^2}{2}(\theta - \sin \theta)$

Exercise 6.1 [Page 322]

1. (2, 4)

3. (−6, −3)

5. (0, −3)

7. (6, 2)

9. (−4, −11)

11. (4, −3)

13. $x'^2 + y'^2 = 9$

15. $x'^2 + y'^2 = 5$

17. $x'^2 + y'^2 = 8$

19. $3x' + 2y' = 0$

21. $(x + 3)^2 + (y + 1)^2 = 25$

23. $(x - 4)^2 + (y + 3)^2 = 1$

25. $2x - 3y - 10 = 0$

27. $3x + 5y = 0$

29. $x = x' + 2$, $y = y' + 6$
$3(x' + 2) - (y' + 6) + 2 = 0$
$3x' + 6 - y' - 6 + 2 = 0$
$3x' - y' + 2 = 0$

31. $\dfrac{\sqrt{13}}{13}$

33. $\dfrac{3\sqrt{193}}{193}$

35. $\dfrac{2\sqrt{10}}{5}$

37. 0

39. a

41. $(x + 1)^2 + (y - 3)^2 = \frac{16}{5}$

43. $(x - h)^2 + (y - k)^2$
$= \dfrac{(ah + bk + c)^2}{a^2 + b^2}$

Exercise 6.2 [Page 334]

1. Vertex: (0, 0)
Focus: (1, 0)
Directrix: $x = -1$
Axis: $y = 0$

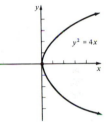

3. Vertex: (0, 0)
Focus: $(-\frac{1}{3}, 0)$
Directrix: $x = \frac{1}{3}$
Axis: $y = 0$

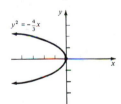

5. Vertex: (0, 0)
Focus: (0, −4)
Directrix: $y = 4$
Axis: $x = 0$

7. Vertex: (0, 0)
  Focus: (0, 7)
  Directrix: $y = -7$
  Axis: $x = 0$

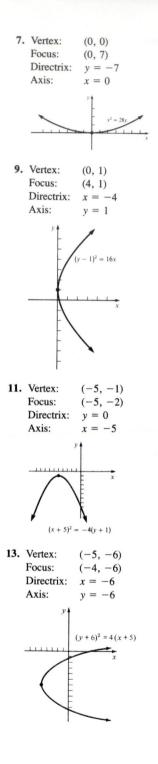

9. Vertex: (0, 1)
  Focus: (4, 1)
  Directrix: $x = -4$
  Axis: $y = 1$

11. Vertex: $(-5, -1)$
  Focus: $(-5, -2)$
  Directrix: $y = 0$
  Axis: $x = -5$

13. Vertex: $(-5, -6)$
  Focus: $(-4, -6)$
  Directrix: $x = -6$
  Axis: $y = -6$

15. Vertex: $\left(-\frac{5}{2}, -1\right)$
  Focus: $\left(-\frac{5}{2}, -\frac{15}{16}\right)$
  Directrix: $y = -\frac{17}{16}$
  Axis: $x = -\frac{5}{2}$

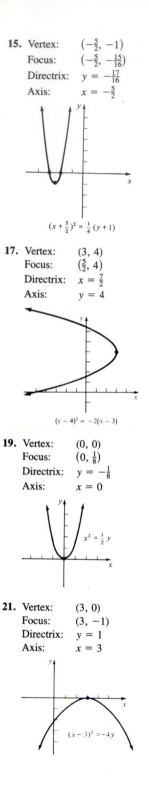

17. Vertex: (3, 4)
  Focus: $\left(\frac{5}{2}, 4\right)$
  Directrix: $x = \frac{7}{2}$
  Axis: $y = 4$

19. Vertex: (0, 0)
  Focus: $\left(0, \frac{1}{8}\right)$
  Directrix: $y = -\frac{1}{8}$
  Axis: $x = 0$

21. Vertex: (3, 0)
  Focus: (3, −1)
  Directrix: $y = 1$
  Axis: $x = 3$

23. Vertex: (−2, 1)
  Focus: (−1, 1)
  Directrix: $x = -3$
  Axis: $y = 1$

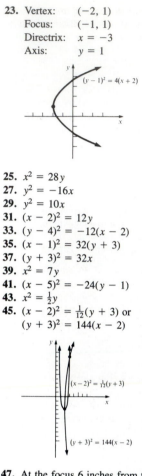

25. $x^2 = 28y$
27. $y^2 = -16x$
29. $y^2 = 10x$
31. $(x - 2)^2 = 12y$
33. $(y - 4)^2 = -12(x - 2)$
35. $(x - 1)^2 = 32(y + 3)$
37. $(y + 3)^2 = 32x$
39. $x^2 = 7y$
41. $(x - 5)^2 = -24(y - 1)$
43. $x^2 = \frac{1}{2}y$
45. $(x - 2)^2 = \frac{1}{12}(y + 3)$ or
  $(y + 3)^2 = 144(x - 2)$

47. At the focus 6 inches from the vertex
49. $y = -2$
51. 27 feet
53. 4.5 feet
57. If we take $V(h, k)$ as the origin of the $x'y'$ system, the equation of the parabola is $y'^2 = 4px'$. From equations (2) in Section 2.5, we obtain $(y - k)^2 = 4p(x - h)$. The directrix $x' = -p$ becomes $x = h - p$.

59. Equating the distance from (x, y) to $(-1, 2)$ with the distance from (x, y) to $2x + 5y - 5 = 0$, we have

$$\sqrt{(x + 1)^2 + (y - 2)^2}$$
$$= \frac{|2x + 5y - 5|}{\sqrt{4 + 25}}.$$

Squaring both sides and simplifying, we obtain $25x^2 - 20xy + 4y^2 + 78x - 66y + 120 = 0$.

61. Without loss of generality we assume that the equation of the parabola is $y^2 = 4px$. Then the focus is at $F(p, 0)$, and the vertex is at the origin. If $P(x, y)$ is a point on the parabola, then the distance between F and P is

$$d(F, P) = \sqrt{(x - p)^2 + y^2}.$$

Since $P(x, y)$ is on the parabola, $y^2 = 4px$ and

$$d(F, P) = \sqrt{(x - p)^2 + 4px}$$
$$= \sqrt{x^2 - 2px + p^2 + 4px}$$
$$= \sqrt{x^2 + 2px + p^2}$$
$$= \sqrt{(x + p)^2}$$
$$= |x + p|.$$

Since x can be any nonnegative real number, $d(F, P)$ will be minimum when $x = 0$. The point on the parabola corresponding to $x = 0$ is $(0, 0)$, which is the vertex.

63. 25 million kilometers

65. x intercept: $\quad 3$
y intercepts: $\quad -2, -6$

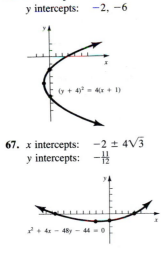

$(y + 4)^2 = 4(x + 1)$

67. x intercepts: $\quad -2 \pm 4\sqrt{3}$
y intercepts: $\quad -\frac{11}{12}$

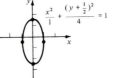

$x^2 + 4x - 48y - 44 = 0$

69. $\left(x + \dfrac{b}{2a}\right)^2 = \dfrac{1}{a}\left(y - \dfrac{4ac - b^2}{4a}\right)$

Vertex: $\quad \left(-\dfrac{b}{2a}, \dfrac{4ac - b^2}{4a}\right)$

Directrix: $\quad y = \dfrac{4ac - b^2 - 1}{4a}$

Focus: $\quad \left(\dfrac{b}{2a}, \dfrac{4ac - b^2 + 1}{4a}\right)$

Axis: $\quad x = -\dfrac{b}{2a}$

71. $(5, -2)$

73. $x^2 + y^2 = \frac{1}{65}$

Exercise 6.3 [Page 346]

1. Center: $\quad (0, 0)$
Foci: $\quad (\pm 4, 0)$
Vertices: $\quad (\pm 5, 0), (0, \pm 3)$

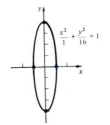

$\frac{x^2}{25} + \frac{y^2}{9} = 1$

3. Center: $\quad (0, 0)$
Foci: $\quad (0, \pm\sqrt{15})$
Vertices: $\quad (0, \pm 4), (\pm 1, 0)$

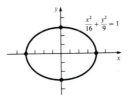

$\frac{x^2}{1} + \frac{y^2}{16} = 1$

5. Center: $\quad (0, 0)$
Foci: $\quad (\pm\sqrt{7}, 0)$
Vertices: $\quad (\pm 4, 0), (0, \pm 3)$

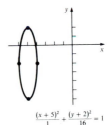

$\frac{x^2}{16} + \frac{y^2}{9} = 1$

7. Center: $\quad (0, 0)$
Foci: $\quad (0, \pm\sqrt{5})$
Vertices: $\quad (0, \pm 3), (\pm 2, 0)$

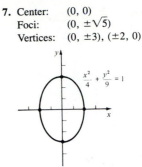

$\frac{x^2}{4} + \frac{y^2}{9} = 1$

9. Center: $\quad (1, 3)$
Foci: $\quad (1 - \sqrt{13}, 3),$
$\quad (1 + \sqrt{13}, 3)$
Vertices: $\quad (-6, 3), (8, 3)$
$\quad (1, -3), (1, 9)$

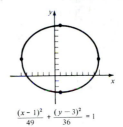

$\frac{(x - 1)^2}{49} + \frac{(y - 3)^2}{36} = 1$

11. Center: $\quad (-5, -2)$
Foci: $\quad (-5, -2 - \sqrt{15}),$
$\quad (-5, -2 + \sqrt{15})$
Vertices: $\quad (-5, -6), (-5, 2)$
$\quad (-6, -2), (-4, -2)$

$\frac{(x + 5)^2}{1} + \frac{(y + 2)^2}{16} = 1$

13. Center: $\quad \left(0, -\frac{1}{2}\right)$
Foci: $\quad \left(0, -\frac{1}{2} - \sqrt{3}\right),$
$\quad \left(0, -\frac{1}{2} + \sqrt{3}\right)$
Vertices: $\quad \left(0, -\frac{5}{2}\right), \left(0, \frac{3}{2}\right)$
$\quad \left(-1, -\frac{1}{2}\right), \left(1, -\frac{1}{2}\right)$

$\frac{x^2}{1} + \frac{\left(y + \frac{1}{2}\right)^2}{4} = 1$

15. Center: $(1, -2)$
Foci: $(1, -2 - \sqrt{6})$, $(1, -2 + \sqrt{6})$
Vertices: $(1, -2 - \sqrt{15})$, $(1, -2 + \sqrt{15})$
$(-2, -2), (4, -2)$

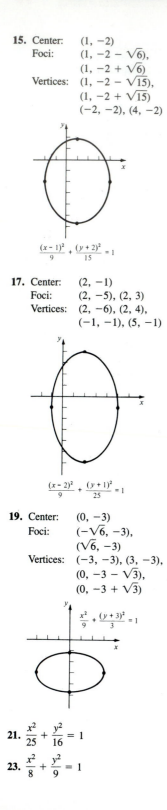

$$\frac{(x-1)^2}{9} + \frac{(y+2)^2}{15} = 1$$

17. Center: $(2, -1)$
Foci: $(2, -5), (2, 3)$
Vertices: $(2, -6), (2, 4)$, $(-1, -1), (5, -1)$

$$\frac{(x-2)^2}{9} + \frac{(y+1)^2}{25} = 1$$

19. Center: $(0, -3)$
Foci: $(-\sqrt{6}, -3)$, $(\sqrt{6}, -3)$
Vertices: $(-3, -3), (3, -3)$, $(0, -3 - \sqrt{3})$, $(0, -3 + \sqrt{3})$

$$\frac{x^2}{9} + \frac{(y+3)^2}{3} = 1$$

21. $\dfrac{x^2}{25} + \dfrac{y^2}{16} = 1$

23. $\dfrac{x^2}{8} + \dfrac{y^2}{9} = 1$

25. $\dfrac{x^2}{1} + \dfrac{y^2}{9} = 1$

27. $\dfrac{(x-1)^2}{16} + \dfrac{(y+3)^2}{4} = 1$

29. $\dfrac{x^2}{9} + \dfrac{y^2}{13} = 1$

31. $\dfrac{x^2}{11} + \dfrac{y^2}{9} = 1$

33. $\dfrac{x^2}{3} + \dfrac{y^2}{12} = 1$

35. $\dfrac{x^2}{16} + \dfrac{(y-1)^2}{4} = 1$

37. $\dfrac{(x-1)^2}{7} + \dfrac{(y-3)^2}{16} = 1$

39. $\dfrac{(x-\frac{15}{2})^2}{\frac{121}{4}} + \dfrac{(y-4)^2}{18} = 1$

41. $\dfrac{x^2}{21,160,000} + \dfrac{y^2}{21,000,000} = 1$

43. 12 feet

45. On the major axis, 12 feet to either side from the center of the room.

47. Using equations (2) in Section 2.5, we rewrite

$$\frac{x'^2}{a^2} + \frac{y'^2}{b^2} = 1$$

as

$$\frac{(x-h)^2}{a^2} + \frac{(y-k)^2}{b^2} = 1.$$

The vertices of this ellipse are at the $x'y'$ points $(\pm a, 0)$ and $(0, \pm b)$. Using $x = x' + h$ and $y = y' + k$, we express these points as $(h \pm a, k)$ and $(h, k \pm b)$. The foci are at the $x'y'$ points $(\pm c, 0)$, which are $(h \pm c, k)$ in the xy system.

49. $5x^2 - 4xy + 8y^2 - 24x - 48y = 0$

51. a. $\frac{8}{3}$
b. The x coordinate of a point on a focal chord is $\pm c$. We substitute $x = c$ into the equation of the ellipse and solve for y:

$$\frac{c^2}{a^2} + \frac{y^2}{b^2} = 1$$

$$y^2 = b^2\left(1 - \frac{c^2}{a^2}\right)$$

$$= \frac{b^2}{a^2}(a^2 - c^2) = \frac{b^2}{a^2}(b^2)$$

$$= \frac{b^4}{a^2}$$

$$y = \pm\frac{b^2}{a}.$$

The endpoints of the right-hand focal chord are thus $(c, b^2/a)$ and $(c, -b^2/a)$. The focal width of the ellipse is then $2b^2/a$.

53. $(-2, -10)$

55. Vertex: $(4, -2)$
Axis: $x = 4$
Focus: $(4, -1)$
Directrix: $y = -3$

Exercise 6.4 [Page 360]

1. Center: $(0, 0)$
Vertices: $(\pm 4, 0)$
Foci: $(\pm\sqrt{41}, 0)$
Asymptotes: $y = \pm\frac{5}{4}x$

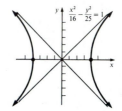

3. Center: $(0, 0)$
Vertices: $(0, \pm 8)$
Foci: $(0, \pm\sqrt{73})$
Asymptotes: $y = \pm\frac{8}{3}x$

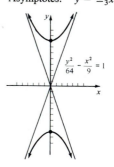

5. Center: $(0, 0)$
Vertices: $(\pm 4, 0)$
Foci: $(\pm\sqrt{20}, 0)$
Asymptotes: $y = \pm\frac{1}{2}x$

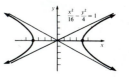

7. Center: $(0, 0)$
Vertices: $(0, \pm 2\sqrt{5})$
Foci: $(0, \pm 2\sqrt{6})$
Asymptotes: $y = \pm\sqrt{5}x$

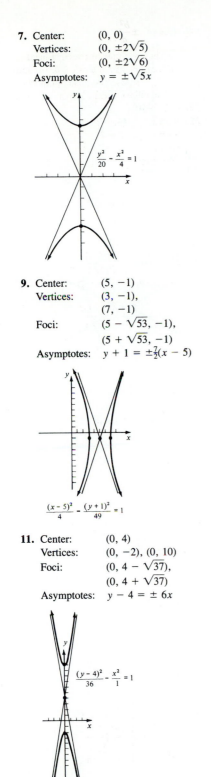

$$\frac{y^2}{20} - \frac{x^2}{4} = 1$$

9. Center: $(5, -1)$
Vertices: $(3, -1),$
    $(7, -1)$
Foci: $(5 - \sqrt{53}, -1),$
    $(5 + \sqrt{53}, -1)$
Asymptotes: $y + 1 = \pm\frac{7}{2}(x - 5)$

$$\frac{(x-5)^2}{4} - \frac{(y+1)^2}{49} = 1$$

11. Center: $(0, 4)$
Vertices: $(0, -2), (0, 10)$
Foci: $(0, 4 - \sqrt{37}),$
    $(0, 4 + \sqrt{37})$
Asymptotes: $y - 4 = \pm 6x$

$$\frac{(y-4)^2}{36} - \frac{x^2}{1} = 1$$

13. Center: $(3, 1)$
Vertices: $(3 - \sqrt{5}, 1),$
    $(3 + \sqrt{5}, 1)$
Foci: $(3 - \sqrt{30}, 1),$
    $(3 + \sqrt{30}, 1)$
Asymptotes: $y - 1 = \pm\sqrt{5}(x - 3)$

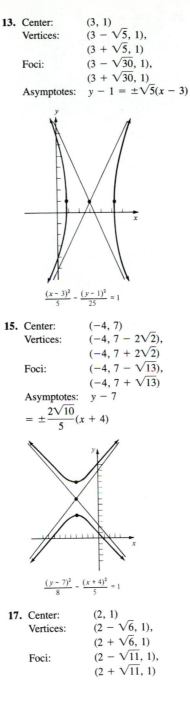

$$\frac{(x-3)^2}{5} - \frac{(y-1)^2}{25} = 1$$

15. Center: $(-4, 7)$
Vertices: $(-4, 7 - 2\sqrt{2}),$
    $(-4, 7 + 2\sqrt{2})$
Foci: $(-4, 7 - \sqrt{13}),$
    $(-4, 7 + \sqrt{13})$
Asymptotes: $y - 7$
    $= \pm\dfrac{2\sqrt{10}}{5}(x + 4)$

$$\frac{(y-7)^2}{8} - \frac{(x+4)^2}{5} = 1$$

17. Center: $(2, 1)$
Vertices: $(2 - \sqrt{6}, 1),$
    $(2 + \sqrt{6}, 1)$
Foci: $(2 - \sqrt{11}, 1),$
    $(2 + \sqrt{11}, 1)$

Asymptotes: $y - 1$
$= \pm\dfrac{\sqrt{30}}{6}(x - 2)$

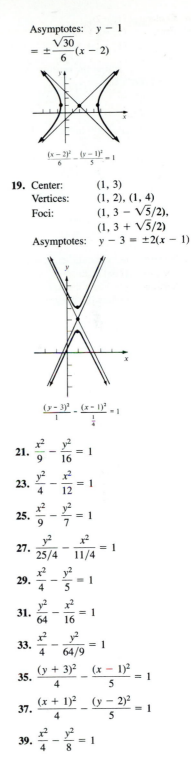

$$\frac{(x-2)^2}{6} - \frac{(y-1)^2}{5} = 1$$

19. Center: $(1, 3)$
Vertices: $(1, 2), (1, 4)$
Foci: $(1, 3 - \sqrt{5}/2),$
    $(1, 3 + \sqrt{5}/2)$
Asymptotes: $y - 3 = \pm 2(x - 1)$

$$\frac{(y-3)^2}{1} - \frac{(x-1)^2}{\frac{1}{4}} = 1$$

21. $\dfrac{x^2}{9} - \dfrac{y^2}{16} = 1$

23. $\dfrac{y^2}{4} - \dfrac{x^2}{12} = 1$

25. $\dfrac{x^2}{9} - \dfrac{y^2}{7} = 1$

27. $\dfrac{y^2}{25/4} - \dfrac{x^2}{11/4} = 1$

29. $\dfrac{x^2}{4} - \dfrac{y^2}{5} = 1$

31. $\dfrac{y^2}{64} - \dfrac{x^2}{16} = 1$

33. $\dfrac{x^2}{4} - \dfrac{y^2}{64/9} = 1$

35. $\dfrac{(y+3)^2}{4} - \dfrac{(x-1)^2}{5} = 1$

37. $\dfrac{(x+1)^2}{4} - \dfrac{(y-2)^2}{5} = 1$

39. $\dfrac{x^2}{4} - \dfrac{y^2}{8} = 1$

41. $\dfrac{(y-3)^2}{1} - \dfrac{(x+1)^2}{4} = 1$

43. $(-7, 12)$

47. In the $x'y'$ system the equation of this hyperbola is

$$\dfrac{y'^2}{a^2} - \dfrac{x'^2}{b^2} = 1.$$

Using equations (2) in Section 2.5, we have

$$\dfrac{(y-k)^2}{a^2} - \dfrac{(x-h)^2}{b^2} = 1.$$

In the $x'y'$ system, the vertices are at $(0, \pm a)$. Using $x = x' + h$ and $y = y' + k$, we may express these points in the xy system as $(h, k \pm a)$.

49. a. 9 **b.** The x coordinate of a point on a focal chord is $\pm c$. We substitute $x = c$ into the equation of the hyperbola and solve for y:

$$\dfrac{c^2}{a^2} - \dfrac{y^2}{b^2} = 1$$

$$y^2 = b^2\left(\dfrac{c^2}{a^2} - 1\right)$$

$$= \dfrac{b^2}{a^2}(c^2 - a^2) = \dfrac{b^4}{a^2}$$

$$y = \pm\dfrac{b^2}{a}.$$

The endpoints of the right-hand focal chord are thus $(c, b^2/a)$ and $(c, -b^2/a)$. The focal width of the hyperbola is $2b^2/a$.

53. Vertex: $(0, 2)$
Focus: $(-\tfrac{1}{4}, 2)$
Directrix: $x = \tfrac{1}{4}$
Axis: $y = 2$

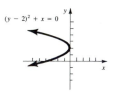

$(y-2)^2 + x = 0$

55. Center: $(1, 0)$
Foci: $(1 - \sqrt{5}, 0),$
$(1 + \sqrt{5}, 0)$
Vertices: $(1 - \sqrt{10}, 0),$
$(1 + \sqrt{10}, 0),$
$(1, \sqrt{5}), (1, -\sqrt{5})$

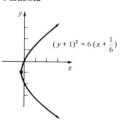

$\dfrac{(x+1)^2}{10} + \dfrac{y^2}{5} = 1$

Exercise 6.5 [Page 365]

1. Parabola

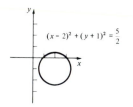

$(y+1)^2 = 6\left(x + \dfrac{1}{6}\right)$

3. Circle

$(x-2)^2 + (y+1)^2 = \dfrac{5}{2}$

5. A point

$\bullet (1, -1)$

7. Ellipse

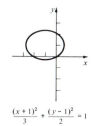

$\dfrac{(x+1)^2}{3} + \dfrac{(y-1)^2}{2} = 1$

9. Ellipse

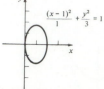

$\dfrac{(x-1)^2}{1} + \dfrac{y^2}{3} = 1$

11. No graph

13. A point

$\bullet (3, 2)$

15. Hyperbola

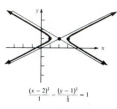

$\dfrac{(x-2)^2}{1} - \dfrac{(y-1)^2}{3} = 1$

17. One straight line and two intersecting straight lines.

19. Suppose that $A > 0$ and $C < 0$ (the case in which $A < 0$ and $C > 0$ is similar). Completing the square, we have

$$A\left(x^2 + \dfrac{D}{A}x + \dfrac{D^2}{4A^2}\right)$$

$$+ C\left(y^2 + \dfrac{E}{C}y + \dfrac{E^2}{4C^2}\right)$$

$$= F + \dfrac{D^2}{4A^2} + \dfrac{E^2}{4C^2}$$

or

$$A\left(x + \dfrac{D}{2A}\right)^2 + C\left(y + \dfrac{E}{2C}\right)^2$$

$$= \dfrac{2A^2C^2F + C^2D^2 + A^2E^2}{4A^2C^2}.$$

Let

$$G = \dfrac{4A^2C^2F + C^2D^2 + A^2E^2}{4A^2C^2}.$$

If $G = 0$, the equation becomes

$$A\left(x + \frac{D}{2A}\right)^2 = C\left(y + \frac{E}{2C}\right)^2$$

or

$$y + \frac{E}{2C} = \pm\sqrt{\frac{A}{C}}\left(x + \frac{D}{2A}\right),$$

which is a pair of intersecting lines. (Since $A \neq 0$, the two lines are distinct and have unequal slopes.) If $G > 0$, the equation may be written

$$\frac{(x + D/2A)^2}{G/A} - \frac{(y + E/2C)^2}{-G/C} = 1,$$

where the denominators are positive as required (since they must be squares). If $G < 0$, we write the equation as

$$\frac{(y + E/2C)^2}{G/C} - \frac{(x + D/2A)^2}{-G/A} = 1.$$

21. $(x - 2)^2 = -28(y + 3)$
23. Vertices: $(-1, 0), (1, 0)$
Foci: $(-\sqrt{10}, 0), (\sqrt{10}, 0)$
Asymptotes: $y = \pm 3$

Exercise 6.6 [Page 371]

1. $\left(\dfrac{2\sqrt{3} + 3}{2}, \dfrac{2 - 3\sqrt{3}}{2}\right)$
 $\approx (3.23, -1.60)$
3. $(-4, 0)$
5. $3.35, 2.97$
7. $(0 -2\sqrt{2}) \approx (0, -2.83)$
9. $(\frac{5}{2}\sqrt{2}, -\frac{5}{2}\sqrt{2}) \approx (3.54, -3.54)$
11. $(2 - 4\sqrt{3}, 2\sqrt{3} + 4)$
 $\approx (-4.93, 7.46)$
13. $x^2 - 2xy + y^2 - 16\sqrt{2}x$
 $+ 16\sqrt{2}y = 0$
15. $xy = -1$
17. $24x^2 + 2\sqrt{3}xy + 22y^2 - 525 = 0$
19. $3x'^2 - 2y'^2 = 48$
21. $x'^2 + 8x' + 8y' = 0$
23. $4x'^2 - y'^2 = 16$
25. $\theta = \arctan\frac{1}{3};\ 13x'^2 + 3y'^2 = 468$
27. $\theta = \dfrac{\pi}{4};\ y'^2 + 4x' = 0$

29. $\dfrac{x'^2}{8/3} + \dfrac{y'^2}{8} = 1$

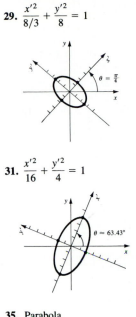

31. $\dfrac{x'^2}{16} + \dfrac{y'^2}{4} = 1$

35. Parabola
37. $\dfrac{(y - 1)^2}{4} - \dfrac{(x - 5)^2}{5} = 1$
39. $(y - 5)^2 = -48(x - 3)$
41. $13x' - 7y' + 21 = 0$

Chapter 6
Review Exercises [Page 375]

1. True
3. False
5. True
7. False
9. False
11. $(10 - \sqrt{2}, -8)$
13. $x - 3y + 2 = 0$
15. $(y + 1)^2 = 12x$
17. Center: $(-1, 3)$
Foci: $(-1, 3 - \sqrt{3}),$
  $(-1, 3 + \sqrt{3})$
Vertices: $(-1, 1), (-1, 5)$
  $(-2, 3), (0, 3)$

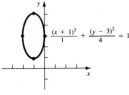

19. $\dfrac{x^2}{4} + (y + 2)^2 = 1$

21. $\dfrac{x^2}{4} - \dfrac{y^2}{16} = 1$

23. Parabola

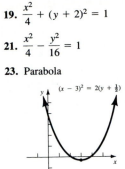

25. Hyperbola

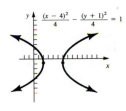

27. 1.95×10^9 m

29.

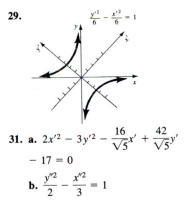

31. a. $2x'^2 - 3y'^2 - \dfrac{16}{\sqrt{5}}x' + \dfrac{42}{\sqrt{5}}y'$
 $- 17 = 0$
 b. $\dfrac{y''^2}{2} - \dfrac{x''^2}{3} = 1$

Exercise A.1 [Page 381]

1. $\{1, 3, 4, 6, 8, 9, 10, 11, 12, 14, 15\}$
3. $\{1, 2, 3, 5, 7, 8, 9, 11, 12, 13, 14\}$
5. $\{1, 8\}$
7. $\{3, 9, 11, 12, 14\}$
9. $\{2, 3, 5, 7, 9, 11, 12, 13, 14\}$

11. $\{3, 4, 6, 9, 10, 11, 12, 15\}$

13. $\{1, 2, 3, 4, 5, 6, 7, 8, 9, 10, 11, 12, 13, 14, 15\}$

15. $\{1, 4, 6, 8, 10, 15\}$

17. $\{\ldots, -5, -3, -1, 1, 3, 5, \ldots\}$

19. $\{(1, 3), (3, 1), (-1, -3), (-3, -1)\}$

21. $\{\ldots, (-4, 1), (0, 0), (4, -1), (8, -2), \ldots\}$

23. $2x^2 - 2x + \dfrac{3}{2} + \dfrac{-\frac{5}{2}}{2x + 1}$

25. $5x^2 + 13x + 40 + \dfrac{114}{x - 3}$

27. $r = \sqrt{13}$

29. $y = \dfrac{\sqrt{3}}{2}$

31. $x = \frac{12}{5}, y = \frac{16}{5}$

Exercise A.2 [Page 388]

1. a. The statement $S(1)$, $2 = (1)^2 + 1$, is true.
b. Assume that $S(k)$,
$2 + 4 + \cdots + 2k = k^2 + k$
is true. Then
$2 + 4 + \cdots + 2k + 2(k + 1)$
$= k^2 + k + 2(k + 1)$
$= k^2 + k + 2k + 2$
$= (k^2 + 2k + 1) + (k + 1)$
$= (k + 1)^2 + (k + 1)$,
and so the statement $S(k + 1)$ is true. Hence, by the principle of mathematical induction, the proof is complete.

3. a. The statement $S(1)$,
$$1^2 = \frac{1(1 + 1)[2(1) + 1]}{6},$$
is true. **b.** Assume that $S(k)$, $1^2 + 2^2 + \cdots + k^2$
$$= \frac{k(k + 1)(2k + 1)}{6}, \text{ is true. Then}$$
$1^2 + 2^2 + \cdots + k^2 + (k + 1)^2$
$$= \frac{k(k + 1)(2k + 1)}{6} + (k + 1)^2$$
$$= \frac{k(k + 1)(2k + 1) + 6(k + 1)^2}{6}$$
$$= \frac{(k + 1)[k(2k + 1) + 6(k + 1)]}{6}$$
$$= \frac{(k + 1)(2k^2 + 7k + 6)}{6}$$
$$= \frac{(k + 1)[(k + 2)(2k + 3)]}{6}$$
$$= \frac{(k + 1)[(k + 1) + 1][2(k + 1) + 1]}{6}$$
and so the statement $S(k + 1)$ is true. Hence, by mathematical induction, the proof is complete.

9. a. The statement $S(1)$, $(1 + a)^1 \geq 1 + (1)a$, for $a \geq -1$ is true.
b. Assume that $S(k)$, $(1 + a)^k \geq 1 + ka$, for $a \geq -1$ is true. Then, for $a \geq -1$,
$(1 + a)^{k+1} = (1 + a)^k(1 + a)$
$\qquad \geq (1 + ka)(1 + a)$
$\qquad = 1 + ka^2 + ka + a$
$\qquad \geq 1 + ka + a$
$\qquad = 1 + (k + 1)a$.
Thus, the statement $S(k + 1)$ is true, and by mathematical induction the proof is complete.

11. Assume that $S(k)$,
$2 + 4 + 6 + \cdots + 2k$
$= k^2 + k + 1$, is true. Then
$2 + 4 + 6 + \cdots + 2k + 2(k + 1)$
$= k^2 + k + 1 + 2(k + 1)$
$= k^2 + k + 1 + 2k + 2$
$= (k^2 + 2k + 1)$
$\quad + (k + 1) + 1$
$= (k + 1)^2 + (k + 1) + 1$,
and so the statement $S(k + 1)$ is true. Note that $S(1)$ is the *false* statement
$2 = (1)^2 + (1) + 1$
$\quad = 3$.

Exercise A.3 [Page 391]

1. $-2\sqrt{3}$

3. $8 - 20i$

5. $6 - 6i$

7. $\frac{2}{29} + \frac{5}{29}i$

9. $-2 + \frac{11}{4}i$

11. $a = \frac{11}{2}, b = 7$

13. $a = \pm 1, b = \pm 1$

Exercise A.4 [Page 397]

1. 1, 0.5955

3. 7, 0.6532

5. 1, 0.3438

7. 0.6362

9. 2.5142

11. 0.6184

13. 289

15. 0.0236

17. 8522

19. 0.1392

21. 0.6111

23. 1.0165

25. 0.2717

27. 0.9900

29. 3.0921

31. −0.7771

33. 0.2675

35. 1.8508

Index

Trigonometric Functions of a Real Number

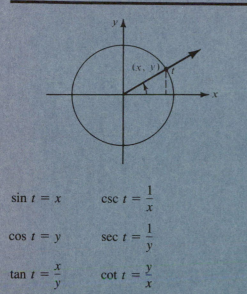

$$\sin t = x \qquad \csc t = \frac{1}{x}$$

$$\cos t = y \qquad \sec t = \frac{1}{y}$$

$$\tan t = \frac{x}{y} \qquad \cot t = \frac{y}{x}$$

Trigonometric Functions of an Acute Angle

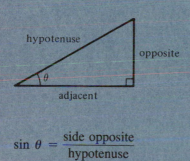

$$\sin \theta = \frac{\text{side opposite}}{\text{hypotenuse}}$$

$$\cos \theta = \frac{\text{side adjacent}}{\text{hypotenuse}}$$

$$\tan \theta = \frac{\text{side opposite}}{\text{side adjacent}}$$

$$\cot \theta = \frac{\text{side adjacent}}{\text{side opposite}}$$

$$\sec \theta = \frac{\text{hypotenuse}}{\text{side adjacent}}$$

$$\csc \theta = \frac{\text{hypotenuse}}{\text{side opposite}}$$

Signs of the Trigonometric Functions

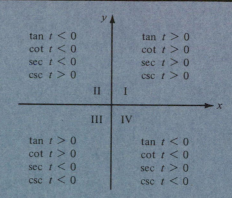

Basic Identities

$$\tan t = \frac{\sin t}{\cos t} \qquad\qquad \sec t = \frac{1}{\cos t}$$

$$\cot t = \frac{\cos t}{\sin t} \qquad\qquad \csc t = \frac{1}{\sin t}$$

$$\cos^2 t + \sin^2 t = 1$$

$$1 + \tan^2 t = \sec^2 t$$

$$1 + \cot^2 t = \csc^2 t$$

Addition Formulas

$$\sin(u + v) = \sin u \cos v + \cos u \sin v$$

$$\sin(u - v) = \sin u \cos v - \cos u \sin v$$

$$\cos(u + v) = \cos u \cos v - \sin u \sin v$$

$$\cos(u - v) = \cos u \cos v + \sin u \sin v$$

$$\tan(u + v) = \frac{\tan u + \tan v}{1 - \tan u \tan v}$$

$$\tan(u - v) = \frac{\tan u - \tan v}{1 + \tan u \tan v}$$